I0065688

Encyclopedia of Marine Biology

Volume I

Encyclopedia of Marine Biology
Volume I

Edited by **Simon Oakenfold**

R CALLISTO REFERENCE

New York

Published by Callisto Reference,
106 Park Avenue, Suite 200,
New York, NY 10016, USA
www.callistoreference.com

Encyclopedia of Marine Biology: Volume I
Edited by Simon Oakenfold

© 2015 Callisto Reference

International Standard Book Number: 978-1-63239-267-1 (Hardback)

This book contains information obtained from authentic and highly regarded sources. Copyright for all individual chapters remain with the respective authors as indicated. A wide variety of references are listed. Permission and sources are indicated; for detailed attributions, please refer to the permissions page. Reasonable efforts have been made to publish reliable data and information, but the authors, editors and publisher cannot assume any responsibility for the validity of all materials or the consequences of their use.

The publisher's policy is to use permanent paper from mills that operate a sustainable forestry policy. Furthermore, the publisher ensures that the text paper and cover boards used have met acceptable environmental accreditation standards.

Trademark Notice: Registered trademark of products or corporate names are used only for explanation and identification without intent to infringe.

Printed in the United States of America.

Contents

Preface

This ever mysterious field of science which is still considered to be known least on this planet is marine biology. It includes the study of the behavior of marine organisms and their interaction with the environment. Thus, marine biology is considered the most encompassing fields of oceanography. In order to understand the behavior of marine organisms, it is very important for biologists to have a basic understanding of all the aspects of oceanography, which includes geological oceanography, physical oceanography as well as chemical oceanography. Therefore, for marine oceanographers marine biology is the key to their research.

Because of a vast number of topics that fall under the field of marine biology, most of the researchers pick up a very specific field of their interest and specialize in it. Now if we consider the specialization, it can be done on the basis of particular organism, ecosystem, species, behavior, etc. Most of the marine biologists select one species and study all the sub-species or natives in all possible climates and regions. Recently marine biologists and researchers are experimenting the ways to learn more about diseased population of farm raised fishes and the drugs to deal with it.

This book gives a close insight into the fields of marine biology and the various aspects related to the recent advancement of marine oceanography. Though it was very difficult to contain this vast field in one book, still it is our high hope that this book will be helpful for those who want to learn and grow their knowledge about marine biology. This book contains chapters that are written by experts. Some of them even study the impact of interaction of human and aquatic animals.

I especially wish to acknowledge the contributing authors. Without them, a work of this magnitude would simply not be feasible. I thank them for allocating much of their valuable time to this project. Not only do I appreciate their participation, but also their adherence as a group to the time parameters set for this publication.

Editor

Summary of Reported Whale-Vessel Collisions in Alaskan Waters

Janet L. Neilson,[1] Christine M. Gabriele,[1] Aleria S. Jensen,[2] Kaili Jackson,[2] and Janice M. Straley[3]

[1] *Division of Resource Management, Glacier Bay National Park and Preserve, P.O. Box 140, Gustavus, AK 99826, USA*
[2] *Office of Protected Resources, National Marine Fisheries Service, P.O. Box 21668, Juneau, AK 99802, USA*
[3] *Department of Biology, University of Alaska Southeast Sitka Campus, 1332 Seward Avenue, Sitka, AK 99835, USA*

Correspondence should be addressed to Janet L. Neilson, janet_neilson@nps.gov

Academic Editor: Frances Gulland

Here we summarize 108 reported whale-vessel collisions in Alaska from 1978–2011, of which 25 are known to have resulted in the whale's death. We found 89 definite and 19 possible/probable strikes based on standard criteria we created for this study. Most strikes involved humpback whales (86%) with six other species documented. Small vessel strikes were most common (<15 m, 60%), but medium (15–79 m, 27%) and large (≥80 m, 13%) vessels also struck whales. Among the 25 mortalities, vessel length was known in seven cases (190–294 m) and vessel speed was known in three cases (12–19 kn). In 36 cases, human injury or property damage resulted from the collision, and at least 15 people were thrown into the water. In 15 cases humpback whales struck anchored or drifting vessels, suggesting the whales did not detect the vessels. Documenting collisions in Alaska will remain challenging due to remoteness and resource limitations. For a better understanding of the factors contributing to lethal collisions, we recommend (1) systematic documentation of collisions, including vessel size and speed; (2) greater efforts to necropsy stranded whales; (3) using experienced teams focused on determining cause of death; (4) using standard criteria for validating collision reports, such as those presented in this paper.

1. Introduction

Ship strikes are a source of injury and mortality for whales worldwide but documenting these events and their outcomes is a significant challenge. The rate at which whale-vessel collisions occur, the types of vessels involved, and the extent to which they affect particular populations of whales are largely unknown, especially in remote areas such as Alaska. Accurate documentation of whale-vessel collisions is difficult for several reasons, ranging from cases where vessel operators are unaware that collisions should be reported, or operators who do not report for fear of reprisal, to incomplete data gathering on the details surrounding the collision and difficulties inherent in accurately assessing a free-ranging whale's condition following a collision. In addition, a large ship may strike a whale and the crew may be unaware that the collision occurred. Determining that a stranded whale died from a collision is especially difficult in Alaska because of the logistical challenges of performing complete necropsies

(e.g., [1]) on stranded animals. These challenges include the remote location of most carcasses, frequent inclement weather, large tides, concerns for human safety when bears are present, limited daylight at some times of the year, and a lack of personnel trained in identifying ship strike injuries.

An overarching challenge in accurately estimating the rate of ship strikes not only in Alaska, but globally, is that there are no universal, standardized criteria for evaluating eyewitness collision reports or stranded whales to determine which cases represent *bona fide* collisions and which reports should be rejected due to a lack of certainty. Other investigators have compiled accounts of ship strikes regionally and worldwide using variable criteria, terminology, and types of evidence [2–20].

Informed management of whale stocks relies upon accurate estimates of the rate of serious injuries and mortalities from ship strikes. In the United States, the Marine Mammal Protection Act (MMPA) [21] defines a serious injury as any injury that will likely result in mortality. The

National Oceanic and Atmospheric Administration (NOAA) is responsible for marine mammal stock assessment reports for all species of cetaceans and all species of pinnipeds except walrus, including an estimate of the annual human-caused mortality and serious injury of each stock by source (e.g., commercial fishing, ship strike, etc.). Internationally, the International Whaling Commission (IWC) considers the number of mortalities from ship strikes with estimates of fisheries bycatch in developing recommendations for large whale conservation. The need for a standardized quality control system to validate collision reports has been recognized by the IWC Vessel Strike Data Standardization Group (VSDG), which formed in 2005 to examine the issue of ship strikes with cetaceans. Since 2007, the group has been developing a global ship strike database that aims, among other things, to identify the level of uncertainty associated with individual records based on strandings and eyewitness collision accounts [22]. The IWC database classifies collision reports into six categories (definite ship strike, probable ship strike, possible ship strike, not a ship strike, whale initiated collision, and rejected report); however, these categories do not yet have standardized definitions (D. Mattila, pers. comm.). Each report is reviewed by the VSDG, and an incident is only classified as a "definite ship strike" if all members are unanimous.

At the individual level, the MMPA contains a general prohibition on the "take" of marine mammals, defined as "to harass, hunt, capture, or kill, or attempt to harass, hunt, capture, or kill any marine mammal." NOAA regulations implementing the MMPA further describe the term "take" to include "the negligent or intentional operation of an aircraft or vessel, or the doing of any other negligent or intentional act which results in disturbing or molesting a marine mammal" [23]. In 2001, NOAA implemented regulations in Alaska limiting approaches to humpback whales to minimize disturbance that could adversely affect individual animals and to manage the threat to these animals caused by increasing vessel traffic and a growing whale watch industry in Alaska. These regulations prohibit vessels from approaching within 91 m (100 yards) of a humpback whale and require vessels to operate at a "slow, safe speed" near humpback whales [24]. This speed is not specified beyond the definition for "safe speed" in 33 US Code 2006, "every vessel shall at all times proceed at a safe speed so that she can take proper and effective action to avoid collision and be stopped within a distance appropriate to the prevailing circumstances and conditions." In addition, since 1979, more protective regulations have existed in Glacier Bay National Park in southeastern Alaska to reduce the risk of humpback whale-vessel collisions and disturbance in park waters. These regulations include limits on the number of vessels allowed to enter Glacier Bay, a 463 m (one-quarter nautical mile) approach limit to humpback whales, and vessel speed and course restrictions in areas where whales are concentrated [25]. For species other than humpback whales, no specific regulations exist in Alaska, although vessel operators are advised to follow a general marine mammal viewing "Code of Conduct" which recommends remaining at least 91 m (100 yards) from marine mammals and avoiding

excessive speeds. These guidelines are intended to prevent mariners from accidentally harassing or injuring whales in violation of the MMPA and US Endangered Species Act (ESA).

Vessel strikes are a significant concern from other perspectives as well. In Alaska, recovering whale populations and increasing vessel traffic are creating a persistent problem. Collisions are costly and dangerous to humans and they can harm mariners' reputations. From commercial whale watching to subsistence whaling, whales are economically and culturally valuable to Alaska residents and visitors. In addition, one can argue that we have an ethical obligation to address the ship strike issue. For example, in 2007 an injured humpback whale (*Megaptera novaeangliae*) with a grossly inflated tongue and deformed head was observed alive in southeastern Alaska for three days before dying. A necropsy revealed that the probable cause of death was blunt trauma [26]. From an animal welfare perspective, it is our human responsibility to learn how to mitigate our actions—in this case, prevent collisions—such that whales are not subject to extended periods of suffering before dying from ship strike injuries.

Vessel traffic in Alaska overlaps with 14 whale species known to occur in waters around the state: humpback whale[1], fin whale (*Balaenoptera physalus*)[1], gray whale (*Eschrichtius robustus*), bowhead whale (*Balaena mysticetus*)[1], minke whale (*Balaenoptera acutorostrata*), blue whale (*Balaenoptera musculus*)[1], sei whale (*Balaenoptera borealis*)[1], North Pacific right whale (*Eubalaena japonica*)[1], sperm whale (*Physeter macrocephalus*)[1], beluga whale (*Delphinapterus leucas*)[2], killer whale (*Orcinus orca*), Cuvier's beaked whale (*Ziphius cavirostris*), Stejneger's beaked whale (*Mesoplodon stejnegeri*), and Baird's beaked whale (*Berardius bairdii*) [27, 28]. Population estimates are not available for most of these species in Alaska waters; however, most of the baleen whale stocks are known or thought to be recovering following the end of commercial whaling in the North Pacific in the 1960s and 1970s. For example, stocks of humpback, fin, bowhead, and gray whales are estimated to be increasing at 3–7% per year [29–32]. A notable exception is the North Pacific right whale, which remains extremely rare with a current population estimate of 31 animals (95% CL 23–54; [33]) and an unknown population trend [28]. The majority of right whale detections have occurred in the southeastern Bering Sea [33], with a smaller number of detections in the Gulf of Alaska south of Kodiak Island [34]. On the other end of the spectrum, a minimum of 12,000 humpback whales are found in high densities in spring through fall in southeastern Alaska, the eastern Aleutian Islands, along the Bering Sea continental shelf edge and break and in the Gulf of Alaska (primarily near the Shumagin Islands, Kodiak Island and from the Barren Islands through Prince William Sound) [28, 31].

The whale strike risk for various vessel types relies on a number of factors, including the number of vessels on the water and their geographic overlap with each whale species. Much of the vessel traffic in Alaskan waters is highly seasonal and concentrated in coastal areas of southeastern and south central Alaska during the summer months, where private and commercial recreational vessels (e.g., charter vessels, commercial whale watch vessels, tour boats, and cruise ships) are prevalent. Other types of vessel traffic in Alaskan waters are more likely to occur year-round and/or over broader geographic areas, including both near shore and offshore waters (e.g., commercial fishing vessels, freighters/tankers, passenger ferries, etc.), where they may overlap with a variety of near shore and offshore species [28]. In general, there is less vessel traffic off western and northern Alaska compared to other parts of the state, although these trends are already changing with climate change-driven decreases in sea ice in the Bering, Chukchi, and Beaufort seas.

Vessel speed and size appear to be important factors in predicting whale-vessel collisions and their outcomes. For example, the probability of a cruise ship having a close encounter with a humpback whale increases with the speed of the ship (especially at speeds >11.8 kn) [35], and Silber et al. [36] demonstrated that during close encounters, reduced ship speeds may reduce the probability of a collision. Further evidence comes from an analysis of worldwide collision records with large whales, in which Laist et al. [8] found that most lethal and severe injuries involve ships traveling 14 kn or faster and ships 80 m or longer. Likewise, Vanderlaan and Taggart [37] analyzed collision records, modeled the probability of lethal injury to a large whale based on vessel speed, and concluded that the chances of a lethal injury exceed 50% at speeds higher than 11.8 kn.

The summary reported here represents the most comprehensive compilation of whale-vessel collision records in Alaska that has yet been assembled [38, 39]. All records included here were evaluated using our newly developed standardized system for classifying collision records (witnessed at sea or based on strandings) into four confidence categories (definite strike, probable strike, possible strike and rejected report). Our primary goals were to (1) summarize the circumstances surrounding whale-collisions in Alaska, (2) recommend ways to improve data collection and validation, and (3) identify measures to help reduce collision risk.

2. Methods

Our study area included all waters of Alaska. We considered records that involved any species of cetacean within 370 km (200 nautical miles) of Alaska except for dolphins and porpoises. Reports of whale-vessel collisions originated from a variety of sources, including NOAA, the US Coast Guard (USCG), vessel owners, tour operators, the media, and anecdotal accounts. These reports were collected opportunistically by the National Park Service (NPS) and the University of Alaska Southeast (UAS) since 1978 and systematically by NOAA since the Alaska Marine Mammal Stranding Network was formed in 1985 [40]. We evaluated records where the whale species was uncertain or unknown on a case-by-case basis. If the species was reported as uncertain but "likely" or "probable" species X and the report was plausible given the seasonal and geographic distribution of species X, then we attributed the report to species X. We counted all other reports where the species was unknown as "unidentified

species." We rejected reports when there was insufficient information to verify that an actual strike occurred.

To analyze seasonal occurrence in collisions, we assigned a month to each record based on when the strike occurred or the carcass was found. Similarly, we assigned a year to each record based on when the strike occurred or when the carcass was found. We assigned one record from the "late 1980s" to the year 1989. We used linear regression to examine the trend in the number of reports over time and log-linear regression to estimate the average annual rate of increase in reports.

2.1. Ship Strike Confidence Categories. The reports were based on (1) collisions witnessed at sea and (2) strandings in which a dead whale was found with evidence of collision injuries. We did not consider reports of whales striking vessels after being shot or harpooned because these collisions are atypical and including them in our analysis would not contribute to our understanding of typical whale-vessel collisions. We error-checked each record against all available documentation and entered the records into a relational database. To avoid potential duplicate reports, we did not include sightings of live whales with visible propeller scars unless the collision that caused the propeller injuries was witnessed. We assigned each record to one of four confidence categories: definite ship strike, probable ship strike, possible ship strike, or rejected report (Table 1).

2.2. Sex and Age Class of Struck Whales. We determined the sex of stranded whales from necropsy reports. It was not possible to determine the sex of live animals; however, in two cases, we knew that individually identified humpback whales were female because we had documented them in previous years with calves (NPS and UAS unpublished data). In one case, we knew that an individually identified humpback whale was male based on genetic analysis (NPS and UAS unpublished data).

We assigned the whale in each report to one of the following age classes: calf, juvenile, adult, or unknown. We based most of our assessments on empirical measurements of dead whales' lengths using guidelines from the scientific literature for each species [41–45]. For humpback whales, we defined calves as <1 year old and juveniles as whales ≥1 year old but <5 years old [46]. We determined that one dead individually identified humpback whale was an adult based on its ≥5-year sighting history (UAS unpublished data). We used the following guidelines to classify dead humpback whales based on body length: calves are typically 4–4.5 m in length at birth [47, 48], grow to 7-8 m in length by late summer [49], attain body lengths of 8–10 m at independence [48], and reach sexual maturity (adulthood) at approximately 12 m in length [50]. We classified an 8.2 m humpback whale that was found dead on March 13, 2005 as a juvenile, even though its length fell within the typical range for calves because it was too big to be a calf based on the date it was found. Also, anisakid nematode parasites were found in the whale's small intestines, indicating that it was feeding on fish, not milk ([51]; F. Gulland, pers. comm.).

TABLE 1: Ship strike confidence categories.

Confidence category	Definition
Definite strike	*There is evidence that a strike occurred beyond a reasonable doubt. For example:* Strike was witnessed by the vessel operator/crew or by the operator/crew of a nearby vessel *or* Strike was not witnessed but whale has massive blunt impact trauma (defined by disarticulated vertebrae or fractures of one or more heavy bones including skull, mandible, scapula, vertebra or adult rib, and a focal area of severe hemorrhaging) *or* Strike was not witnessed but carcass has apparent propeller wounds[1] (i.e., deep parallel slashes or cuts into the blubber) on the dorsal aspect *or* Strike was not witnessed but carcass has propeller wounds on the ventral and/or lateral aspect which a necropsy confirms were produced ante mortem *or* Strike was not witnessed but carcass has an amputated appendage (e.g., fluke or flipper) which a necropsy confirms occurred ante mortem due to a sudden and traumatic laceration (versus an entanglement injury causing a slow, ischemic loss of the appendage) *or* Strike was not witnessed but evidence of a collision was found on the vessel (e.g., whale skin or tissue) *or* Whale was found on the bow of a ship .... >Subcategory: Whale struck stationary vessel Vessel was stationary at the time of the collision (i.e., anchored or drifting)[2]
Probable strike	*The report is likely to be true; having more evidence for than against, but some evidence is lacking. For example:* Vessel operator/crew or operator/crew of a nearby vessel believes that a strike occurred but cannot confirm the strike with absolute certainty *or* Strike was not witnessed, and the whale is a calf with smaller broken bones (e.g., ribs) that could have been fractured by another animal rather than by a vessel *or* Strike was not witnessed and the whale shows partial evidence of a collision other than as defined under definite strike. For example: (i) Whale has a focal area of severe hemorrhaging but no known broken bones; therefore, it is possible the trauma was caused by another animal rather than by a vessel; (ii) Carcass has propeller wounds on the ventral and/or lateral aspect; however, the necropsy is not able to determine if they were produced ante mortem
Possible strike	*The report may be true; however, a majority of evidence is lacking. For example:* Vessel operator/crew or operator/crew of a nearby vessel believes that a strike may have occurred but there is significant uncertainty *or* Vessel operator/crew or operator/crew of a nearby vessel believes that a strike occurred, while the vessel operator/crew or operator/crew of a nearby vessel believes that a strike did not occur

TABLE 1: Continued.

Confidence category	Definition
	or
	Strike was not witnessed, and the whale shows partial evidence of a collision other than as defined under definite or probable strike, such as damage to an appendage or skin, but the necropsy is incomplete or there is no close examination of the whale (e.g., whale is viewed from a distance only)
Rejected report	*The report is not credible. For example:*
	Third-hand report
	or
	No credible eyewitnesses[3]
	or
	Lacking sufficient detail or documentation to be credible
	or
	Necropsy determines an alternate cause of death

[1]We only included whales with propeller wounds where there was evidence that the strike occurred in Alaska (i.e., the propeller wounds had to be from a strike that was witnessed and/or the propeller wounds had to be fresh (bleeding) or assessed to be fresh by a trained observer.)

[2]We counted collisions involving kayaks and canoes under this subcategory unless the kayak/canoe was known to be traveling at >0 kn.

[3]The credibility of the eyewitness(es) was assessed on a case-by-case basis. The most credible eyewitness is someone who had "something to lose" in reporting the collision (e.g., the captain and/or the crew of the vessel that struck the whale) because it is presumed they would not risk reporting the collision if it had not occurred. The least credible eyewitness is a passenger on a commercial vessel (e.g., whale watch vessel, cruise ship, etc.) who reports a collision, but there is no supporting evidence (photos, observation of wound, blood, etc.) or other eyewitnesses. In these cases, the report was rejected unless the passenger was an experienced observer and/or additional eyewitnesses were available to corroborate the report (assessed on a case-by-case basis).

Most observations of live whales were classified as age class unknown; however, we classified two live sightings of humpback whales made by knowledgeable observers as calves based on their close, consistent affiliation with an adult whale, presumed to be the mother (after [52]; J. Neilson, pers. obs.; commercial whale watch captain, pers. obs.). Similarly, we classified one live sighting of a humpback whale as a juvenile based on the animal's very small body size (J. Neilson, pers. obs.). We determined that three live individually identified humpback whales were adults based on their ≥5 year sighting histories (NPS, UAS, and Kewalo Basin Marine Mammal Laboratory unpublished data).

2.3. Vessel Characteristics. We assigned each report to one of the following vessel categories: private recreational, non-motorized recreational (e.g., kayaks and canoes), commercial recreational (e.g., charter vessels, tour boats, and commercial whale watch vessels), cruise ship, cargo (e.g., oil tankers, container ships, and landing craft), commercial fishing, research, USCG cutter, state ferry, or unknown. After Laist et al. [8], we classified vessel lengths as small (<15 m), medium (15–79 m), large (≥80 m), or unknown. We searched the USCG's Port State Information Exchange (PSIX) online database [53] and commercial vessel operator's websites to fill in missing vessel lengths when the vessel name was reported.

We evaluated the vessel's activity *prior to the collision* by assigning each record to one of the following categories: anchored or drifting with engine off, slow travel (<12 kn), fast travel (≥12 kn), travel at unknown speed, whale watching, intentionally approaching whales (e.g., whale research), intentionally ramming whales, commercial longline fishing and unknown. Similarly, we evaluated the vessel's activity

at the time of the collision by assigning each record to one of the following categories: anchored or drifting silently, slow travel (<12 kn), fast travel (≥12 kn), decelerating from fast travel, decelerating from unknown speed, travel at unknown speed, and unknown. We classified vessel speed at the time of the collision as anchored or drifting, 1–11 kn, ≥12 kn, or unknown. Separating vessel activity into these two components allowed us to link particular vessel behavior with collision risk and to assess the outcome of the collision with some knowledge of the force with which the whale was struck.

2.4. Fate of Whales. We evaluated the fate of the whale after the collision by assigning each report to one of the following categories: minor injury (presumably not life threatening—e.g., no blood reported in water), severe injury (potentially life threatening—e.g., blood reported in water), dead, or unknown. We described dead whales' injuries as unknown, blunt trauma, or sharp trauma [54, 55].

2.5. Human Toll and Property Damage from Collisions. We assessed the human toll and/or property damage resulting from each collision by counting the number of reports in which passengers onboard the vessel were knocked down, injured, or thrown into the water. To avoid double-counting reports, passengers who were knocked down and injured were only counted as injured. However, passengers who were injured and thrown into the water were counted in both categories because we were interested in the frequency of both of these two outcomes. We also counted the number of reports in which there was significant damage to the vessel or the vessel sank. We defined significant damage as that which required repairs for continued use of the vessel.

2.6. Collision Hotspots. We used the kernel density analysis tool in ArcGIS 10.0 (ESRI Inc., Redlands, CA, USA) to identify potential high risk areas for whale-vessel collisions in southeastern Alaska. Only collisions that were witnessed at sea were included in the analysis. Dead whales where no collision was reported (including bow-caught whales where the collision was not witnessed), were excluded because the location where they were found may not be the same as the location where they were struck [8]. We set the output raster cell size to be 100 m and the search radius (kernel bandwidth) to be 20 km. For clarity, raster cell values representing extremely low collision densities (<0.0025 collisions per km^2) were excluded from the map. The remaining raster cell values (range 0.0025–0.0211 collisions per km^2) were manually divided into 32 equal classes and displayed in colors ranging from yellow (moderate collision risk) to red (higher collision risk).

3. Results

We verified 108 and rejected 11 reports of whale-vessel collisions in Alaska waters between 1978 and 2011. The 11 rejected reports were not included in further analyses. Most strikes ($n = 93$, 86%) involved humpback whales, although six other species were documented (Table 2, Appendix 1 in supplementary material) (Supplementary Material will be available online at doi:10.1155/2012/106282). In eight reports (7%) the species was uncertain; however, we assigned seven of these records to humpback whales and one record to a Cuvier's beaked whale. In one report, a pair of humpback whales, thought to be a cow and calf, were involved in a collision but it was unknown which animal was hit; we counted this as one strike, not two, with the sex and age class of the struck whale unknown.

We found a significant increase in the number of reports over time between 1978 and 2011 (regression, $r^2 = 0.6999$, df $= 32$, $P < 0.001$). Most strikes ($n = 98$, 91%) occurred in May through September and there were no reports from December or January. The majority of strikes ($n = 82$, 76%) were reported in southeastern Alaska (Figure 1), where the number of humpback whale collisions increased 5.8% annually from 1978 to 2011.

Most reports ($n = 86$, 80%) were based on collisions witnessed at sea, while the remaining 22 reports (20%) were based on dead whales where no collision was reported. The geographic location of the 22 dead whales and the dates when they were found did not correlate with any of the witnessed collisions; therefore, we do not believe we double-counted any of these reports. Three of the collisions witnessed at sea are known to have resulted in mortalities, for a total of 25 dead whales.

3.1. Ship Strike Confidence Categories. The majority of reports ($n = 89$, 82%) were assessed to be definite strikes and in 15 (17%) of these cases, a whale struck a stationary vessel. Seventy-nine (89%) of the 89 definite strikes were based on witnessed collisions, and 10 reports (11%) were based on dead whales where no collision was reported. Two (22%) of the nine probable strikes were based on witnessed collisions,

and seven reports (78%) were based on dead whales where no collision was reported. Five (50%) of the 10 possible strikes were based on witnessed collisions, and five reports (50%) were based on dead whales where no collision was reported.

Two of the nine probable strikes were thoroughly investigated, but seven reports were not and may have been upgraded to definite strikes with more complete follow-up (e.g., complete necropsies). In one of the two probable strikes that were witnessed, a dead humpback whale washed ashore within 3 km of where a 190 m cruise ship transiting at an unknown speed reported striking what they believed to be a whale three days earlier; however, there was no close examination of the whale [56]. Similarly, three of the 10 possible strikes were thoroughly investigated, but seven reports were not and may have been upgraded to definite strikes with more complete follow-up. For example, two of the vessel operators involved in witnessed collisions were not interviewed, and four of the five dead whales were not necropsied or examined closely. The fifth dead whale was necropsied; however, the necropsy did not get down to bone to look for fractures diagnostic of a collision.

3.2. Sex and Age Class of Struck Whales. Nine of the 25 dead whales were female, nine were male, and seven were of unknown sex (Table 3). In addition, we documented three live individually identified humpback whales (two females and one male) for a total, of 21 whales of known sex (10 males and 11 females).

There were 25 whales of known age involved in collisions: seven calves, seven juveniles, and 11 adults. Five dead whales were calves, six were juveniles, eight were adults, and six were of unknown age (Table 3). In addition, six humpback whales in witnessed collisions were assigned to age classes (two calves, one juvenile, and three adults). Six adult female humpback whales are known to have died from collisions and four of these mortalities occurred in southeastern Alaska between 2001 and 2011.

3.3. Vessel Characteristics

3.3.1. Vessel Type. In 19 cases, the type of vessel involved in the collision was unknown (18 were dead whales where no collision was reported, but one was a witnessed collision where the type of vessel was not recorded.) In the 89 cases where the vessel type was known, 35% ($n = 31$) were private recreational, 35% ($n = 31$) were commercial recreational, 8% ($n = 7$) were cruise ships, 7% ($n = 6$) were commercial fishing vessels, 4% ($n = 4$) were USCG cutters, 3% ($n = 3$) were cargo, 3% ($n = 3$) were nonmotorized recreational, 3% ($n = 3$) were research, and 1% ($n = 1$) was a state ferry. The three cargo vessels were a 254-m oil tanker, a 216-m container ship, and a 10-m landing craft. The seven cases where the vessel type was known and the whale died involved large cruise ships ($n = 5$) or cargo vessels ($n = 2$; one container ship and one oil tanker). All three non-motorized recreational vessel strikes occurred in Glacier Bay.

3.3.2. Vessel Length. In 44 reports (41%) vessel length was not reported; however, in 18 of these cases we were able to

TABLE 2: Summary of whale-vessel collisions reported in Alaska 1978–2011. Rejected reports are not included.

Species	Confidence category			Total	Number of known dead
	Definite strike	Probable strike	Possible strike		
Humpback whale	78	8	7	93 (86.1%)	17
Fin whale	3			3 (2.8%)	2
Gray whale	1			1 (0.9%)	1
Sperm whale	1			1 (0.9%)	
Cuvier's beaked whale		1	1	2 (1.9%)	2
Stejneger's beaked whale			1	1 (0.9%)	1
Beluga whale			1	1 (0.9%)	1
Unidentified whale	6			6 (5.6%)	1
Total	89 (82.4%)	9 (8.3%)	10 (9.3%)	108 (100%)	25

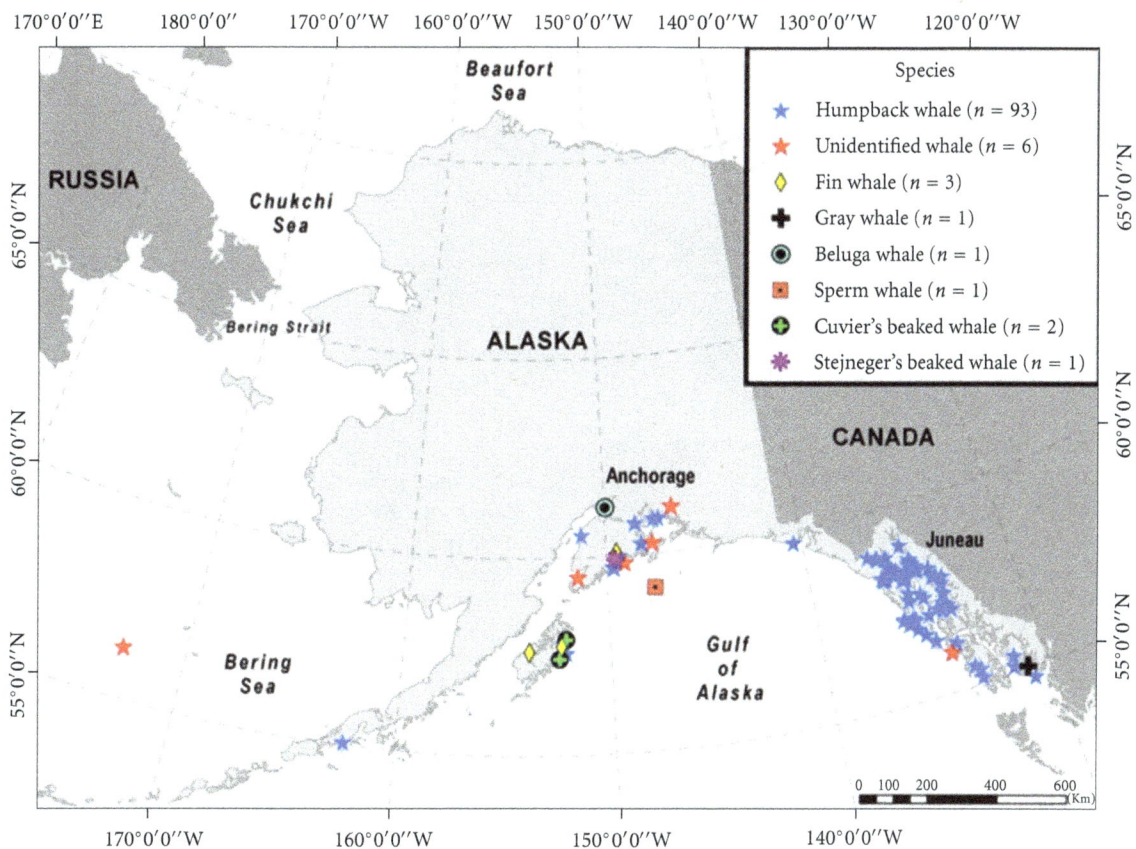

FIGURE 1: Location of whale-vessel collision reports in Alaska by species 1978–2011 ($n = 108$). Rejected reports are not included.

infer the vessel's length category based on the vessel type ($n = 4$, e.g., kayak and Zodiac) or look up the vessel length online using the vessel's name ($n = 14$). This left 26 reports where vessel length was unknown. Eighteen of these cases were dead whales where no collision was reported; however, eight were witnessed collisions, of which five were reported to and/or investigated by federal law enforcement officials.

In the 82 reports where vessel length was known (range 5–294 m), small (<15 m) vessels were the most commonly reported ($n = 49$, 60%), followed by medium (15–79 m) vessels ($n = 22$, 27%), and large (≥80 m) vessels ($n = 11$, 13%). The difference in the number of reports in each vessel length category is significant (chi-square test for goodness of fit, $\chi^2 = 27.976$, df = 2, $P < 0.001$).

TABLE 3: Sex and age classes of the 25 whales known to have been killed by vessels.

Species	Male			Female			Sex unknown		Total
	Calf	Juvenile	Adult	Calf	Juvenile	Adult	Adult	Age class unknown	
Humpback whale	4	2	1	1		6		3	17
Fin whale		1			1				2
Gray whale					1				1
Sperm whale									
Cuvier's beaked whale		1						1	2
Stejneger's beaked whale							1		1
Beluga whale								1	1
Unidentified whale								1	1
Total	4	4	1	1	2	6	1	6	25

3.3.3. Vessel Activity Prior to Collision. In 18 reports (17%) the vessel's activity prior to the collision was unknown or not reported. In the 90 reports where the vessel's activity was known, 44% ($n = 40$) were engaged in fast travel, 16% ($n = 14$) were anchored or drifting silently, 14% ($n = 13$) were engaged in slow travel, 12% ($n = 11$) were traveling at an unknown speed, 7% ($n = 6$) were whale watching, 3% ($n = 3$) were intentionally approaching whales, 2% ($n = 2$) were intentionally ramming whales, and 1% ($n = 1$) were commercial fishing. Note that whale watching vessels that were traveling prior to the collision were classified under one of the traveling vessel activity categories. The difference in the number of reports in each vessel activity category is significant (chi-square test for goodness of fit, $\chi^2 = 99.867$, df $= 7$, $P < 0.001$).

3.3.4. Vessel Activity at Time of Collision. In 19 reports (18%) the vessel's activity at the time of the collision was unknown or not reported. In the 89 reports where the vessel's activity was known, 33 (37%) were engaged in fast travel, 19 (21%) were engaged in slow travel, 15 (17%) were anchored or drifting silently, 12 (13%) were traveling at an unknown speed, 9 (10%) were decelerating from fast travel, and one (1%) was decelerating from slow travel. The 10 vessels that reported decelerating did so in response to seeing the whale just prior to the collision, thus in some cases, their speed at the time of the collision (below) is lower (i.e., 1–11 kn versus ≥12 kn). The difference in the number of reports in each vessel activity category is significant (chi-square test for goodness of fit, $\chi^2 = 39.157$, df $= 5, P < 0.001$). All 15 of the cases where a whale struck a stationary vessel involved humpback whales hitting vessels that were anchored or drifting with their engine off.

3.3.5. Vessel Speed at Time of Collision. In 47 reports (44%) vessel speed at the time of the collision was unknown or not reported; however, in 14 of these cases we were able to infer the vessel's speed based on other information in the report (e.g., "sailboat under power" was classified as 1–11 kn). This resulted in 75 reports (69%) where vessel speed was known (range 0–35 kn), with vessels, traveling at ≥12 kn the most commonly reported ($n = 37$, 49%), followed by vessels traveling at 1–11 kn ($n = 23$, 31%), and anchored or

drifting vessels ($n = 15$, 20%). The difference in the number of reports in each vessel speed category is significant (chi-square test for goodness of fit, $\chi^2 = 9.92$, df $= 2$, $P < 0.05$).

Twenty-two of the 33 cases (67%) where vessel speed was unknown were dead whales where no collision was reported; however, 11 (33%) were witnessed collisions in which speed was not recorded. The maximum speed reported (35 kn) was a 10 m jet boat whose operator intentionally rammed a pair of humpback whales thought to be a cow and calf [57].

3.4. Fate of Whales. In most cases ($n = 78$, 72%), the fate of the whale following the collision was unknown, but 25 cases (23%) were known mortalities, and in five cases (5%) the whale was documented alive in subsequent months or years using individual identification techniques (NPS and UAS unpublished data).

3.4.1. Minor Injuries. In 11 cases (10%) the whale was observed with either a presumably minor injury or no visible injuries (all were humpback whales). Five of these whales are known to have survived; however, the fate of the other six whales is unknown. The five surviving whales (one calf, three adults, and one age unknown) were hit by vessels <20 m in length (range 7–19.8 m) traveling at 5 kn ($n = 1$), 10 kn ($n = 2$), 25 kn ($n = 1$), and an unknown speed ($n = 1$). The latter vessel was whale watching and was therefore likely traveling at 1–11 kn. Three of the whales had blunt trauma injuries after being struck by the bows of vessels and two had sharp trauma injuries from propellers. The collision that occurred at 25 kn was reported in 2008 by the captain of a 10 m aluminium tour boat after he struck a humpback whale as the whale came up to breathe [58]. The captain believed that the whale he hit was an individual with a uniquely marked dorsal fin that was well known to tour boat captains in the area. The speed of the vessel decreased approximately 3-4 kn after the strike, and he did not see the whale come up again, but it is unknown how long the vessel stayed on scene. Later that day, this uniquely marked adult whale was documented behaving normally and lunge feeding nearby (NOAA unpublished data), and it was observed as recently as 2011 with no visible injuries (NPS unpublished data). The calf that was struck was documented alive with its mother 75 days after being hit by an 18 m commercial fishing vessel transiting at 10 kn (NOAA unpublished data). The other three whales that are known to have survived have been documented for a minimum of six years post-collision (NPS unpublished data).

3.4.2. Severe Injuries. In five cases (5%) the whale was observed after the collision with a severe injury (three humpback whales and two unidentified large whales); however, the fate of these whales is unknown. In four of these cases, blood was reported in the water. Three of these whales had sharp trauma injuries from propellers, while the type of injury sustained by the fourth whale was unknown. In the fifth case, a humpback whale punched a 1.5 m hole through the hull of an anchored 22 m wooden sailboat, sinking the vessel and leaving six plates of baleen measuring approximately 0.3 m in length held together by torn flesh inside the splintered hull.

3.4.3. Mortalities. In 25 cases (23%) the whale is known to have died, but vessel length and speed were known in only three of these cases. Two of the 86 collisions witnessed at sea that are known to have caused mortalities (both were adult humpback whales) involved 232 m and 243 m cruise ships traveling 14 kn and 19 kn, respectively. In a third case, a dead humpback whale was found on the bow of a 216 m container ship, and the vessel's speed at the time of the collision is unknown; however, its typical transit speed was 12–19 kn. Statewide, humpback whale vessel-strike mortalities peaked in 2010 ($n = 4$) and we found an increasing trend in the number of humpback whales killed between 1978 and 2011 (regression, $r^2 = 0.1193$, df $= 32$, $P < 0.05$).

Thirteen (52%) of the 25 dead whales were first reported floating: five were towed to shore for examination, five are known to have washed ashore on their own, and three were not towed and floated away. Seven (28%) of the dead whales were first reported beach-cast. Five of the dead whales (20%) were caught on the bulbous bows of large ships (three humpback whales, one fin whale and one unidentified large baleen whale in 2009 that appeared to be a fin, blue, or sei whale). One of the humpback whales slipped off a 243 m cruise ship's bow and sank when the ship slowed down, the other four bow-caught whales remained pinned to the ships' bows (288 m cruise ship, 294 m cruise ship, 254 m oil tanker, and 216 m container ship) until they came into port or stopped. The state of decomposition and point of collision impact on two of the whales is unknown. However, the fin whale and two of the humpback whale carcasses were fresh (not bloated) and appeared to have been struck on the dorsal side of their bodies, indicating that the whales were alive when they were hit [9, 54]. This is inferred because most large whales (except right and bowhead whales) sink when they die and then rise to the surface, ventral side up, as decomposition gases inflate the abdomen (assuming the abdominal cavity is intact and the carcass is in relatively shallow water) [18, 59, 60]. Depending on blubber thickness, some whales may float immediately upon death; in these cases, they typically will float ventral side up within approximately 24 hours as decomposition gases inflate the abdomen (F. Gulland, pers. comm.). Therefore, collision injuries on the dorsal side of a

TABLE 4: Types of injuries sustained by the 25 whales known to have been killed by vessels.

Species	Blunt trauma	Sharp trauma	Unknown injuries
Humpback whale	12	2	3
Fin whale	2		
Cuvier's beaked whale	1	1	
Stejneger's beaked whale	1		
Gray whale		1	
Beluga whale		1	
Unidentified whale			1
Total	16 (64%)	5 (20%)	4 (16%)

whale provide indirect evidence that the whale was alive (or extremely recently dead) when it was struck, otherwise the point of collision impact would be expected on the whale's ventral or lateral side ([54], F. Gulland, pers. comm.).

The first whale necropsy conducted in Alaska with a veterinarian trained in assessing ship strike injuries occurred in 2001. Since then, numerous veterinarians, stranding team members and other personnel have gained experience in assessing ship strike injuries and 13 more necropsies have found evidence that whales died from collisions. However, six of these necropsies were incomplete, meaning that the carcass was not flensed down to the bone to look for fractures. In several cases, the necropsy team ran out of time as the incoming tide covered the carcass. Overall, 11 humpback whales, two fin whales, and one Cuvier's beaked whale with ship strike injuries have been necropsied since 2001.

Most of the 25 dead whales ($n = 16$, 64%) had blunt trauma injuries, five (20%) had sharp trauma injuries, and four (16%) had unknown injuries because they were not necropsied (Table 4); however, at least three (two humpback whales and one unidentified large baleen whale that appeared to be a fin, blue, or sei whale) likely suffered from blunt trauma because they were found pinned to ships' bows. The fourth whale stranded in 1978 after a cruise ship reported striking what they believed to be a whale; however, there was no close examination of this humpback whale, and it is unknown if the ship's bow or propeller(s) struck the whale.

The necropsy of an adult female humpback whale found on the bow of a 288 m cruise ship in 2010 revealed a potentially complicated history [61]. A necropsy was conducted, and both gross and internal assessments of the carcass were made; however, the necropsy was limited by an incoming tide. Though it was not possible to strip the carcass entirely to the bone, the animal was found to have a sharp trauma injury (amputated pectoral flipper cut cleanly at 0.8 m in diameter), acute degenerative myopathy in several muscle tissues (indicating severe ante mortem stress and muscle exertion), and a large area of missing inframandibular tissue, indicating that the whale may have been fed on by killer whales. Elevated saxitoxin levels were also detected, which could have caused the whale to behave abnormally, making it more vulnerable to being struck. It has been proposed that the whale may have been struck initially by a different large vessel, shearing off the pectoral fin and causing debility and/or death, followed by possible predation by killer whales, and eventual postmortem entrapment on the bow of the cruise ship. However, in initial photos of the carcass on the ship's bow, the whale does not appear to be bloated and the point of collision impact is on the dorsal thorax, indicating that the whale may have been alive when it was struck. We include the details of this particular report to illustrate the complexities involved in piecing together case histories and determining cause of death.

3.5. Human Toll and Property Damage from Collisions. In 37 reports (34%) the passengers and vessel were not affected by the collision, in 36 reports (33%) there was some kind of human toll and/or property damage resulting from the collision, and in 35 reports (32%) the outcome of the collision for the passengers and vessel is unknown. There were 19 reports in which passengers were knocked down (affecting a minimum of 41 people), 10 reports in which passengers were injured (affecting a minimum of 18 people), 9 reports in which passengers were thrown into the water (affecting a minimum of 15 people), 20 reports of significant property damage, and three reports of private recreational vessels sinking. Two of the vessels that sank (a 10 m fiberglass sailboat and a 22 m wooden sailboat) were anchored or drifting with their engine off when they were rammed by humpback whales. The third case involved a 8 m polyethylene powerboat that sank after striking an unidentified large whale while transiting at 19 kn.

3.6. Collision Hotspots. We identified several high risk areas for whale-vessel collisions in southeastern Alaska (Figure 2). All of the high risk areas were located in the northern portion of southeastern Alaska. The areas with the highest collision densities centered around Point Adolphus in Icy Strait and around North Pass in lower Lynn Canal, both popular whale watching destinations. Medium-risk areas centered around the Inian Islands in Cross Sound and in Sitka Sound. Other areas where we identified a collision risk included eastern Icy Strait near Hoonah, the lower West Arm of Glacier Bay, upper Stephens Passage, and eastern Frederick Sound.

4. Discussion

The great majority of ship strikes in Alaska occur with humpback whales in southeastern Alaska. This area is primarily comprised of protected waters and supports a genetically distinct feeding aggregation of 3,000–5,000 humpback whales [31]. The number of humpback whale collisions detected in this region increased by 5.8% annually from 1978 to 2011, which closely matches the 6.8% annual growth rate of the humpback whale population in southeastern Alaska between 1986 and 2008 [62]. Although the problem at present may not be resulting in population level impacts, a collision with a large whale is considered a "take" under the MMPA and is therefore a cause for concern, as are other considerations such as human safety. Our results showing

an increase over time in whale and vessel collisions are susceptible to several biases inherent in the dataset, yet we believe that this conclusion is valid based on the seasonal overlap of high densities of humpback whales and vessels and an increasing whale population trend in southeastern Alaska.

4.1. Reporting Biases. Although we attempted to capture all whale-vessel collisions throughout Alaska, the number we report here represents a minimum level of occurrence due to under-reporting of witnessed collisions and the significant challenges involved in investigating cause of death in whale mortalities in a large and remote state. We know that under-reporting of witnessed collisions occurs; for example, a survey of recreational boaters in southeastern Alaska documented that at least three out of four whale-vessel collisions in this region were not reported (J. Straley, pers. comm.), and similar rates of under-reporting have been found among professional mariners in Hawaii [63]. This lack of reporting could be due to fear of possible repercussions or simple ignorance that collisions should be reported to NOAA. In 2009, NOAA implemented a toll-free Marine Mammal Stranding Hotline in Alaska, which increased public awareness about the existence of a stranding network and the agency's interest in collecting ship strike information and may have led to an increase in reports in recent years. One only has to engage in casual conversation with nearly any Alaskan boater to hear anecdotal stories of whale strikes that happened to them or someone they know. Most of these reports lack so many critical details such as vessel speed, location, and the fate of whale that although they would contribute to a better understanding of the true frequency of whale-vessel collisions, they might not advance our knowledge of the specific factors leading to collisions or their outcomes.

We documented collisions with seven of the 14 whale species known to occur in Alaska, with 86% of the reports involving humpback whales and none involving bowhead, minke, blue, sei, North Pacific right, Baird's beaked, or killer whales. We recognize that the records compiled here may be biased towards humpback whales because the authors are based in southeastern Alaska; however, the overwhelming number of live and dead reports involving humpback whales indicates that they are the most heavily impacted species, at least in terms of absolute numbers. The seasonal trend in collisions, with 91% of reports occurring in May through September, is not surprising because these are the months when humpback whales, which migrate in winter to lower latitudes, are most common in Alaska. The number of humpback whales that are known to have died from collisions in Alaska ($n = 17$) is much higher than in Washington from 1980–2006 ($n = 1$) [18] or British Columbia from 1995–2007 ($n = 0$) [64], despite both areas being important summer habitat for this species. The reason for this difference is unknown, but Douglas et al. [18] were surprised by the virtual absence of dead ship-struck humpback whales in Washington.

When a dead whale is reported in Alaska, there are limited resources and personnel to respond and conduct a

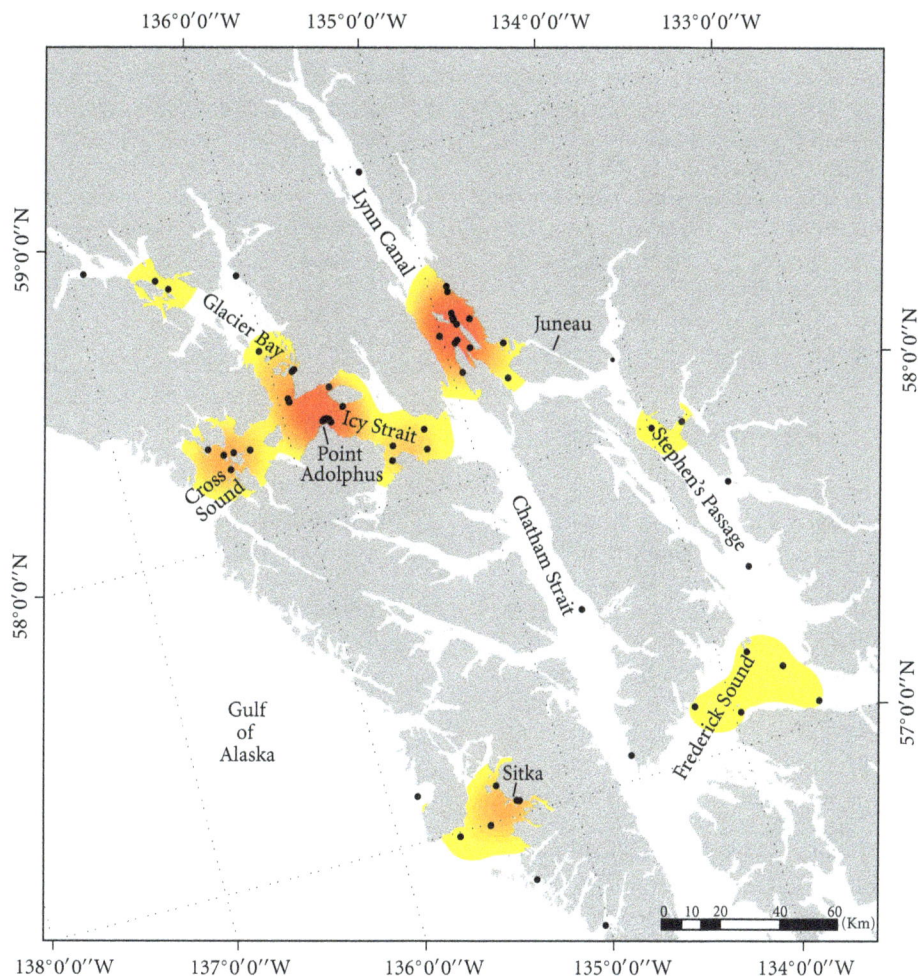

FIGURE 2: Whale-vessel collision hotspots in southeastern Alaska based on kernel density estimation. Yellow indicates moderate collision risk and red indicates higher collision risk. The locations of the collision reports used to create the map are displayed.

necropsy. Whether a necropsy is pursued or not depends on a variety of factors, including the condition of the carcass (ranging from fresh to skeletal), location and accessibility, safety, weather, available expertise, and whether the carcass is secured (such that it will not be washed away by the tide before a team can respond). Priority may be given to species listed under the ESA, species which are rarely encountered or for which little data exists (i.e., beaked whales), or incidents where there is a likelihood of human interaction (e.g., suspected ship strike, entanglement, shooting, etc.).

We were surprised to find so few collisions with fin whales ($n = 3$) given their abundance and widespread distribution in many parts of Alaska [28], especially compared to other studies, which have found them to be the most common species struck by vessels [8, 9, 18]. One reason so few fin whale collisions were observed is that fin whales are rare in the inside waters of southeastern Alaska frequented by vessels and occur more commonly offshore where a dead whale is less likely to be noticed. Collisions with gray whales ($n = 1$), sperm whales ($n = 1$), and killer whales ($n = 0$) were also rare compared to trends documented elsewhere [8, 18, 64].

Although vessel collisions with beaked whales have been documented in other areas [15, 17, 65], we were surprised to find three fatal strikes involving two Cuvier's beaked whales and one Stejneger's beaked whale because these species are observed rarely and typically inhabit offshore waters [27, 28].

We acknowledge several geographic biases in the records we compiled. The records are likely biased towards southeastern Alaska and there is a statewide bias towards human population centers (e.g., Juneau, Anchorage, Seward, and Kodiak) where there are more observers on the water. However, this goes hand-in-hand with more vessels on the water, so these areas probably do have a higher collision risk. Our dataset, like other ship strike datasets [9], is biased towards species that inhabit near shore waters, such as humpback whales, because carcasses of near shore species are more likely to be found (and subsequently examined) compared to offshore species. Furthermore, carcasses reported floating far offshore are unlikely candidates for towing to shore for necropsies given the long distances involved. In addition, whales that die offshore in water depths greater than 1,000 m may not float to the surface because the hydrostatic pressure at these depths limits the generation of buoyant decomposition gases [60]. There is generally more vessel traffic in near shore areas compared to offshore areas, which likely puts near shore species at a higher risk for collisions. We propose that a better understanding of the geographic extent of ship strikes in Alaska could be obtained by an effort to actively solicit information about past events from resource managers, law enforcement officers, the media, and the maritime community throughout the state. Maintaining NOAA's current focus on systematic data collection about ship strikes as they occur will also help facilitate equal representation of all parts of Alaska.

4.2. Ship Strike Confidence Categories.

We recommend that the standardized system we developed to classify collision records into four confidence categories (definite strike, probable strike, possible strike, and rejected report), or a similar detailed system, be universally adopted to reduce uncertainty in interpreting ship strike data. Other investigators have employed similar tiered classification systems for dead whales with evidence of ship strike injuries (e.g., [10, 13, 15, 16, 18]); however, our definitions contain a higher level of detail which we feel makes our approach more useful as a classification tool. Also, unlike other classification systems, ours includes criteria for classifying eyewitness collision reports. We recognize that our definitions err on the side of classifying strikes as definite when it is possible that some of these collisions occurred postmortem; however, we surmise that postmortem strikes of large whales are unlikely and consequently rare, given that floating carcasses are, in most cases, significantly bloated and therefore highly visible to the naked eye and radar ([15]; F. Gulland, pers. comm.). To rule out misclassifying postmortem collisions, we recommend that whenever possible, samples from stranded whales be collected and analyzed using histochemical techniques that can detect fat emboli diagnostic of ante mortem bone fractures and severe soft-tissue damage [66, 67].

We acknowledge that including possible and probable strikes in our analyses positively biased the number of ship strike records; however, like Van Waerebeek et al. [15], we chose to include these reports in our analysis because (a) we are trying to quantify a problem that we know is under-reported and (b) we feel that the confidence codes are generally conservative, meaning that the majority of the probable and possible strikes are likely to have been genuine collisions but incomplete followup and/or necropsies precluded classifying many of them as definite strikes.

Along with a standardized system to evaluate the level of certainty associated with individual collision reports, we also recommend that a universal standardized reporting form for collisions witnessed at sea would improve the quality of ship strike data by reducing the number of reports lacking key information such as vessel size and speed at the time of the collision. An outreach campaign to the public and to the resource protection agency personnel likely to respond to reported collisions will help ensure systematic reporting of the salient details of collisions. Currently, the NOAA National Marine Mammal Stranding Database does not accept ship strike records, thus recording and cataloguing methods vary among NOAA regions across the country. Regions typically record collision reports on a general marine mammal stranding reporting form rather than using a specialized ship strike reporting form that prompts for key details. The latter approach is needed both nationally and internationally to ensure more systematic documentation of collisions and their outcomes.

4.3. Sex and Age Class of Struck Whales.

We did not detect any difference in the collision risk for male versus female whales, but did find that calves and juveniles appear to be at higher risk of collisions than adult whales, which is consistent with other studies [3, 7, 8, 13, 14, 17, 18]. Our age class data are biased towards dead whales; therefore, it is unknown if young animals are overall more likely than adults to be struck (based on differences in their behavior, sightability, or other factors). It is also plausible that young animals are more likely to die from collisions because of their smaller body size.

4.4. Vessel Characteristics.

All types and sizes of vessels collided with whales; however, small (<15 m) recreational vessels were the most common. This result contrasts with other studies that have concluded that small vessels are less likely than larger vessels to strike whales [8]. We found more private recreational vessel strikes and fewer commercial recreational vessel strikes than in Hawaii, where the majority of recorded collisions with humpback whales involved commercial whale watch vessels [14], but this result might be biased by different numbers of private versus commercial whale watch vessels in the two areas. The number of large vessels that we documented is presumably an underestimate because compared to smaller vessels, the crews of large vessels may be less likely to see collisions when they occur due to limited visibility around their bows, and the impact of a collision is less likely to be felt in larger vessels [8]. Undetected collisions with large vessels presumably account for some of the 22 cases where dead whales were found but no collision was reported. Alternatively, these collisions may have been witnessed but not reported. We recognize that the majority of records were based on witnessed collisions and that our conclusions regarding the types of vessels that hit whales are likely to be biased by different reporting and detection rates between vessel types. For example, some user groups may be more wary of reporting collisions to federal officials, and overall some user groups are more aware than others that collisions should be reported. For instance, in recent years, there has been a high level of awareness in the Alaska cruise ship industry about whale collision avoidance and reporting, but other user groups may not be as aware of the issue, leading to under-reporting.

It is notable that all 15 stationary vessels that were struck by humpback whales were drifting with their engine off or anchored. This suggests that the whales did not detect the vessels and that being in a silent vessel may increase the risk of a collision. Further evidence comes from a study of sailing vessel collisions with cetaceans, which found that 79%

of collisions occurred when the vessels were under sail, as opposed to motoring [19]. Many boaters erroneously assume that whales are aware of their presence and location at all times. Increasing public outreach and education programs that emphasize that sperm whales are the only large whale species that uses echolocation could be beneficial in reducing collision risk.

4.5. Fate of Whales.
Our data support previous findings that collisions are more likely to be lethal when they involve large ships and higher vessel speeds [8, 37]. In the three mortalities where both vessel length and speed were known, the ships ranged from 216–243 m in length and were traveling 12–19 kn. Four other mortalities involved 190–294 m ships traveling at unknown speeds. In addition to these seven mortalities, there were eight more dead whales whose massive injuries (e.g., fractured skulls) indicate that they were likely struck by large ships in collisions that were either not detected or witnessed but not reported. Conversely, four of the five cases in which struck humpback whales were known to have survived provide evidence that collisions with smaller, slower moving vessels are less likely to inflict serious or fatal injuries [8, 37]. We know of at least 23 other humpback whales in southeastern Alaska that have survived collisions based on live sightings of 15 different whales with healed propeller wounds and eight whales with deep gashes and other wounds that appear to be from vessel collisions (NPS, UAS, NOAA, and Alaska Whale Foundation unpublished data). The vessel types, sizes, and speeds involved in these nonfatal collisions are unknown, but all of the propeller wounds appear to be from relatively small vessels based on the size and close spacing of the propeller scars.

The majority (80%) of the collision records were based on strikes witnessed at sea, with the fate of the whale unknown in most (72%) cases. However, over half (49%) of the witnessed collisions occurred at vessel speeds ≥ 12 kn, and therefore some of these collisions may have been fatal, though the smaller size of most of the vessels presumably means that lethal collisions were less likely [8, 37]. We found that vessel operators are often exceeding a "slow, safe speed" near humpback whales as required in Alaska [24] and that overall, vessels engaged in fast travel are at a greater risk of striking a whale. In the majority of cases, the collisions were accidental, with little or no time for evasive action. In a few cases ($n = 10$), vessel operators reported decelerating just prior to hitting the whale.

4.5.1. Mortalities.
We found blunt trauma injuries (e.g., broken bones and a focal area of hemorrhaging) to be more than three times as common as sharp trauma injuries (e.g., propeller wounds) in whales that died from ship strikes in Alaska, whereas propeller injuries dominate among dead ship struck right whales along the US Atlantic and South African coasts [8] and gray whales in Washington [18]. Blunt trauma injuries were prevalent in ship struck balaenopterids examined in Washington [18] and in ship struck fin, blue, and sei whales along the US Atlantic and French coasts

[8]. Models indicate that whales at the water's surface are more likely to be hit by the bows of ships than whales submerged near the surface, which are more likely to suffer propeller strikes [36]. The majority (12 of 16) of the blunt trauma injuries in our sample were sustained by humpback whales. Humpback whales in Alaska typically make short, shallow dives [68] and spend a relatively high proportion of their time feeding, socializing, and resting at the surface (NPS unpublished data). This behavior pattern may make humpbacks more susceptible to bow strikes than propeller strikes, explaining why we found more blunt trauma injuries than sharp trauma injuries. In contrast, North Atlantic right whales spend the majority of their time submerged 0.5–2.5 m below the water's surface, which may explain why vessel collisions in general, and propeller injuries in particular, are so common in this species [69]. Douglas et al. [18] proposed two other possible explanations for the greater percentage of blunt traumas found in some species: (1) deep propeller wounds may open the body cavity and make the whale more likely to sink and not be recovered; (2) bow-caught whales (i.e., blunt trauma cases) are more likely to be transported to coastal waters where they can be recovered and examined. Both of these hypotheses may apply to our observations, but neither fully explains our findings. Note that in our dataset, 15 of the 16 whales with blunt trauma injuries were found floating or beach-cast, not bow-caught. However, some of these whales may have been bow-caught originally but then slipped off after the ships slowed down or stopped. Ships displacing 1600 or more gross tons are required to test their forward/astern propulsion within 12 hours of entering or getting underway in US waters [70], which could increase the chances of a bow-caught whale slipping off before it is detected.

A total of five dead whales were reported caught on the bulbous bows of large ships (three humpback whales, one fin whale, and one unidentified large baleen whale that appeared to be a fin, blue, or sei whale). Previously, stocky whale species such as humpback whales were not thought to be susceptible to being pinned to the bows of ships compared to longer, sleeker rorquals such as fin whales [8]. This conclusion was based on a single known case from Alaska of a humpback whale draped over a cruise ship's bulbous bow, and this whale slipped off the bow and sank when the ship slowed down [8, 71]. A second case, reported to have occurred in Alaska in 2006 and cited by Van Waerebeek et al. [15], was misidentified in the media as a bow-caught humpback whale, but this was actually a fin whale [72]. In addition to the single bow-caught humpback whale case already reported in Laist et al. [8], we documented two new verified cases in which humpback whales were caught on the bows of ships. In both cases, the whales did not slip off when the ships stopped; in fact, in one case, it was difficult to dislodge the whale from the bow [61].

The 25 whales that we concluded had died from ship strikes from 1978–2011 represent the minimal number of whale mortalities from ship strikes in Alaska during this time period. Over the same time span, 516 large whales (i.e., baleen whales and sperm whales) were reported dead in Alaska (NOAA Alaska Region Stranding Database unpublished data). Thirty-two (6%) of these carcasses were necropsied, with 13 of the whales classified as ship strikes in this study. Excluding two bow-caught whales (because they are not representative of the typical floating or beach-cast dead whale), 37% (11 of 30) of the large whales necropsied in Alaska since 1978 have died from ship strikes. Similar high rates of ship strike mortalities have been found along the U.S. East Coast in some whale species (e.g., one-third of stranded northern right whales and fin whales) [8]. It is unknown how many more dead sank whales in Alaska were scavenged, floated offshore, and/or sunk without being located, but considering the remoteness of the state's coastline and offshore areas, 516 dead whales presumably represents a small fraction of the true number of dead whales over this 34-year period. Studies in the Gulf of Mexico suggest that on average, only 2% (range 0–6.2%) of cetacean carcasses are recovered [73], and low detection rates (range <1%–17%) have also been documented in several other cetacean species in other areas [74–77]. The high rate of ship strike mortalities in Alaska, as indicated by the available necropsy data (37%), suggests that many ship strike mortalities are likely going undetected in floating and beach-cast whales that are not examined.

In recent years, there has been improvement in the investigation of cause of death in whale stranding mortalities in Alaska, due to increased resources and expertise within the state, from sources such as the Prescott Marine Mammal Stranding Grant Program, the Alaska SeaLife Center, and additional veterinary support within the Alaska Marine Mammal Stranding Network. These improvements may explain some of the apparent increase in humpback whale ship strike mortalities over time. For example, 72% (24 of 32) of the large whale necropsies conducted in Alaska since 1978 occurred between 2001 and 2011 (NOAA Alaska Region Stranding Database unpublished data), which reflects NOAA's increased commitment to necropsy whales over the past decade. Despite these improvements, limited resources and personnel, combined with the logistical challenges of responding to remote carcasses, continue to result in missed opportunities to investigate the cause of death in many whale strandings. While federal resource agencies in Alaska strive to promote and facilitate necropsies led by experienced teams, ideally veterinarians, to investigate cases of whale mortality, additional resources are recommended to increase capacity and infrastructure in necropsy response to improve cause of death investigations. For instance, establishing a statewide network of vessels that are available to tow floating whale carcasses to shore would reduce the number of missed opportunities for necropsies. In many cases, multiday necropsies may be needed to flense a carcass down to bone to examine the skeleton for fractures, especially because necropsy sites in Alaska are generally too remote for heavy equipment to assist with maneuvering large carcasses [1]. It may be beneficial to involve northern Alaska Eskimo subsistence whalers, who are highly skilled in flensing whales without the aid of machines, in large whale necropsy teams. Alternatively, returning to inspect carcasses over time to look for newly exposed broken bones may be helpful, although postmortem damage to bones on weather-beaten shores may confuse matters. Responding to whale strandings in Alaska will always be more challenging than in less remote areas where necropsy rates may be as high as 69% [18], but continuing to increase efforts to perform complete necropsies (e.g., down to bone to examine for fractures, Table 1) using experienced teams focused on determining cause of death [1] is needed to allow for a more accurate determination of the rate of ship strike mortality in Alaska.

Performing full necropsies on ship struck whales is also important because they can reveal underlying factors such as disease, biotoxins, parasites, prior injuries, and entanglements in fishing gear that may have compromised a whale and predisposed it to being hit by a vessel [8, 15]. Researchers investigating northern sea otter (*Enhydra lutris*) mortalities from vessel collisions in Alaska have found that many of the struck otters had underlying health issues such as bacterial infections and biotoxins that may have made them more susceptible to being hit (V. Gill, pers. comm.). In our sample, one adult humpback whale was found to have elevated saxitoxin levels that may have caused it to behave abnormally, which could have made it more vulnerable to being struck [61]. Systematic sample collection in all necropsies to test for an array of underlying factors is needed to gain a better understanding of how often these other stressors may be contributing to collisions. Recognizing that the pathology results from necropsies are often not available for weeks or months after the stranding, detecting the proximate and ultimate causes of vessel strikes will require stranding network personnel ensure that these results are systematically entered into the main record for each stranding in such a way that meta-analyses are possible. Storing these data in a usable fashion may require modifications to the national stranding database structure.

4.6. Human Toll and Property Damage from Collisions. The discovery that one-third of collisions resulted in some kind of human toll and/or property damage highlights that whale-vessel collisions are a human safety issue. To date, there have been no confirmed human fatalities from collisions in Alaska, although in one of the reports we rejected, a 5 m skiff reportedly struck a gray whale, and the operator died after falling into the water [78]. The human fatality was confirmed but we could not confirm that the accident was caused by a collision with a whale. Threats to human safety posed by collisions have been documented elsewhere [8, 17, 19, 79], but the frequency of human injuries and property damage we documented may be positively biased because presumably these cases are more likely to be reported than other collisions. Increased attention to systematic documentation of human injuries and/or property damage in all collision reports is needed to allow for a more quantitative assessment of the problem. Regardless, the number of documented incidents indicates that boaters in Alaska, especially those operating small open vessels where the likelihood of being thrown into the water from a collision is high, would benefit from public outreach and education programs that raise awareness of the risks posed by collisions and how these risks can be minimized (e.g., slow down, keep a sharp lookout for whales, always wear a life-jacket, etc.).

4.7. Management Recommendations. As we have shown in our analyses, the problem of whale-vessel collisions is clearly one that can be detrimental to whales and humans. Conversely, avoiding whale-vessel collisions is mutually beneficial, but the challenge is to understand how best to reach and advise each user group, given the tangle of human factors that influence vessel operators' decisions. These factors include, but are not limited to: economics, convenience, knowledge and tolerance of risk, and whether they are professional or recreational vessel operators. For the professional mariner, the recently published International Whaling Commission and International Maritime Organization collision avoidance leaflet [80] gives practical advice (e.g., pay attention, avoid areas where you know there are whales, and slow down) in an appealing and respectful format. Available on the internet in six languages, this leaflet also highlights the importance of reporting collisions to foster an understanding that will help avoid future incidents. Wide distribution of this leaflet in the international maritime industry will highlight the issue and create an ongoing dialog on whale avoidance in the industry that seems likely to alleviate some collision risk.

For recreational boaters, we suggest that the most effective approach for raising awareness of the issue would occur in nonregulatory settings using contemporary modes of communication including social networking, to inform people how to avoid collisions, and the need to report incidents when they occur. A key message for operators of small boats in Alaska is that the likelihood of colliding with a whale is increasing, and that people can get hurt, costly vessel damage can occur and the whale can be injured or killed. Simple but specific preventive measures that encourage vigilance and the willingness to use slow speeds in high-density whale areas should be made widely available in a sound-byte format that is easy to digest. Creating and distributing these messages is a step toward creating a culture where people understand the risks and will do what they can to avoid collisions with whales.

Collision hotspots (Figure 2) are areas that warrant special attention in the form of vessel speed limits, public service announcements, increased law enforcement presence or other measures. The map we created for this paper is the first regional look at the geography of collisions in Alaska, and may be a useful approach for analysis of other collision datasets outside Alaska. High-risk areas need to be closely examined and coupled with predictive modeling to assess areas where conservation action (e.g., vessel speed limits) may be targeted to prevent future vessel collisions with whales in Alaska. For example, a recommendation to reduce speed at night in known hotspot areas may be particularly relevant for large ships (such as cruise ships) which routinely transit at night. Commercial vessels may want to consider marketing "whale friendly" voyages by advertising and adhering to lowered speeds as part of their standard operations, along with increased care and attentiveness in hotspot areas. Reduced speeds have been used successfully in Glacier Bay National Park for many years (termed "whale waters"), where the park superintendent implements vessel course and speed restrictions in areas where whale concentrations have been detected [25]. Protective measures applied to relatively small areas with reliably high whale densities may yield a disproportionately large reduction in collision risk for humpback whales in southeastern Alaska and presumably impact fewer vessel operators compared to other mitigation measures [81]. As whale populations and vessel traffic continue to change throughout the state, improved data collection and validation of collision reports will enhance our understanding of collisions, with the ultimate goal of reducing the frequency of whale-vessel collisions in Alaska.

Acknowledgments

The authors gratefully acknowledge the many organizations and individuals who have reported and collected data on whale-vessel collisions over the years including members of the Alaska Marine Mammal Stranding Network; the US Coast Guard; NOAA Enforcement; the Alaska Department of Fish & Game; the Alaska State Troopers; tour operators; vessel captains, pilots, and crew; harbormasters; fishermen; recreational boaters; Charles Jurasz; and C. Scott Baker. They thank John Sease, Linda Shaw, and Kaja Brix for developing the Alaska Marine Mammal Stranding Network with limited resources; Mary Sternfeld, who shepherded the NOAA Alaska Region Stranding Database through its infancy; Doug DeMaster for initiating the first paper on this topic for the IWC in 2007 and for providing valuable comments on this paper. Special credit goes to Dr. Frances Gulland from The Marine Mammal Center for leading necropsies in Alaska, training local responders in ship strike necropsy methods, and contributing her expertise to this paper. They extend sincere thanks to the Alaskan marine mammal veterinarians (Dr. Kathy Burek, Dr. Rachel Dziuba, Dr. Carrie Goertz, Dr. Kate Savage, and Dr. Pam Tuomi) and volunteers who have conducted and participated in whale necropsies. They are indebted to Jen Cedarleaf (UAS) for her expert fluke matching skills which allowed them to identify several of the dead humpback whales in this study. They thank John Moran (NOAA), Fred Sharpe (Alaska Whale Foundation), and Erin Falcone (Cascadia Research Collective) for sharing photos of live whales with collision injuries. They are grateful to David Mattila (IWC), Ed Lyman (NOAA), and Jerry Dzugan (Alaska Marine Safety Education Association) for contributing to and supporting this study. They thank Whitney Rapp and Greg Ambrose for their help developing the hotspot map. They thank two anonymous reviewers for their valuable comments on this paper. Necropsies on endangered whales were conducted under National Marine Fisheries Service (NMFS) permits 932-1489 and 932-1905.

Endnotes

1. Listed as endangered under the ESA.

2. Listed as endangered under the ESA (Cook Inlet stock only).

References

[1] W. McLellan, S. Rommel, M. Moore, and D. A. Pabst, "Right whale necropsy protocol," Final Report to NOAA Fisheries for Contract #40AANF112525, 2004.

[2] J. George, L. Philo, K. Hazard, D. Withrow, G. Carroll, and R. Suydam, "Frequency of killer whale (*Orcinus orca*) attacks and ship collisions based on scarring on bowhead whales (*Balaena mysticetus*) of the Bering-Chukchi-Beaufort Seas stock," *Arctic*, vol. 47, no. 4, pp. 247–255, 1994.

[3] D. N. Wiley, R. A. Asmutis, T. D. Pitchford, and D. P. Gannon, "Stranding and mortality of humpback whales, *Megaptera novaeangliae*, in the mid-Atlantic and southeast United States, 1985–1992," *Fishery Bulletin*, vol. 93, pp. 196–205, 1995.

[4] I. N. Visser, "Propeller scars on and known home range of two orca (*Orcinus orca*) in New Zealand waters," *New Zealand Journal of Marine and Freshwater Research*, vol. 33, pp. 635–642, 1999.

[5] P. B. Best, V. M. Peddemors, V. G. Cockcroft, and N. Rice, "Mortalities of right whales and related anthropogenic factors in South African waters, 1963–1998," *Journal of Cetacean Research and Management*, vol. 2, no. 2, pp. 171–176, 2001.

[6] J. Capella, L. Flórez-González, and P. Falk, "Mortality and anthropogenic harassment of humpback whales along the Pacific coast of Colombia," *Memoirs of the Queensland Museum*, vol. 47, no. 2, pp. 547–553, 2001.

[7] A. R. Knowlton and S. D. Kraus, "Mortality and serious injury of northern right whales (*Eubalaena glacialis*) in the western North Atlantic Ocean," *Journal of Cetacean Research and Management*, vol. 2, pp. 1–15, 2001.

[8] D. W. Laist, A. R. Knowlton, J. G. Mead, A. S. Collet, and M. Podesta, "Collisions between ships and whales," *Marine Mammal Science*, vol. 17, no. 1, pp. 35–75, 2001.

[9] A. S. Jensen and G. K. Silber, "Large whale ship strike database," NOAA Technical Memorandum NMFS-OPR-25, U.S. Department of Commerce, Washington, DC, USA, 2003.

[10] M. J. Moore, A. R. Knowlton, S. D. Kraus, W. A. McLellan, and R. K. Bonde, "Morphometry, gross morphology and available histopathology in North Atlantic right whale (*Eubalaena glacialis*) mortalities (1970–2002)," *Journal of Cetacean Research and Management*, vol. 6, no. 3, pp. 199–214, 2004.

[11] M. Weinrich, "A review of worldwide collisions between whales and fast ferries," Paper SC/56/BC9, International Whaling Commission Scientific Committee, Cambridge, UK, 2005, IWC Secretariat, http://www.iwcoffice.org/publications/doclist.htm.

[12] M. Weinrich, "A review of collisions between whales and whale watch boats," Paper SC/57/WW8, International Whaling Commission Scientific Committee, Cambridge, UK, 2005, IWC Secretariat, http://www.iwcoffice.org/publications/doclist.htm.

[13] S. Panigada, G. Pesante, M. Zanardelli, F. Capoulade, A. Gannier, and M. T. Weinrich, "Mediterranean fin whales at risk from fatal ship strikes," *Marine Pollution Bulletin*, vol. 52, no. 10, pp. 1287–1298, 2006.

[14] M. O. Lammers, A. A. Pack, and L. Davis, "Trends in whale/vessel collisions in Hawaiian waters," Paper SC/59/BC14, International Whaling Commission Scientific Committee, Cambridge, UK, 2007, IWC Secretariat, http://www.iwcoffice.org/publications/doclist.htm.

[15] K. Van Waerebeek, A. N. Baker, F. Félix et al., "Vessel collisions with small cetaceans worldwide and with large whales in the Southern Hemisphere, an initial assessment," *Latin American Journal of Aquatic Mammals*, vol. 6, no. 1, pp. 43–69, 2007.

[16] S. Behrens and R. Constantine, "Large whale and vessel collisions in northern New Zealand," Paper SC/60/BC9, International Whaling Commission Scientific Committee, Cambridge, UK, 2008, IWC Secretariat, http://www.iwcoffice.org/publications/doclist.htm.

[17] M. Carrillo and F. Ritter, "Increasing numbers of ship strikes in the Canary Islands: proposals for immediate action to reduce risk of vessel-whale collisions," *Journal of Cetacean Research and Management*, vol. 11, no. 2, pp. 131–138, 2010.

[18] A. B. Douglas, J. Calambokidis, S. Raverty, S. J. Jeffries, D. M. Lambourn, and S. A. Norman, "Incidence of ship strikes of large whales in Washington State," *Journal of the Marine Biological Association of the United Kingdom*, vol. 88, pp. 1121–1132, 2008.

[19] F. Ritter, "Collisions of sailing vessels with cetaceans worldwide: first insights into a seemingly growing problem," Paper SC/61/BC1, International Whaling Commission Scientific Committee, Cambridge, UK, 2009, IWC Secretariat, http://www.iwcoffice.org/publications/doclist.htm.

[20] M. Berman-Kowalewski, F. M. Gulland, S. Wilkin et al., "Association between blue whale (*Balaenoptera musculus*) mortality and ship strikes along the California coast," *Aquatic Mammals*, vol. 36, no. 1, pp. 59–66, 2010.

[21] U.S. Marine Mammal Protection Act of 1972, as amended (16 U.S. Code 1361 *et seq*).

[22] K. van Waerebeek and R. Leaper, "Report from the IWC vessel strike data standardisation group," Paper SC/59/BC12, International Whaling Commission Scientific Committee, Cambridge, UK, 2007, IWC Secretariat, http://www.iwcoffice.org/publications/doclist.htm.

[23] "Definitions," Code of Federal Regulations, Title 50, Part 216.3.

[24] "Special prohibitions for endangered marine mammals," Code of Federal Regulations, Title 50, Part 224.103.

[25] Special Regulations-Glacier Bay National Park and Preserve, Code of Federal Regulations, Title 36, Part 13, Subpart N.

[26] NOAA Alaska Region Stranding Record #2007068, (unpublished, available from National Marine Fisheries Service, Protected Resources Division, PO Box 21668, Juneau, AK, 99802).

[27] K. Wynne, *Guide to Marine Mammals of Alaska*, Alaska Sea Grant College Program, Fairbanks, Alaska, USA, 3rd edition, 2007.

[28] B. M. Allen and R. P. Angliss, "Alaska marine mammal stock assessments 2010," NOAA Technical Memorandum NMFS-AFSC-223, U.S. Department of Commerce, Washington, DC, USA, 2011.

[29] J. C. George, J. Zeh, R. Suydam, and C. Clark, "Abundance and population trend (1978–2001) of western arctic bowhead whales surveyed near Barrow, Alaska," *Marine Mammal Science*, vol. 20, no. 4, pp. 755–773, 2004.

[30] A. N. Zerbini, J. M. Waite, J. L. Laake, and P. R. Wade, "Abundance, trends and distribution of baleen whales off western Alaska and the central Aleutian Islands," *Deep-Sea Research I*, vol. 53, no. 11, pp. 1772–1790, 2006.

[31] J. Calambokidis, E. A. Falcone, T. J. Quinn et al., "SPLASH: structure of populations, levels of abundance and status of humpback whales in the North Pacific," Final Report for Contract AB133F-03-RP-00078, U.S. Department of Commerce, Washington, DC, USA, 2008.

[32] A. E. Punt and P. R. Wade, "Population status of the eastern North Pacific stock of gray whales in 2009," NOAA Technical Memorandum NMFS-AFSC-207, U.S. Department of Commerce, Washington, DC, USA, 2010.

[33] P. R. Wade, A. Kennedy, R. LeDuc et al., "The world's smallest whale population?" *Biology Letters*, vol. 7, no. 1, pp. 83–85, 2011.

[34] P. R. Wade, A. de Robertis, K. R. Hough et al., "Rare detections of North Pacific right whales in the Gulf of Alaska, with observations of their potential prey," *Endangered Species Research*, vol. 13, no. 2, pp. 99–109, 2011.

[35] S. M. Gende, A. N. Hendrix, K. R. Harris, B. Eichenlaub, J. Nielsen, and S. Pyare, "A Bayesian approach for understanding the role of ship speed in whale-ship encounters," *Ecological Applications*, vol. 21, no. 6, pp. 2232–2240, 2011.

[36] G. K. Silber, J. Slutsky, and S. Bettridge, "Hydrodynamics of a ship/whale collision," *Journal of Experimental Marine Biology and Ecology*, vol. 391, no. 1-2, pp. 10–19, 2010.

[37] A. S. M. Vanderlaan and C. T. Taggart, "Vessel collisions with whales: the probability of lethal injury based on vessel speed," *Marine Mammal Science*, vol. 23, no. 1, pp. 144–156, 2007.

[38] C. M. Gabriele, A. Jensen, J. L. Neilson, and J. M. Straley, "Preliminary summary of reported whale-vessel collisions in Alaskan waters: 1978–2006," Paper SC/59/BC16, International Whaling Commission Scientific Committee, Cambridge, UK, 2007, IWC Secretariat, http://www.iwcoffice.org/publications/doclist.htm.

[39] A. S. Jensen, J. L. Neilson, C. M. Gabriele, and J. M. Straley, "Summary of reported whale-vessel collisions in Alaskan waters: 1978–2008," The Alaska Marine Science Symposium, 2010.

[40] S. T. Zimmerman, "A history of marine mammal stranding networks in Alaska, with notes on the distribution of the most commonly stranded cetacean species, 1975–1987," in *Marine Mammal Strandings in the United States—Proceedings of the 2nd Marine Mammal Stranding Workshop*, J. E. Reynolds and D. K. Odell, Eds., pp. 43–53, National Oceanic and Atmospheric Administration Technical Report 98, Washington, DC, USA, 1991.

[41] H. Omura, K. Fujino, and S. Kimura, "Beaked whale Berardius bairdi of Japan, with notes on Ziphius cavirostris," The Scientific Reports of the Whales Research Institute 10, 1955.

[42] C. Lockyer, "The age at sexual maturity of the southern fin whale (*Balaenoptera physalus*) using annual layer counts in the ear plug," *Journal du Conseil, Conseil International pour l'Exploration de la Mer*, vol. 34, pp. 276–294, 1972.

[43] J. G. Mead, "Survey of reproductive data for the beaked whales (*Ziphiidae*)," *Reports of the International Whaling Commission*, no. 6, pp. 91–96, 1984.

[44] J. L. Sumich and J. T. Harvey, "Juvenile mortality in gray whales (*Eschrichtius robustus*)," *Journal of Mammalogy*, vol. 67, pp. 179–182, 1986.

[45] B. M. Allen, R. L. Brownell, and J. G. Mead, "Species review of Cuvier's beaked whale, *Ziphius cavirostris*," Paper SC/63/SM17, International Whaling Commission Scientific Committee, Cambridge, UK, 2011, IWC Secretariat, http://www.iwcoffice.org/publications/doclist.htm.

[46] P. J. Clapham, "Age at attainment of sexual maturity of humpback whales, *Megaptera novaeangliae*," *Canadian Journal of Zoology*, vol. 70, no. 7, pp. 1470–1472, 1992.

[47] R. G. Chittleborough, "The breeding cycle of the female humpback whale, *Megaptera nodosa* (Bonnaterre)," *Australian Journal of Marine & Freshwater Research*, vol. 9, no. 1, pp. 1–18, 1958.

[48] P. J. Clapham, S. E. Wetmore, T. D. Smith, and J. G. Mead, "Length at birth and at independence in humpback whales," *Journal of Cetacean Research and Management*, vol. 1, no. 2, pp. 141–146, 1999.

[49] J. M. Straley, *Seasonal characteristics of humpback whales* (Megaptera novaeangliae) *in southeastern Alaska*, M.S. thesis, University of Alaska, Fairbanks, Alaska, USA, 1994.

[50] D. W. Rice, "Progress report on biological studies of the larger cetaceans in the waters off California," *Norsk Hvalfangst-Tidende*, vol. 52, no. 7, pp. 181–187, 1963.

[51] NOAA Alaska Region Stranding Record #2005003, (unpublished, available from National Marine Fisheries Service, Protected Resources Division, PO Box 21668, Juneau, AK, 99802).

[52] C. M. Gabriele, J. M. Straley, S. A. Mizroch et al., "Estimating the mortality rate of humpback whale calves in the central North Pacific Ocean," *Canadian Journal of Zoology*, vol. 79, no. 4, pp. 589–600, 2001.

[53] U.S. Coast Guard and United States Coast Guard Maritime Information eXchange, "Port State Information eXchange," http://cgmix.uscg.mil/PSIX/PSIXSearch.aspx/2011.

[54] R. Campbell-Malone, S. G. Barco, P. Y. Daoust et al., "Gross and histologic evidence of sharp and blunt trauma in North Atlantic right whales (*Eubalaena glacialis*) killed by vessels," *Journal of Zoo and Wildlife Medicine*, vol. 39, no. 1, pp. 37–55, 2008.

[55] S. Barco and K. Touhey, "Handbook for recognizing, evaluating, and documenting human interaction in stranded cetaceans and pinnipeds," Report from the Virginia Aquarium Stranding Response Program and Cape Cod Stranding Network, 2006.

[56] NOAA Alaska Region Stranding Record #1978050, (unpublished, available from National Marine Fisheries Service, Protected Resources Division, P.O. Box 21668, Juneau, AK, 99802).

[57] NOAA Alaska Region Stranding Record #2008145, (unpublished, available from National Marine Fisheries Service, Protected Resources Division, P.O. Box 21668, Juneau, AK, 99802).

[58] NOAA Alaska Region Stranding Record #2008031, (unpublished, available from National Marine Fisheries Service, Protected Resources Division, P.O. Box 21668, Juneau, AK, 99802).

[59] E. J. lijper, *Whales and Dolphins*, The University of Michigan Press, Ann Arbor, Mich, USA, 1978.

[60] P. A. Allison, C. R. Smith, H. Kukert, J. W. Deming, and B. A. Bennett, "Deep-water taphonomy of vertebrate carcasses: a whale skeleton in the bathyal Santa Catalina Basin," *Paleobiology*, vol. 17, no. 1, pp. 78–89, 1991.

[61] NOAA Alaska Region Stranding Record #2010089, (unpublished, available from National Marine Fisheries Service, Protected Resources Division, P.O. Box 21668, Juneau, AK, 99802).

[62] A. N. Hendrix, J. Straley, C. Gabriele, and S. Gende, "Bayesian estimation of humpback whale (*Megaptera novaeangliae*) population abundance and movement patterns in Southeast Alaska," *Canadian Journal of Fisheries and Aquatic Science*. In press.

[63] International Whaling Commission and Agreement on the Conservation of Cetaceans in the Black Sea, Mediterranean Sea and Contiguous Atlantic Area, "Report of the joint IWC-ACCOBAMS workshop on reducing risk of collisions between vessels and cetaceans," Tech. Rep., International Whaling Commission and Agreement on the Conservation of Cetaceans in the Black Sea, Mediterranean Sea and Contiguous Atlantic Area, Beaulieu-sur-Mer, France, 2010.

[64] R. Williams and P. O'Hara, "Modelling ship strike risk to fin, humpback and killer whales in British Columbia, Canada,"

Journal of Cetacean Research and Management, vol. 11, no. 1, pp. 1–8, 2010.

[65] M. L. Dalebout, K. G. Russell, M. J. Little, and P. Ensor, "Observations of live Gray's beaked whales (*Mesoplodon grayi*) in Mahurangi Harbour, North Island, New Zealand, with a summary of at-sea sightings," *Journal of the Royal Society of New Zealand*, vol. 34, no. 4, Article ID R03012, pp. 347–356, 2004.

[66] A. Fernández, J. F. Edwards, F. Rodriguez et al., "Gas and fat embolic syndrome' involving a mass stranding of beaked whales (*family Ziphiidae*) exposed to anthropogenic sonar signals," *Veterinary Pathology*, vol. 42, no. 4, pp. 446–457, 2005.

[67] M. J. Shkrum and D. A. Ramsay, *Forensic Pathology of Trauma: Common Problems for the Pathologist*, Humana Press, Totowa, NJ, USA, 2007.

[68] W. F. Dolphin, "Ventilation and dive patterns of humpback whales, (*Megaptera novaeangliae*), on their Alaskan feeding grounds," *Canadian Journal of Zoology*, vol. 65, no. 1, pp. 83–90, 1987.

[69] S. E. Parks, J. D. Warren, K. Stamieszkin, C. A. Mayo, and D. Wiley, "Dangerous dining: surface foraging of North Atlantic right whales increases risk of vessel collisions," *Biology Letters*, vol. 8, no. 1, pp. 57–60, 2012.

[70] "Tests before entering or getting underway," Code of Federal Regulations, Title 33, Part 164.25.

[71] NOAA Alaska Region Stranding Record #1999095, (unpublished, available from National Marine Fisheries Service, Protected Resources Division, P.O. Box 21668, Juneau, AK, 99802).

[72] NOAA Alaska Region Stranding Record #2006140, (unpublished, available from National Marine Fisheries Service, Protected Resources Division, P.O. Box 21668, Juneau, AK, 99802).

[73] R. Williams, S. Gero, L. Bejder et al., "Underestimating the damage: interpreting cetacean carcass recoveries in the context of the Deepwater Horizon/BP incident," *Conservation Letters*, vol. 4, no. 3, pp. 228–233, 2011.

[74] J. E. Heyning and M. E. Dahlheim, "Strandings and incidental takes of gray whales," Paper SC/A90/G2, International Whaling Commission Scientific Committee on the Assessment of Gray Whales, Cambridge, UK, 1990, IWC Secretariat, http://www.iwcoffice.org/publications/doclist.htm.

[75] S. D. Kraus, M. W. Brown, H. Caswell et al., "North Atlantic right whales in crisis," *Science*, vol. 309, no. 5734, pp. 561–562, 2005.

[76] J. E. Moore and A. J. Read, "A Bayesian uncertainty analysis of cetacean demography and bycatch mortality using age-at-death data," *Ecological Applications*, vol. 18, no. 8, pp. 1914–1931, 2008.

[77] H. Peltier, W. Dabin, P. Daniel et al., "The significance of stranding data as indicators of cetacean populations at sea: modelling the drift of cetacean carcasses," *Ecological Indicators*, vol. 18, pp. 278–290, 2012.

[78] J. Empire, *State and Local Briefly*, 2000.

[79] R. de Stephanis and E. Urquiola, "Collisions between ships and cetaceans in Spain," Paper SC/58/BC5, International Whaling Commission Scientific Committee, Cambridge, UK, 2006, IWC Secretariat, http://www.iwcoffice.org/publications/doclist.htm.

[80] Belgian Federal Public Service Health, Food Chain Security and Environment, "Whales: avoiding collisions prevents damage to ships, and injuries to passengers, crew and whales," http://iwcoffice.org/_documents/sci_com/shipstrikes/, 2011 English%20whale%20strike%20folder.pdf, 2011.

[81] E. M. Chenoweth, C. M. Gabriele, and D. F. Hill, "Tidal influences on humpback whale habitat selection near headlands," *Marine Ecology Progress Series*, vol. 423, pp. 279–289, 2011.

List of Zooplankton Taxa in the Caspian Sea Waters of Iran

Siamak Bagheri,[1] **Jalil Sabkara,**[1] **Alireza Mirzajani,**[1,2] **Seyed Hojat Khodaparast,**[1] **Esmaeil Yosefzad,**[1] **and Foong Swee Yeok**[3]

[1] *Inland Waters Aquaculture Institute, Iranian Fisheries Research Organization (IFRO), Anzali 66, Iran*
[2] *Faculty of Natural Resources, University of Tehran, P.O. Box 4314, Karaj 31587-77878, Iran*
[3] *School of Biological Sciences, Universiti Sains Malaysia, 11800 Penang, Malaysia*

Correspondence should be addressed to Siamak Bagheri; siamakbp@gmail.com

Academic Editor: E. A. Pakhomov

A total of 61 zooplankton taxa were found in the southwestern Caspian Sea between 1996 and 2010. Thirteen of them were meroplankton taxa and forty-eight were holoplankton taxa. The occurrence of 14 freshwater taxa indicated the influence of the Anzali wetland and river inflows. The decrease in zooplankton taxa was detected since 1996-1997 and continued till 2010. *Pleopis polyphemoides*, the only one out of the nine recorded Cladocera species in 1996-1997, was found after 2001. Similarly, of the five Copepoda species recorded in 1996-1997, only one, *Acartia tonsa*, was found abundant during the 2001–2010 sampling period. It was striking that many species which were abundant in the Caspian Sea in 1996-1997 were not found after 2000. Many reasons could have contributed to the changes in the zooplankton composition of the southern Caspian Sea, notably the serious environmental degradation since the early 1990s. It is also possible that invasive species might play a role in wiping out some sensitive endemic species.

1. Introduction

The Caspian Sea is a large inland water body. It is called a sea even though not being directly connected to any marine system due to its large size [1]. Water is mainly sourced from the big river Volga in the north (almost 76.3% of the total), and other rivers like Kura, Ural, Terek, and Sefidrood [2]. In the Iranian coast, the Sefidrood river is the largest river pouring into the Caspian Sea with 67,000 km^2 of catchment area and an average discharge of 4,037 million m^3 per year [3, 4]. The Caspian Sea has undergone significant ecological alteration during the past 30 years. This could be linked to dramatic changes in the southern Caspian Sea due to increased pollution: sewage, industry effluent, and agricultural waste water discharges into the river as well as deforestation of the river's watersheds [5, 6].

Zooplankton is recognized among the best indicators to be particularly useful to investigate and document environmental changes [7]. Main zooplankton taxa have short life cycle and the community structure is able to reflect real-time scenario as it is less enforced by the stability of individuals

from previous years [8]. Besides this, zooplankton is also the food of choice for many fishes and as such plays a very significant role in pilaring the upper stages of the food chain [9].

Bagirov [10] reported that the number of zooplankton taxa was almost 200 in the northern Caspian Sea with 70 taxa of Protista, 50 taxa of Rotatoria, 30 taxa of Cladocera, and 20 taxa of Copepoda. Meroplankton, represented mainly by larvae of bivalves and crustaceans, contributed to the biodiversity of plankton communities in the North. Whereas in the South of Caspian Sea, Hossieni et al. [11] documented that zooplankton community consisted of 36 taxa, including Cladocera (24 taxa), Copepoda (7 taxa), and meroplankton (2 taxa) along the Iranian coastal area of the Caspian Sea in 1996. Subsequently in 1999, Sabkara et al. [12] reported over 50 zooplankton taxa with holoplankton comprising >80% of the taxa sampled and Rotifera (22 species) dominating. Meroplankton accounted for the remaining diversity. Recently, Roohi et al. [13] noted that only 18 zooplankton taxa were observed in the southern Caspian Sea, of which five taxa belonged to holoplankton (four Copepoda and one Cladocera) and 13 to meroplankton.

TABLE 1: "×" denotes sampling done based on season in the Caspian Sea waters of Iran during 1996–2010.

Year	Season			
	Winter	Spring	Summer	Autumn
1996		×	×	×
1997	×			
2001	×		×	×
2002	×	×	×	×
2003	×			×
2004	×	×	×	×
2005	×	×	×	×
2006	×		×	×
2008	×	×	×	×
2009			×	×
2010	×	×		×

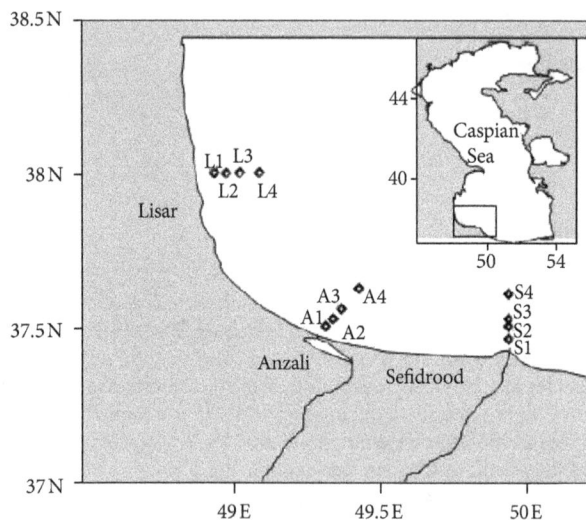

FIGURE 1: Study sites sampled 1996 to 2010 in the Caspian Sea waters of Iran. L = Lisar, A = Anzali, S = Sefidrood, 5 m (L1, A1, S1), 10 m (L2, A2, S2), 20 m (L3, A3, S3), and 50 m (L4, A4, S4).

In recent years, few studies have been conducted on the distribution and abundance of zooplankton in the southern Caspian Sea [11–14], while there was hardly any work done to look at the composition of zooplankton taxa. In order to investigate the state of ecological condition of the Caspian Sea, an eleven-year study of the zooplankton community in the Caspian Sea waters of Iran was compiled between 1996 and 2010.

2. Materials and Methods

2.1. Study Area. The area under investigation is located at the southwestern corner of the Caspian Sea. The identification of zooplankton taxa was carried out along three transects along the mouth of the Anzali wetland, Lisar and Sefidrood rivers. The sampling was performed at 5 m, 10 m, 20 m, and 50 m depths during 1996–2010 (Figure 1). Table 1 shows the seasonal sample collection done in between 1996–2010. In

some years, sampling was unable to be carried out due to logistic problems. In total, a collection of 132 samples were carried out during these periods (Table 1).

2.2. Data Collection. Zooplankton was sampled using a Juday net (opening diameter: 36 cm, mesh size: 100 μm). At every station, a vertical haul with a Juday net was carried out from bottom to surface using a handle pulley for heaving the net. Zooplankton samples were preserved in neutral 4% formaldehyde and analyzed in the laboratory. Samples were divided into subsamples using a 1 mL Hensen-Stempel pipette and transferred to a Bogorov chamber for identification. At least 100 individuals were counted per sample and identified to species, and life-cycle stages were determined using an inverted microscope [15]. Zooplankton taxonomic classification was performed based on Birshtain et al. [16], Kasimov [17], and James and Covich [18].

3. Results

A total of 61 zooplankton taxa belonging to 48 taxa of holozooplankton were found in the area of investigation (Tables 2 and 3). Fourteen of them, *Pleroxus trigenellus* O.F. Mueller, *Chydorus* sp. Leach, *Moina* sp. Baird, all Cladocera, the copepod *Cyclops* sp. Risso, *Cyclops* sp. nauplii and the rotifers *Filinia* sp. Vincent, *Keratella cochlearis*, *Lecane* sp. Nitzsch, *Notholca acuminate* Ehrenberg, *Philodina* sp. Ehrenberg, *Polyarthera dolichoptera* Idelson, *Testudinella patina* Hermann, *Trichocerca* sp. Lamarck, and *Trichocerca caspia* Tschugunoff are freshwater species in the area.

Thirteen meroplankton taxa consisted of individuals of water spiders (Arachnida), crab (*Rhithropanopeus harrisii* Gould), Foraminifera, Ostracoda larvae, *Asteromeyenia* sp. Weltner, *Pseudocuma* sp. G.O. Sars, and larvae of Bivalvia, Nematoda, Polychaeta (*Hediste diversicolor* O.F. Muller, *Hypania* sp. Ostroumouw), Pisces (egg and larvae), and multitudinous Cirripedia (*Balanus* sp. Costa) represented by nauplii and cypris larvae (Table 2).

Thirty-four holoplankton taxa belong to Ctenophora (*Mnemiopsis leidyi* A. Agassiz), Ciliata (*Codonella* sp. Haeckel, *Codonella relicta* Minikiewich, *Tintinnopsis* sp. Lamarck,

TABLE 2: Checklist of zooplankton taxa in the Caspian Sea waters of Iran during 1996–2010.

Holoplankton taxa	After appearance of *Mnemiopsis leidyi*								
	1996-1997	2001	2002	2003	2004	2005	2006	2008	2009-2010
Macrozooplankton									
Ctenophora:									
Mnemiopsis leidyi A. Agassiz, 1865	−	+	+	+	+	+	+	+	+
Mesozooplankton									
Cladocera:									
Cercopagis pengoi Ostroumov, 1891	+	−	−	−	−	−	−	−	−
Cercopagis prolongata G.O. Sara, 1897	+	−	−	−	−	−	−	−	−
Chydorus Leach, 1843	−	−	−	−	−	−	−	+	−
Evaden anonyx G.O. Sars, 1897	+	−	−	−	−	−	−	−	−
Moina Baird, 1850	−	−	−	−	−	−	−	+	−
Pleroxus trigonellus O.F. Muller, 1785	−	−	−	−	−	−	−	+	−
Podonevadne Gibitz, 1922	−	−	−	−	−	−	−	+	−
Podonevaden angusta G.O. Sars, 1902	+	−	−	−	−	−	−	−	−
Podonevaden camptonyx G.O. Sars, 1897	+	−	−	−	−	−	−	−	−
Podonevaden trigona G.O. Sars, 1897	+	−	−	−	−	−	−	−	−
Podon intermedius Lilljeborg, 1853	+	−	−	−	−	−	−	−	−
Pleopis polyphemoides Leuckart, 1859	+	−	+	−	+	+	+	+	+
Polyphemus exiguus G.O. Sars, 1897	+	−	−	−	−	−	−	−	−
Copepoda:									
Acartia tonsa nauplii	+	+	+	+	+	+	+	+	+
Acartia tonsa Dana, 1849	+	+	+	+	+	+	+	+	+
Calanipeda aquae dulcis nauplii	+	−	−	−	−	−	−	−	−
Calanipeda aquae dulcis Kritchagin, 1873	+	−	−	−	−	−	−	−	−
Cyclops nauplii	−	−	−	−	−	−	−	+	−
Cyclops Risso, 1826	+	+	−	+	+	+	−	+	+
Eurytemora grimmi nauplii	+	−	−	−	−	−	−	−	−
Eurytemora grimmi G.O. Sars, 1897	+	−	−	−	−	−	−	−	−
Halicyclops sarsi Akatova, 1935	+	−	+	−	−	−	−	−	+
Ectinosoma concinnum Akatova, 1935	+	−	−	−	−	−	−	+	−
Limnocalanus grimaldii grimaldii nauplii	+	−	−	−	−	−	−	−	−
Limnocalanus grimaldii grimaldii Guerne, 1886	+	−	−	−	−	−	−	−	−
Rotifera:									
Brachionus Pallas, 1766	−	−	−	−	+	+	−	−	−
Brachionus angularis Gosse, 1851	−	+	+	−	−	−	−	−	−
Brachionus calyciflorus Pallas, 1766	−	−	+	−	+	+	+	−	+
Brachionus plicatilis Muller, 1786	−	+	+	−	−	−	−	+	+
Filinia Bory de St. Vincent, 1824	−	−	+	−	−	−	−	−	−
Keratella cochlearis Gosse, 1851	−	+	+	+	−	−	−	+	+
Lecane Nitzsch, 1827	−	+	−	−	−	−	−	−	−
Notholca acuminata Ehrenberg, 1832	−	−	+	−	−	−	−	−	−
Philodina Ehrenberg, 1830	−	−	−	−	−	−	−	−	+
Polyarthera dolichoptera Idelson, 1925	−	−	+	−	−	−	−	−	+
Synchaeta Ehrenberg, 1832	+	+	+	+	+	+	+	−	+
Synchaeta stylata Wierzejski, 1893	+	−	−	−	−	−	−	−	−
Synchaeta vorax Rousselet, 1902	+	−	−	−	−	−	+	−	−
Testudinella patina Hermann, 1783	−	−	−	−	−	−	−	−	+
Trichocerca Lamarck, 1801	−	−	+	−	−	−	−	−	−
Trichocerca capica Tschugunoff, 1921	−	−	+	−	−	−	−	−	−
Microzooplankton									
Ciliata:									
Codonella Haeckel, 1873	−	−	+	−	−	−	−	−	−

TABLE 2: Continued.

Holoplankton taxa	After appearance of *Mnemiopsis leidyi*								
	1996-1997	2001	2002	2003	2004	2005	2006	2008	2009-2010
Codonella relicta Minikiewich, 1905	−	+	+	−	−	−	−	−	+
Tintinnopsis Stein 1867	−	+	+	−	−	−	−	−	−
Tintinnopsis karajacensis Brandt, 1896	−	+	−	−	−	−	−	−	−
Tintinnopsis tubulosa Levander, 1900	−	+	+	+	+	+	−	+	+
Meroplankton taxa (Mesozooplankton)									
Arachnida Larvae Cuvier, 1812	+	−	−	−	−	−	−	+	+
Cirripedia:									
Balanus sp. cypris Costa, 1778	+	+	+	+	+	+	+	+	+
Balanus sp. nauplii Costa, 1778	+	+	+	+	+	+	+	+	+
Crustacea:									
Rhithropanopeus harrisii Gould, 1841	+	−	−	−	−	−	−	−	−
Foraminifera larvae	+	+	+	−	+	+	−	−	+
Bivalvia larvae Linnaeus, 1758	+	+	+	+	+	+	+	+	−
Nematoda larvae	−	−	−	−	+	+	+	+	+
Ostracoda larvae Latreille, 1802	−	−	+	−	−	−	−	−	−
Pisces larvae	+	+	+	−	+	+	+	+	+
Pisces ovae	+	+	−	−	+	+	+	+	+
Polychaeta:									
Hypania Ostroumouw, 1897	−	−	−	−	−	−	−	+	−
Hediste diversicolor O.F. Muller, 1776	−	+	+	+	+	+	+	+	+
Porifera:									
Asteromeyenia Weltner, 1913	−	−	−	−	−	−	−	+	−
Cumacea:									
Pseudocuma G.O. Sars, 1865	+	−	−	−	−	−	−	−	−

*Alien taxa; bold font: freshwater taxa indicated the influence by river inflow.

TABLE 3: Zooplankton number of taxa (holomeroplankton) in the Caspian Sea waters of Iran during 1996–2010.

Zooplankton taxa	After appearance of *Mnemiopsis leidyi*									Number of taxa
	1996-1997	2001	2002	2003	2004	2005	2006	2008	2009-2010	
Ctenophora	0	1	1	1	1	1	1	1	1	1
Ciliata	0	4	4	1	1	1	0	1	2	5
Cladocera	9	0	1	0	1	1	1	5	1	13
Copepoda	11	3	3	3	3	3	2	5	4	12
Rotifera	3	5	10	2	3	3	3	2	7	17
Total holoplankton	*23*	*13*	*19*	*7*	*9*	*9*	*7*	*14*	*15*	*48*
Arachnida	1	0	0	0	0	0	0	1	1	1
Cirripedia	2	2	2	2	2	2	2	2	2	2
Crustacea	1	0	0	0	0	0	0	0	0	1
Foraminifera	1	1	1	0	1	1	0	0	1	1
Bivalvia	1	1	1	1	1	1	1	1	0	1
Nematoda	0	0	0	0	1	1	1	1	1	1
Ostracoda	0	0	1	0	0	0	0	0	0	1
Pisces	2	2	1	0	2	2	2	2	2	2
Polychaeta	0	1	1	1	1	1	1	2	1	1
Porifera	0	0	0	0	0	0	0	1	0	1
Pseudocuma	1	0	0	0	0	0	0	0	0	1
Total meroplankton	*9*	*7*	*7*	*4*	*8*	*8*	*7*	*10*	*8*	*13*
Total zooplankton	**32**	**20**	**26**	**11**	**17**	**17**	**14**	**24**	**23**	**61**

TABLE 4: Seasonal list of zooplankton taxa in the Caspian Sea waters of Iran during 1996–2010.

Group	Zooplankton taxa	Season			
		Winter	Spring	Summer	Autumn
Arachnida	Arachnida larvae	+	−	−	+
Bivalvia	Bivalvia larvae	+	+	+	+
Ciliata	Codonella sp.	+	−	−	−
Ciliata	Codonella relicta	+	−	−	−
Ciliata	Tintinnopsis sp.	+	−	−	−
Ciliata	Tintinnopsis karajacensis	+	−	−	+
Ciliata	Tintinnopsis tubulosa	+	−	+	+
Cirripedia	Balanus sp. cypris	+	+	+	+
Cirripedia	Balanus sp. nauplii	+	+	+	+
Cladocera	Cercopagis pengoi	−	−	−	+
Cladocera	Cercopagis prolongata	−	−	−	+
Cladocera	**Chydorus sp.**	+	−	−	−
Cladocera	Evaden anonyx	−	+	−	+
Cladocera	**Moina sp.**	−	−	−	+
Cladocera	**Pleroxus trigonellus**	−	−	−	+
Cladocera	Podonevadne sp.	−	+	−	−
Cladocera	Podonevaden angusta	+	−	−	+
Cladocera	Podonevaden camptonyx	−	+	−	−
Cladocera	Podonevaden trigona	+	+	+	+
Cladocera	Podon intermedius	+	−	−	−
Cladocera	Pleopis polyphemoides	+	+	−	+
Cladocera	Polyphemus exiguus	+	+	+	+
Copepoda	Acartia tonsa nauplii	+	+	+	+
Copepoda	Acartia tonsa	+	+	+	+
Copepoda	Calanipeda aquae dulcis nauplii	+	+	+	+
Copepoda	Calanipeda aquae dulcis	+	+	+	+
Copepoda	**Cyclops nauplii**	+	−	−	+
Copepoda	**Cyclops sp.**	+	−	−	+
Copepoda	Eurytemora grimmi nauplii	+	+	+	+
Copepoda	Eurytemora grimmi	+	+	+	+
Copepoda	Halicyclops sarsi	+	+	+	+
Copepoda	Ectinosoma concinnum	+	+	+	+
Copepoda	Limnocalanus grimaldii grimaldii nauplii	+	+	+	+
Copepoda	Limnocalanus grimaldii grimaldii	+	+	+	+
Ctenophora	Mnemiopsis leidyi	+	+	+	+
Crustacea	Rhithropanopeus harrisii	+	−	−	+
Cumacea	Pseudocuma sp.	−	−	−	+
Foraminifera	Foraminifera larvae	−	+	+	−
Nematoda	Nematoda larvae	+	+	−	+
Ostracoda	Ostracoda larvae	+	−	−	−
Pisces	Pisces larvae	+	−	+	−
Pisces	Pisces ovae	+	+	−	−
Polychaeta	Hypania sp.	−	−	+	+
Polychaeta	Hediste diversicolor	+	+	+	+
Porifera	Asteromeyenia sp.	−	+	−	+
Rotifera	Brachionus sp.	+	+	−	+
Rotifera	Brachionus angularis	+	+	−	+
Rotifera	Brachionus calyciflorus	+	+	−	+
Rotifera	Brachionus plicatilis	+	+	−	+
Rotifera	**Filinia sp.**	+	−	−	−
Rotifera	**Keratella cochlearis**	+	−	−	−
Rotifera	**Lecane sp.**	+	−	−	−

TABLE 4: Continued.

Group	Zooplankton taxa	Season			
		Winter	Spring	Summer	Autumn
Rotifera	***Notholca acuminata***	+	−	−	−
Rotifera	***Philodina* sp.**	−	+	−	−
Rotifera	***Polyarthera dolichoptera***	−	+	−	−
Rotifera	*Synchaeta* sp.	+	+	−	−
Rotifera	*Synchaeta stylata*	+	+	−	+
Rotifera	*Synchaeta vorax*	+	+	−	+
Rotifera	***Testudinella patina***	−	−	−	+
Rotifera	***Trichocerca* sp.**	+	+	−	−
Rotifera	***Trichocerca capica***	+	−	−	−

Bold font: freshwater taxa indicated the influence by river inflow.

Tintinnopsis karajacensis Brandt, and *Tintinnopsis tubulosa* Levander), Rotifera (*Brachionus* sp. Pallas, *Brachionus angularis* Gosse, *Brachionus calyciflorus* Pallas, *Brachionus plicatilis* Mueller, *Synchaeta* sp. Ehrenberg, *Synchaeta stylata* Wierzejski, and *Synchaeta vorax* Rousselet), Cladocera (*Cercopagis pengoi* Ostroumov, *Cercopagis prolongata* G.O. Sars, *Evaden anonyx* Sars, *Podonevadne* sp. Gibitz, *Podonevaden angusta* G.O. Sars, *Podonevaden camptonyx* G.O. Sars, *Podonevaden trigona* G.O. Sars, *Podon intermedius* Lilljeborg, *Pleopis polyphemoides* Leuckart, and *Polyphemus exiguus* G.O. Sars), and Copepoda (*Acartia tonsa* Dana, *Calanipeda aquae dulcis* Kritchagin, *Eurytemora grimmi* G.O. Sars, *Halicyclops sarsi* Akatova, *Ectinosoma concinnum* Akatova, and *Limnocalanus grimaldii grimaldii* Guerne).

The Copepoda (*Eurytemora grimmi*, *Limnocalanus grimaldii grimaldii*, *Acartia tonsa*, *Ectinosoma concinnum*, and *Halicyclops sarsi*), Cladocera (*Podonevaden trigona*, *Pleopis polyphemoides*, *Polyphemos exiguus*), Cirripedia (*Balanus* sp.), Polychaeta (*Hediste diversicolor*), Bivalvia larvae, and Ctenophora (*Mnemiopsis leidyi*) dominated the zooplankton taxa in all seasons in the southwestern Caspian Sea (Table 4).

4. Discussion

In comparison with earlier surveys by Hossieni et al. [11] and Roohi et al. [13] which were carried out in the southern Caspian Sea, major changes in zooplankton community became obvious after 2000 (Tables 2 and 3). During the present study, 26 taxa including *Codonella* sp. and *Codonella relicta* (Ciliata); *Tintinnopsis* sp., *Tintinnopsis karajacensis*, and *Tintinnopsis tubulosa* (Ciliata); *Brachionus* sp., *Brachionus angularis*, *Brachionus calyciflorus*, and *Brachionus plicatilis* (Rotifera); *Filinia* sp. (Rotifera), *Keratella cochlearis* (Rotifera), *Lecane* (Rotifera), *Notholca acuminata* (Rotifera), *Philodina* (Rotifera), *Polyarthera dolichoptera* (Rotifera), *Synchaeta* sp. (Rotifera), *Synchaeta stylata*, *Synchaeta vorax* (Rotifera), *Testudinella patina* (Rotifera), *Trichocerca* (Rotifera), and *Trichocerca capica*; *Podonevadne* sp. (Cladocera); *Rhithropanopeus harrisii* (Crustacea); *Asteromeyenia* sp. (Porifera); *Pseudocuma* sp. (Cumacea) and *Hypania* larvae (Polychaeta) were documented. Hossieni et al. [11] and Roohi et al. [13] listed none of these taxa.

Hossieni et al. [11] did not list the following species, which were sampled in subsequent years: Ostracoda larvae, Arachnida larvae, Harpacticoida (*Ectinosoma concinnum*), Nematoda larvae, and *Hediste diversicolor*. Roohi et al. [13] listed Chironomidae, Nematidae, and Mysidacaea (*n* = 2 in 2006), which were not observed in this study.

Hossieni et al. [11] listed 24 Cladocera species in the southern Caspian Sea (cited by Roohi et al., [13]). Three of these taxa were repeated twice under different names as *Apagis cylindrata* (*Cercopagis cylindrata*), *Apagis longicaudata* (*Cercopagis longicaudata*), and *Apagis ossiani* (*Cercopagis ossiani*). Additionally fourteen taxa cited were not recognized as validly published taxa by CSBP (Caspian Sea Biodiversity Project), ITIS (Integrated Taxonomic Information System), MarBEF (Marine Biodiversity and Ecosystem Functioning), and WoRNS (World Register of Marine Species), while there were nine Cladocera listed in the present study in 1996-1997 and only two of them (*Pleopis polyphemoides* and *Podonevadne* sp.) could be found after 2000. Five Copepoda species were present in 1996 [11] and the present study, but they were either absent or found in very low numbers during 2001 to 2010 (Table 2): *Calanipeda aquae dulcis*, *Limnocalanus grimaldii*, *Eurytemora minor*, and *Eurytemora grimmi*.

It was striking that many species which were abundant in the Caspian Sea since 1996-1997 were not found after 2000 (Tables 3 and 4). However, a similar drastic decline in species numbers was not observed in other invaded seas [19]. Possible reasons could be twofold: the endemic Caspian Sea fauna is very sensitive to disruptions of invaders [5, 20], or it a results of serious environmental degradation which started since the beginning of the 1990s [5]. Rodionov [21], Bilio and Niermann [22], and Polonskii et al. [23] theorized that hydrobiological changes in the Caspian Sea, Black Sea, and Baltic Sea during the 1990s and 2000s could definitely be correlated to these changes. Recent remarks on other seas indicated that the shifting plankton communities can also be related to climatic inconsistency [7, 24–26].

The survey helped to characterize the composition of the zooplankton taxa in the southwestern Caspian Sea between 1996 and 2010. In comparison with previous publications, the multiyear composition of zooplankton taxa was estimated for the first time. We call for a standardization in the result presentation during future studies in this area. A comprehensive database including international participation should be launched to facilitate long-term comparisons of planktonic species to monitor anthropogenic and climatic effects on the Caspian Sea ecosystem.

Acknowledgments

The authors are grateful to Peter Boyce for improving the English of the draft paper. They would like to thank the Inland Waters Aquaculture Institute and Iranian Fisheries Research Organization (IFRO) for financially supporting this project. They deeply appreciate the assistance received from F. Maddadi, Y. Zahmatkesh, M. Sayad-Rahim, A. Sedaghat-Kish, H. Norouzi, and M. Iran-Pour in this study.

References

[1] V. A. Putans, L. R. Merklin, and O. V. Levchenko, "Sediment waves and other forms as evidence of geohazards in Caspian Sea," *International Journal of Offshore and Polar Engineering*, vol. 20, no. 4, pp. 241–246, 2010.

[2] H. J. Dumont, "The Caspian Lake: history, biota, structure, and function," *Limnology and Oceanography*, vol. 43, no. 1, pp. 44–52, 1998.

[3] H. A. Lahijani, V. Tavakoli, and A. H. Amini, "South Caspian river mouth configuration under human impact and sea level fluctuations," *Environmental Sciences*, vol. 5, pp. 65–86, 2008.

[4] S. Bagheri, M. Mansor, M. Turkoglu, W. O. Wan Maznah, and H. Babaei, "Temporal distribution of phytoplankton in the southwestern Caspian Sea during 2009-2010: a comparison with previous surveys," *Journal of the Marine Biological Association of the United Kingdom*, vol. 92, pp. 1243–1255, 2012.

[5] H. Dumont, "Ecocide in the Caspian Sea," *Nature*, vol. 377, no. 6551, pp. 673–674, 1995.

[6] S. Bagheri, M. Mansor, M. Marzieh et al., "Fluctuations of phytoplankton community in the coastal waters of Caspian Sea in 2006," *American Journal of Applied Sciences*, vol. 8, no. 12, pp. 1328–1336, 2011.

[7] C. Sipkay, K. T. Kiss, C. Vadadi-Fülöp, and L. Hufnagel, "Trends in research on the possible effects of climate change concerning aquatic ecosystems with special emphasis on the modelling approach," *Applied Ecology and Environmental Research*, vol. 17, no. 2, pp. 171–198, 2009.

[8] A. J. Richardson, "In hot water: zooplankton and climate change," *ICES Journal of Marine Science*, vol. 65, no. 3, pp. 279–295, 2008.

[9] A. H. Taylor, J. I. Allen, and P. A. Clark, "Extraction of a weak climatic signal by an ecosystem," *Nature*, vol. 416, no. 6881, pp. 629–632, 2002.

[10] R. M. Bagirov, *The Azov and Black Sea species introduced to the Caspian benthos and biofouling [Ph.D. thesis]*, University of Baku, 1989.

[11] A. Hossieni, A. Roohi, K. A. Ganjian et al., "Hydrology and hydrobiology of the southern Caspian Sea," Registration 96.132, Agricultural Research and Education Organization, 1998.

[12] J. Sabkara, S. Bagheri, and M. Makaremi, "Identification of Cladocera in the Caspian Sea," *Journal of Fisheries Sciences*, vol. 5, pp. 61–76, 2011.

[13] A. Roohi, Z. Yasin, A. E. Kideys, A. T. S. Hwai, A. G. Khanari, and E. Eker-Develi, "Impact of a new invasive ctenophore (*Mnemiopsis leidyi*) on the zooplankton community of the Southern Caspian sea," *Marine Ecology*, vol. 29, no. 4, pp. 421–434, 2008.

[14] A. Ganjian, W. O. Wan Maznah, K. Yahya et al., "Seasonal and regional distribution of phytoplankton in the southern part of the Caspian Sea," *Iranian Journal of Fisheries Sciences*, vol. 9, no. 3, pp. 382–402, 2010.

[15] R. P. Harris, P. H. Wiebe, J. Lenz, and H. R. Skjoldal, *Zooplankton Methodology Manual*, Academic, 2000.

[16] Y. A. Birshtain, L. G. Vinogradova, N. N. Kondakov, M. S. Koon, T. V. Astakhova, and N. N. Romanova, *Invertebrate Atlas Caspian Sea*, Industry Food, Moscow, Russia, 1968.

[17] A. Kasimov, *Methods of Monitoring in Caspian Sea*, QAPP-POLIQRAF, Azerbaijan, 2000.

[18] H. T. James and A. P. Covich, *Ecology and Classification of North American Freshwater Invertebrates*, Academic press, 2nd edition, 2001.

[19] J. E. Purcell, T. A. Shiganova, M. B. Decker, and E. D. Houde, "The ctenophore Mnemiopsis in native and exotic habitats: U.S. estuaries versus the Black Sea basin," *Hydrobiologia*, vol. 451, pp. 145–176, 2001.

[20] S. Bagheri, *Ecological assessment of plankton community and effects of alien species in the southwestern Caspian Sea [Ph.D. thesis]*, Universiti Sains Malaysia, 2012.

[21] S. N. Rodionov, *Global and Regional Climate Interaction: The Caspian Sea Experience*, Kluwer Academic, Dordrecht, The Netherlands, 1994.

[22] M. Bilio and U. Niermann, "Is the comb jelly really to blame for it all? Mnemiopsis leidyi and the ecological concerns about the Caspian Sea," *Marine Ecology Progress Series*, vol. 269, pp. 173–183, 2004.

[23] A. B. Polonskii, D. V. Basharin, E. N. Voskresenskaya, and S. Worley, "North atlantic oscillation: description, mechanisms, and influence on the eurasian climate," *Physical Oceanography*, vol. 14, no. 2, pp. 96–113, 2004.

[24] S. Bagheri, M. Mansor, M. Turkoglu, M. Marzieh, W. O. Wan Maznah, and H. Negaresatan, "Phytoplankton composition and abundance in the Southwestern Caspian Sea," *Ekoloji*, vol. 21, pp. 32–43, 2012.

[25] U. Sommer and A. Lewandowska, "Climate change and the phytoplankton spring bloom: warming and overwintering zooplankton have similar effects on phytoplankton," *Global Change Biology*, vol. 17, no. 1, pp. 154–162, 2011.

[26] S. Bagheri, U. Niermann, J. Sabkara, A. Mirzajani, and H. Babaei, "State of Mnemiopsis leidyi (Ctenophora: Lobata) and mesozooplankton in Iranian waters of the Caspian Sea during 2008 in comparison with previous surveys," *Iranian Journal of Fisheries Sciences*, vol. 11, no. 4, pp. 732–754, 2012.

3

High-Speed Vessel Noises in West Hong Kong Waters and Their Contributions Relative to Indo-Pacific Humpback Dolphins (*Sousa chinensis*)

Paul Q. Sims,[1] Samuel K. Hung,[2] and Bernd Würsig[3]

[1] *Department of Biology, McGill University, 1205 Avenue Docteur Penfield, Montreal, QC, Canada H3A 1B1*
[2] *Hong Kong Cetacean Research Project, Flat C 22/F., Block 13, Sceneway Garden, Lam Tin, Kowloon, Hong Kong*
[3] *Department of Marine Biology, Texas A&M University at Galveston, 200 SeaWolf Pkwy, Galveston, TX 77553, USA*

Correspondence should be addressed to Paul Q. Sims, paul.q.sims@gmail.com

Academic Editor: Nobuyuki Miyazaki

The waters of West Hong Kong are home to a population of Indo-Pacific humpback dolphins (*Sousa chinensis*) that use a variety of sounds to communicate. This area is also dominated by intense vessel traffic that is believed to be behaviorally and acoustically disruptive to dolphins. While behavioral changes have been documented, acoustic disturbance has yet to be shown. We compared the relative sound contributions of various high-speed vessels to nearby ambient noise and dolphin social sounds. Ambient noise levels were also compared between areas of high and low traffic. We found large differences in sound pressure levels between high traffic and no traffic areas, suggesting that vessels are the main contributors to these discrepancies. Vessel sounds were well within the audible range of dolphins, with sounds from 315–45,000 Hz. Additionally, vessel sounds at distances ≥100 m exceeded those of dolphin sounds at closer distances. Our results reaffirm earlier studies that vessels have large sound contributions to dolphin habitats, and we suspect that they may be inducing masking effects of dolphin sounds at close distances. Further research on dolphin behavior and acoustics in relation to vessels is needed to clarify impacts.

1. Introduction

Natural and anthropogenic sounds are part of the ocean environment. Natural sound is produced by physical (e.g., sea state, wind speed, precipitation, earthquakes) and biological (marine mammal vocalizations, fish communication, and snapping shrimp) sources ([1, 2] provide summaries). Anthropogenic sound, often termed "noise," is caused by human activities such as explosives, seismic exploration, sonar, ships, industrial activities, and acoustic deterrent and harassment devices [1, 3]. Some of these noises affect marine mammal communication sounds, including Indo-Pacific humpback dolphins [4–7]. Additionally, short-term behavioral changes can occur in cetaceans due to noise (e.g., changes in surfacing, diving, and movement patterns [8–10]), but long-term or physiological impacts have been less well explored. Chronic sources of noise pollution have been hypothesized to contribute to population differences in the sound repertoire of various species, but this evidence is observational in nature and does not exclude intrinsic population differences, such as subspeciation [11–13]. Thus, a detailed investigation of the potential impacts of these local chronic noises may help to clarify the nature of population differences in marine mammal sounds and deepen our understanding of the potential effects of chronic noise exposure.

Hong Kong waters are particularly busy, with many sources of anthropogenic disturbance (e.g., land reclamation, construction, dredging, heavy vessel traffic, chemical pollution, piling, dolphin tourism, etc.) throughout the area [14, 15]. These waters are also home to Indo-Pacific humpback dolphins (*Sousa chinensis*) and Indo-Pacific finless porpoises (*Neophocaena phocaenoides*) and are considered important habitat for these cetaceans [16]. Previous research [8, 17] focused on various anthropogenic disturbances that impacted local marine mammals, but only a few researchers

High-Speed Vessel Noises in West Hong Kong Waters and Their Contributions Relative to Indo-Pacific
Humpback Dolphins (Sousa chinensis)

27

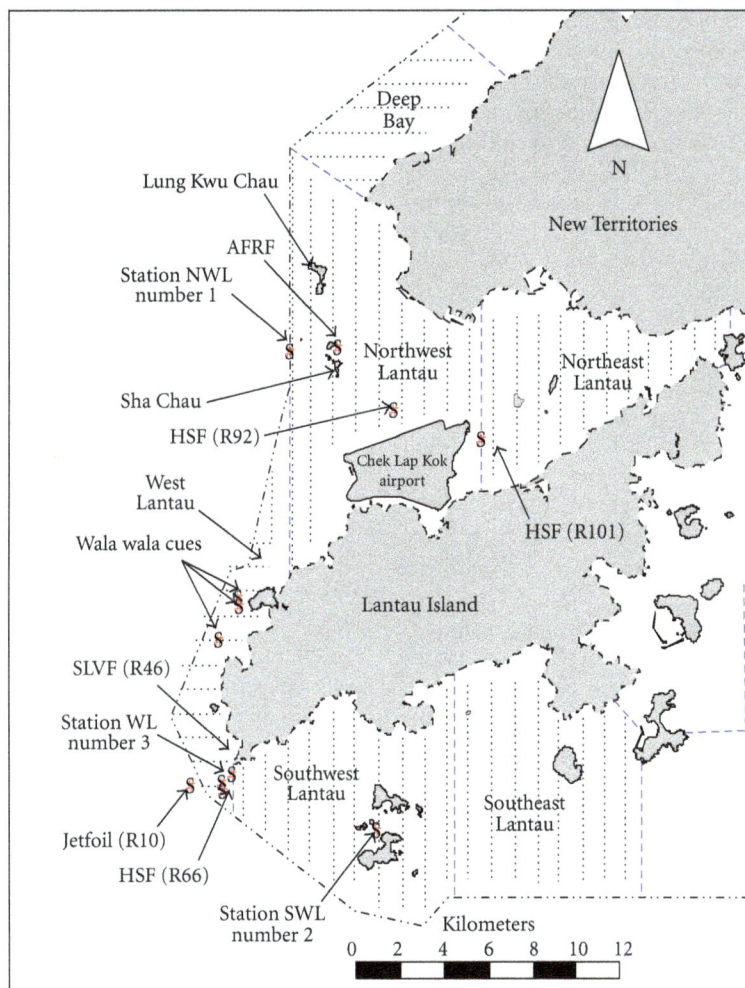

FIGURE 1: A map of Lantau Island in Hong Kong with recording stations, recording locations, and transects. The letter "R" and the following number correspond to the individual recording number associated with the location name. "HSF" stands for an unidentified high-speed ferry, "Jetfoil" is the name of an identified high-speed ferry, and "Wala wala" is an arbitrary name associated with the small tour boats in the area. The "SLVF" is the South Lantau Vessel Fairway and the "AFRF" is the retired Aviation Fuel Receiving Facility.

have studied the effects of noise pollution on these local species [18, 19]. Würsig and Greene [19] documented sound pressure level (SPL) relationships to different frequencies associated with tankers and tugs either offloading, approaching, or departing the Aviation Fuel Receiving Facility (AFRF, Figure 1). Their findings showed that North Lantau waters are relatively noisy, but the vessels in question still meet airport authority requirements; however, they also noted that the effects of these sound disturbances to the cetaceans (almost exclusively humpback dolphins in North Lantau waters) inhabiting the area are yet to be documented.

Our objectives of the current study were to broaden the scope of Würsig and Greene's [19] research by additionally examining the sound contributions of abundant high-speed vessels in the area and better quantify their various contributions to the high levels of background noise. Recent data indicate that the various activities of these vessels may be partially related to recent declines of Indo-Pacific humpback dolphins in Hong Kong waters [20]. Thus, we provide a summary of select high-speed vessel sounds

relative to background sound levels and dolphin hearing and communication sounds. Understanding the various sound contributions of these vessels will be useful in determining their effects on marine mammals in the area and provide data for potential attempts at mitigation. Additionally, this research may also generate further insights into whether differences in sound repertoires between the Hong Kong and Australian humpback dolphin populations can be attributed to ambient noise.

2. Materials and Methods

2.1. Field Methods. We sampled vessel, ambient, and Indo-Pacific humpback dolphin sounds at various area stations (Figure 1) in the waters surrounding Lantau Island in Hong Kong (latitude 22°15′00″, longitude 113°55′00″), from April to October 2010 and from February to August 2011. Samples were taken in conjunction with a long-term sound monitoring program conducted by the Hong Kong Cetacean Research Project. This program annually conducts line transects throughout the Hong Kong Special Administrative Region, which is divided into twelve different survey areas, with line-transect surveys conducted among six of these areas (i.e., Northwest (NWL), Northeast (NEL), West (WL), Southwest (SWL), Southeast Lantau (SEL), and Deep Bay (DB) Figure 1). We recorded vessel, ambient, and Indo-Pacific humpback dolphin sounds from the stern of a 15-m diesel vessel, the "standard 31516", with vessel noise off and the vessel drifting. We used a Cetacean Research Technology spot-calibrated hydrophone (model: CR1; sensitivity: 197.69 dB, re. V/μPa; linear frequency range listed as: 0.0002 kHz–48 kHz $\pm$ 3 dB; usable frequency range listed as: 0.00004 kHz–68kHz $\pm$ 3/–20 dB, only analyzing sounds up to 48 kHz due to our linear frequency range) to record sounds, and a Fostex digital recorder (model: FR-2; frequency response: 20 Hz–80 kHz $\pm$ 3 dB) with a preamplified signal conditioner (model: PC200-ICP; precision gain: x0.1–x100; frequency range: >100 kHz; system response: 1 Hz–100 kHz $\pm$ 0.25 dB) to prevent overloading. The hydrophone, suspended by a 2 m spar buoy, was lowered into the water at 3 to 7 m depths and recorded (sampling rate: 24-bit at 192 kHz) various durations in Broadcast Wave Format, ranging from 21 s to 15 min and 16 s. The spar buoy acted to prevent excessive hydrophone movement from wave and boat motion. During each sampling event, we recorded vessel type, distance from the recording vessel at cue time, vessel activity, and dolphin presence. The distance to vessels was noted using Bushnell laser rangefinding binoculars (distance accuracy $\pm$0.5 m up to 700–800 m). We also recorded the date, start and end times, hydrophone and water depths, Beaufort sea state, area, start and end location, gain, event, and any additional notes for each sampling event. We determined locations using a Garmin eTrex Legend H GPS unit. A total of 219 recordings were taken over 2010-2011, both with and without the presence of various vessel types; however, many recordings took place in the presence of multiple vessels.

2.2. Data Analysis. We selected and analyzed recordings of high-speed ferries, small tour boats, dolphins, and the ambient noise of various sites using SpectraLAB software (version 4.32) on a Lenovo ThinkPad T400 7417-PLU notebook PC. We divided vessel selections into two categories of solitary and multiple vessels present during the recording. We defined vessels as solitary if there were no other vessels present within 2 km from the recording vessel throughout the duration of the recording. Recordings in which there were two or more vessels within 2 km of each other in the study area were classified as having multiple vessels. We analyzed solitary vessel selections at specific cue times that described vessel distance and direction. These selections were analyzed over 5 s segments, $\pm$2.5 s of the cue time to accurately capture their sound pressure level without averaging out their sounds. We computed 1/3 octave band sound pressure levels and narrowband spectra in 1 Hz bands using SpectraLAB's "compute average spectrum" analysis for solitary vessel selections. We used a 1/3 octave bandwidth because of its general approximation to cetacean auditory bands [21] and narrowband spectra for finer scale detection

of vessel tonal signatures. These two measurements (1/3 octave band sound pressure levels and narrowband spectra) allowed us to describe the sound pressure levels and tonal sound signature, respectively, of the individual vessel at specific distances, relative to dolphin hearing and sounds. The multiple vessel recordings were used to gauge the relative contribution of the individual vessels to ambient sounds when multiple ships were present.

For ambient noise measurements, we took 10 s nonoverlapping section measurements throughout the recording starting at the beginning. Most recording times were not a multiple of 10, and we only measured the full 10 s clips for these. To avoid sound selection bias, we also repeated our measurements starting from the end of the recording. Furthermore, we randomly selected 18 of these selections and averaged them for each ambient sound recording to compute 1/3 octave band ambient sound pressure levels. To reduce geographic or nearby traffic differences between ambient sites and individual recordings, we selected sites near the individual vessel recording for ambient sound comparisons as recordings of the site with no vessels present were not available. These ambient sounds were used to assess individual vessel sound contributions relative to the natural background sounds (i.e., without ships).

For dolphin sounds, we played back spectrograms (fast Fourier transform (FFT): 8192; smoothing window: hanning; postprocess: 192 kHz; bandwidth: 1 Hz; FFT window overlap: 50%) to visually select sounds of interest, particularly sounds associated with social communication, for example, burst pulses and whistles [22]. We only used recordings of dolphin sounds when no ships were present. Dolphin proximity during recordings varied and ranged from around 50 to 200 m of our vessel during sound recordings. We analyzed selected dolphin sounds as narrowband spectra to compare to vessel spectral sounds. We compared these sounds to assess potential vessel sound masking effects on dolphin social communication. While only one audiogram is available for Indo-Pacific humpback dolphins [23], several exist for common bottlenose dolphins (*Tursiops truncatus*, hereafter simply "bottlenose dolphins") [24, 25]. Popov et al. [25] observed variation amongst individual bottlenose dolphin audiograms; as such, the single audiogram available for Indo-Pacific humpback dolphins may not accurately represent the mean hearing sensitivity of the species. Past research on humpback dolphin communication frequencies [12, 22] indicates that they share similarities in repertoire and frequency range to bottlenose dolphins, suggesting that these two species may also share similar audiograms. Therefore, we used published audiograms of both bottlenose dolphins [24, 25] and the single audiogram of an Indo-Pacific humpback dolphin [23] for comparison with the received sound pressure levels. For the bottlenose dolphin audiograms, we used the average sound pressure level for each frequency band since both audiograms gave multiple sound pressure level thresholds per frequency unit. For the Johnson [24] audiogram, we converted dB re 1 μbar to dB re 1 μPa by adding 100 to the recorded sound pressure level [21].

The data for all sound selections were saved to Microsoft Excel 2007 and subsequently plotted using R statistical software 2.13.1 [26].

3. Results

Of the 219 available recordings, only four were available for analysis of solitary High-Speed Ferries (HSF), specifically the Jetfoil and three different unclassified High-Speed Ferries. We also examined three recordings of a small speed boat ("Wala wala") escorting tourists to watch dolphins, in isolation. We measured the ambient sounds of four areas: Southwest Lantau Station number 2 (SWL number 2, Figure 1), South Lantau Vessel Fairway (SLVF, Figure 1), Northwest Lantau Station number 1 (NWL number 1, Figure 1), and West Lantau Station number 3 (WL number 3, Figure 1); see Table 1 for site details. We used SWL number 2 and NWL number 1 for comparisons to natural ambient sounds, while WL number 3 was used as a comparison to a usually busy traffic area with only one HSF present. Lastly, we used SLVF near Fan Lau (the southwest tip of Lantau Island; Figure 1), for an ambient sound recording of a generally busy traffic area with moderate vessel traffic (i.e., the presence of a shrimp trawler and several HSFs; Table 1).

3.1. Ambient Noise. A comparison of ambient sound levels for the four sites revealed several notable differences. The SLVF ambient sound levels were markedly higher throughout most of the frequency range of SWL number 2, (i.e., 50–10,000 Hz; Figure 2). SLVF ambient sound levels were also higher than parts of NWL number 1 and WL number 3's frequency ranges, particularly frequencies 316–20,000 Hz for NWL number 1 and both 50–500 Hz and 3162–25,000 for WL number 3. However, the differences in ambient sound levels between SLVF and WL number 3 were progressively less pronounced when approaching both above and below a frequency of 1,000 Hz. In fact, WL number 3 sound pressure levels slightly exceeded SLVF ambient sounds around 1,000 Hz. The relatively high sound pressure levels associated with SLVF correspond with the presence of several ships, a shrimp trawler, and three HSFs; as such, it is labeled as a busy traffic area.

In contrast, WL number 3, also considered a busy traffic area, had only one vessel present (a HSF away at 702 m) during the recording. For the lower frequencies (50–316 Hz), WL number 3 had the lowest sound pressure levels; however, sound pressure levels rapidly rose from 316–1,000 Hz and gradually declined to an equilibrium around 10,000 Hz at 100 dB re 1 μPa (Figure 2).

One of the areas considered to have quiet background sound levels, NWL number 1, maintained relatively low sound pressure levels throughout the frequency range, except in the lower frequencies, 50–400 Hz (Figure 2). In this short frequency range, NWL number 1 sound pressure levels remained near 95 dB re 1 μPa, which were the highest sound pressure levels of the four ambient sound recordings, until SLVF levels exceeded NWL number 1 around 160 Hz. Additionally, NWL number 1 displayed a brief spike in sound

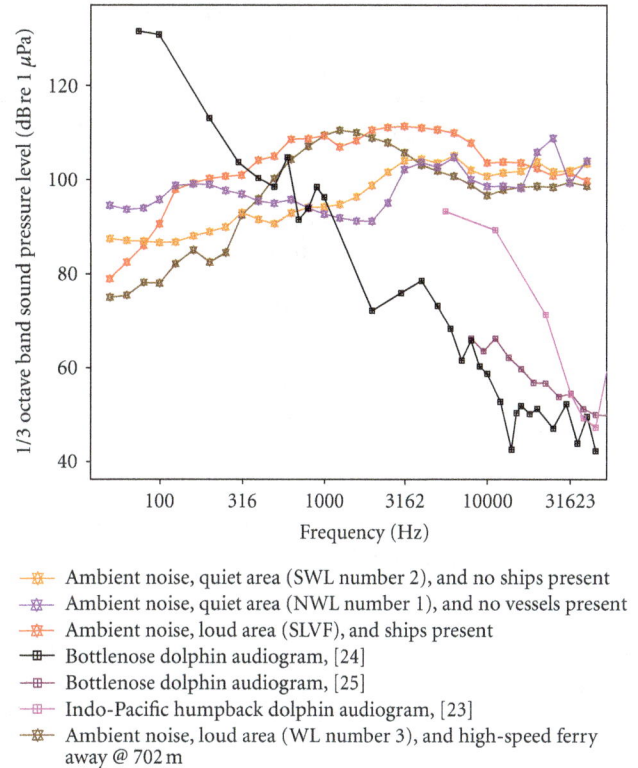

FIGURE 2: Ambient sounds of the four areas varying in general ship traffic and types of ships present. Bottlenose and humpback dolphin audiograms show the magnitudinal difference between vessel sounds and minimum audible levels for the dolphins. Southwest Lantau number 2 (SWL number 2) is between the Soko Islands with very little boat traffic and a Beaufort sea state (Bss) of 3. Likewise, Northwest Lantau number 1 (NWL number 1) is described as having very little boat traffic, located within the Sha Chau and Lung Kwu Chau Marine Park, recorded during a Bss of 3. West Lantau number 3 (WL number 3) is within the very busy shipping route at South Lantau Vessel Fairway with a high-speed ferry away at 702 m, Bss 3. South Lantau Vessel Fairway (SLVF) was recorded with several ferries and a shrimp trawler present during a Beaufort sea state of 0.

pressure levels to 108 dB re 1 μPa at 25,000 Hz, the peak sound pressure level for that site.

Lastly, SWL number 2, also considered to be a quiet area, was consistently in the relatively lower range of sound pressure levels to about 6,300 Hz, where it slightly exceeded the sound levels from two other sites by a few decibels (Figure 2). Sound pressure levels gradually began to increase at 160 Hz and peaked around 6,300 Hz, ranging in sound pressure levels from 93 to 104 dB re 1 μPa.

We also compared the dolphin audiograms to the ambient sounds to describe the audibility of the average background sound levels. The Johnson [24] bottlenose dolphin audiogram extended above all ambient sound levels to around 400 Hz, where it dropped below the ambient sounds of SLVF (Figure 2). All audiograms for the dolphins (bottlenose and humpback) followed a declining pattern as frequency increased, thereby augmenting the difference

TABLE 1: Descriptions of ambient noise recordings.

Location	Site description	Beaufort sea state	Vessel(s) present	Recording date
West Lantau Station no. 3 (WL no. 3)	Within the very busy shipping route at South Lantau Vessel Fairway	3	High-speed ferry	June 30, 2010
South Lantau Vessel Fairway (SLVF)	A busy area which experiences much traffic, particularly from ferries	0	Shrimp trawler, high-speed ferries: turbo jet, NWT, and zuhai	May 30, 2010
Southwest Lantau Station no. 2 (SWL no. 2)	Between the Soko Islands with very little boat traffic	3	None	September 1, 2010
Northwest Lantau Station no. 1 (NWL no. 1)	Within the Sha Chau and Lung Kwu Chau Marine Park with very little boat traffic	3	None	August 13, 2010

in sound pressure levels between dolphin hearing thresholds and the various ambient sounds. However, we found interspecific variation in the magnitude of the difference between dolphin hearing thresholds and sound pressure. Notably, the difference between the humpback dolphin audiogram and SLVF sound pressure levels was smaller as compared to the bottlenose dolphin audiogram and SLVF sound pressure levels. For example, near 5,600 Hz the difference for humpback dolphins was ~17 dB re 1 μPa compared to ~37 dB re 1 μPa for bottlenose dolphins. However, this interspecific difference between audiograms and SLVF rapidly decreased as frequencies increased, with the two species converging around 31 kHz. While our study did not extend to frequencies above 48 kHz, it should be noted that the humpback audiogram diverged from the bottlenose audiogram and increased in sound pressure levels following frequencies above 48 kHz. We also observed intraspecific variation in hearing thresholds between the bottlenose audiograms. The Popov et al. [25] audiogram declined at a slower rate as compared to the Johnson [24] audiogram. Additionally, the Popov et al. [25] audiogram was an average of 13 bottlenose dolphin subjects and may be a more accurate representation of a bottlenose audiogram. We were limited by the existence of only one available audiogram from a single humpback dolphin, and individual variation may potentially bias our observed differences. While both bottlenose audiograms show a clear continuing trend of decline, the Popov et al. [25] audiogram appeared to begin leveling out around the end of the frequency range, that is, 48,000 Hz. The data from Johnson [24] audiogram did not extend beyond an upper frequency limit of 45,000 Hz; likewise, Popov et al. [25] did not record responses to frequencies below 8,000 Hz. Because of these data gaps, we display both audiograms for better clarity in frequency and sound pressure auditory thresholds in bottlenose dolphins.

3.2. High-Speed Ferry and Small Tour Boat Sounds. At most distances, the HSF and small tour boat (also referred to as "Wala wala") sounds were much louder when compared to the corresponding natural ambient sound levels from either SWL number 2 or NWL number 1 (Figures 3 and

4, but see Figure 5). These higher sound pressure levels were consistent throughout the frequency range, though they usually declined to levels similar to those of the natural ambient sound in the upper frequencies (e.g., 4,000–10,000 Hz). HSF and some small tour boat sound pressure levels generally extended beyond the SLVF ambient sounds, but this tended to be at the closer distances, for example, between 100 and 400 m of apparent sound sources (Figures 3 and 4, but see Figure 5). The frequency ranges varied for sound pressure levels exceeding the SLVF ambient sounds. Some HSF sounds stayed above the SLVF ambient sounds throughout the frequency range (Figures 3(c), 4(c)) while others only exceeded the SLVF ambient sound levels across select frequencies (Figures 3(a), 4(a), and 5(a)). Sound pressure levels also tended to peak between 100–3,000 Hz, although the exact peak varied between vessels. The highest sound pressure levels peaked around 120 dB, except for one HSF (Figure 4(c)). These peaks were associated with a range of distances, from <100–556 m. The direction of the vessel (i.e., approaching or away) may have affected some of the received sounds. In some cases, received sounds were higher from distances away than from approaching (Figures 4(a), 4(c), and 5(a)). Additionally, in a few cases, sound levels were generally higher for stopped rather than moving tour boats (Figure 3(a), at 43 m from 1,500–4,000 Hz).

We found differences between the vessel generated sounds and the dolphin audiograms similar to those described for ambient sounds. However, most vessel sounds exceeded the ambient sound levels, increasing the differences in sound levels between vessels and dolphin audiograms. This increased difference was readily apparent in Figures 3(a), from frequencies 1,000 to 10,000 Hz, 4(a), from frequencies 316 to 2,500 Hz, and 4(c), from frequencies 316 to 45,000 Hz. While ambient sounds did not appear to be audible to bottlenose dolphins around frequencies ≤400 Hz, vessel sound pressure levels reached or exceeded the dolphin auditory threshold at lower frequencies, from 200–300 Hz (Figures 3(a), 3(c), 4(a), and 4(c)). Our humpback dolphin audiogram did not extend below 5,600 Hz, so we were unable to determine if humpback dolphins show similar decreases in hearing sensitivity to bottlenose dolphins in the lower frequencies.

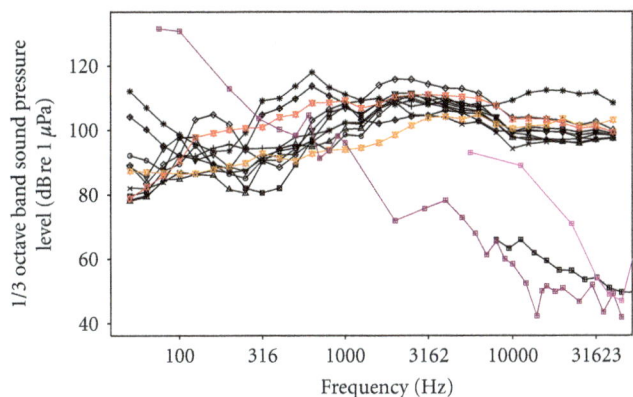

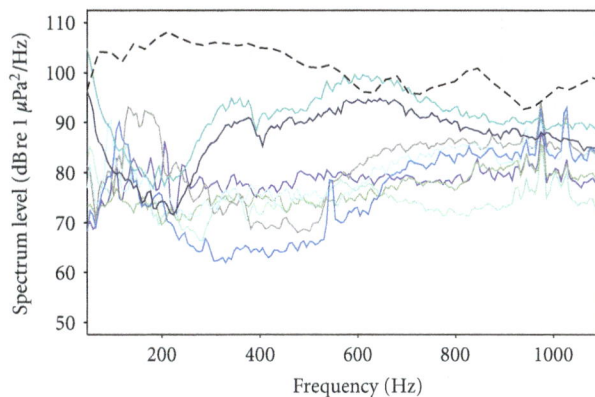

(a)

(b)

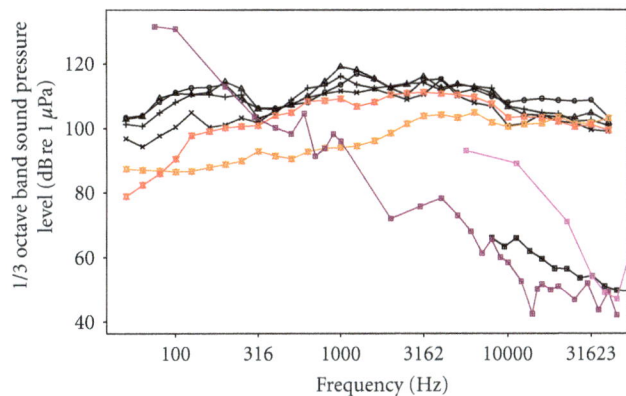

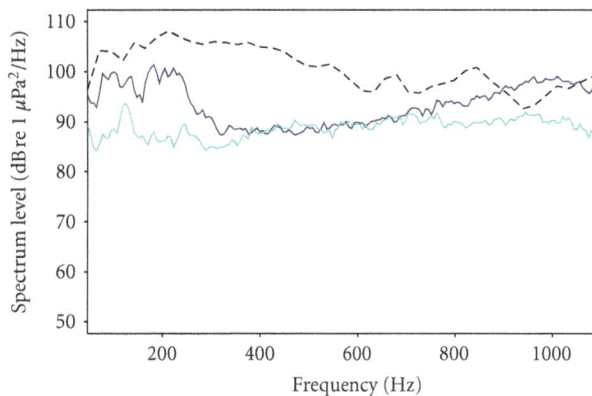

(c)

(d)

FIGURE 3: 1/3 octave band sound pressure levels for a small tour boat (referred to as "Wala wala") in West Lantau (a), Beaufort sea state of 1. The orange line indicates the ambient sounds of Southwest Lantau number 2, located between the Soko Islands with very little boat traffic and no vessels present during recording. The red line indicates South Lantau Vessel Fairway, a busy traffic area, especially for ferries. South Lantau Vessel Fairway was recorded with several ferries and a shrimp trawler present. Selected sound spectra for the small tour boat and humpback sounds are displayed below (b). (c) shows the distribution of sound pressure levels for the Jetfoil high-speed ferry at varying distances, with a Beaufort sea state of 4 at West Lantau Station number 3. The ambient sound levels are the same as those for the small tour boat. (d) shows selected sound spectra for the Jetfoil compared to those of humpback dolphin communication sounds at a distance ≤100 m.

(a)

(b)

(c)

(d)

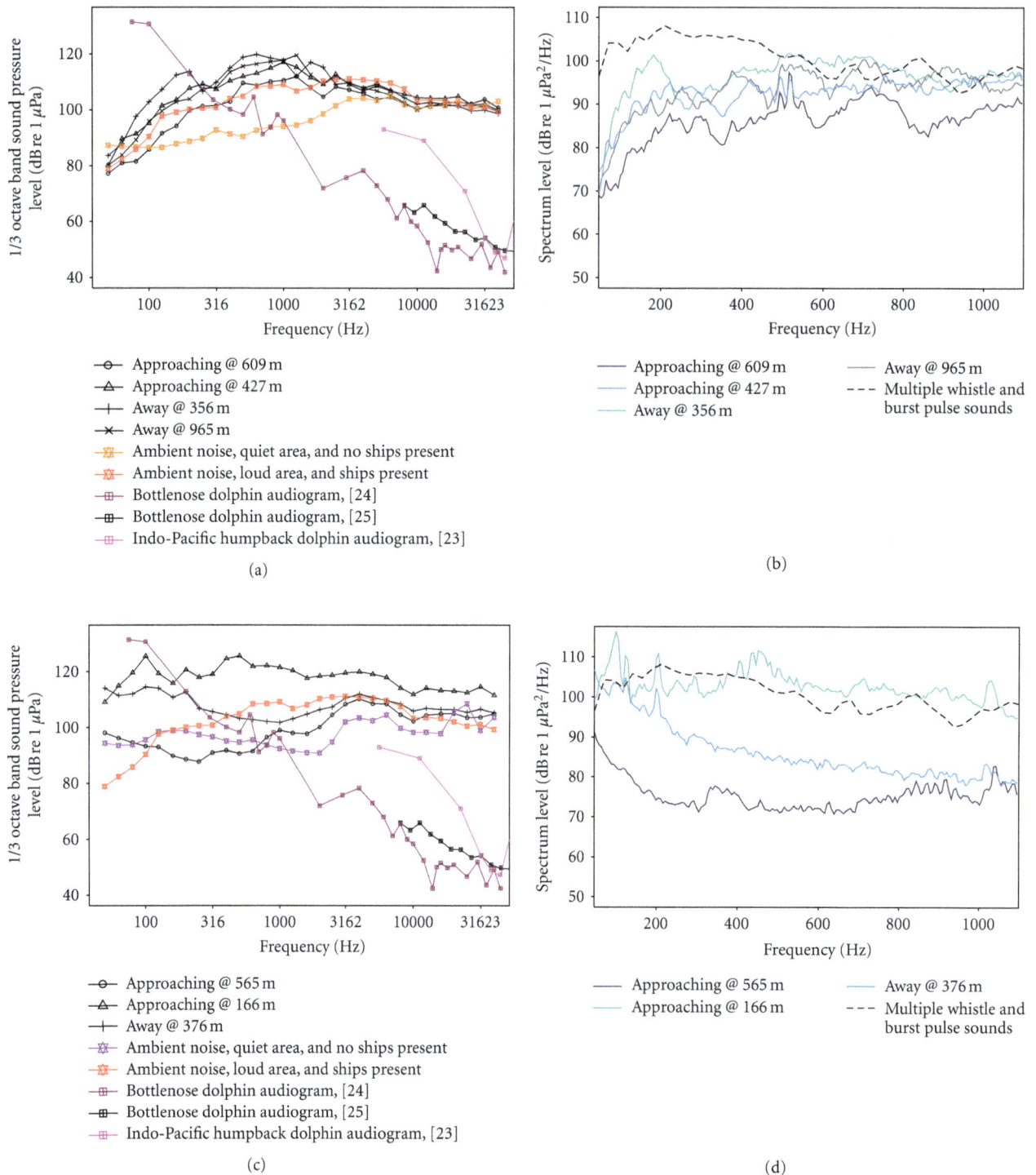

FIGURE 4: The various sound pressure level contributions of a high-speed ferry (R66; a), at West Lantau number 3 with a Beaufort sea state of 2. The orange line indicates the ambient sounds of Southwest Lantau number 2, located between the Soko Islands with very little boat traffic and no vessels present during recording. The red line indicates South Lantau Vessel Fairway, a busy traffic area, especially for ferries. South Lantau Vessel Fairway was recorded with several ferries and a shrimp trawler present. (b) shows selected sound spectra of the above high-speed ferry in comparison to the spectra of humpback dolphin communication sounds at a distance of ≤100 m. (c) The various sound pressure level contributions of a high-speed ferry (R92), at Northwest Lantau number 5 with a Beaufort sea state of 4. The ambient noise level (represented in purple) was taken from Northwest Lantau number 1, an area with very little boat traffic, located within the Sha Chau and Lung Kwu Chau Marine Park. The red line is the same as that in (a). Sound spectra are represented below (d) for the high-speed ferry, including the same humpback dolphin spectra as (b).

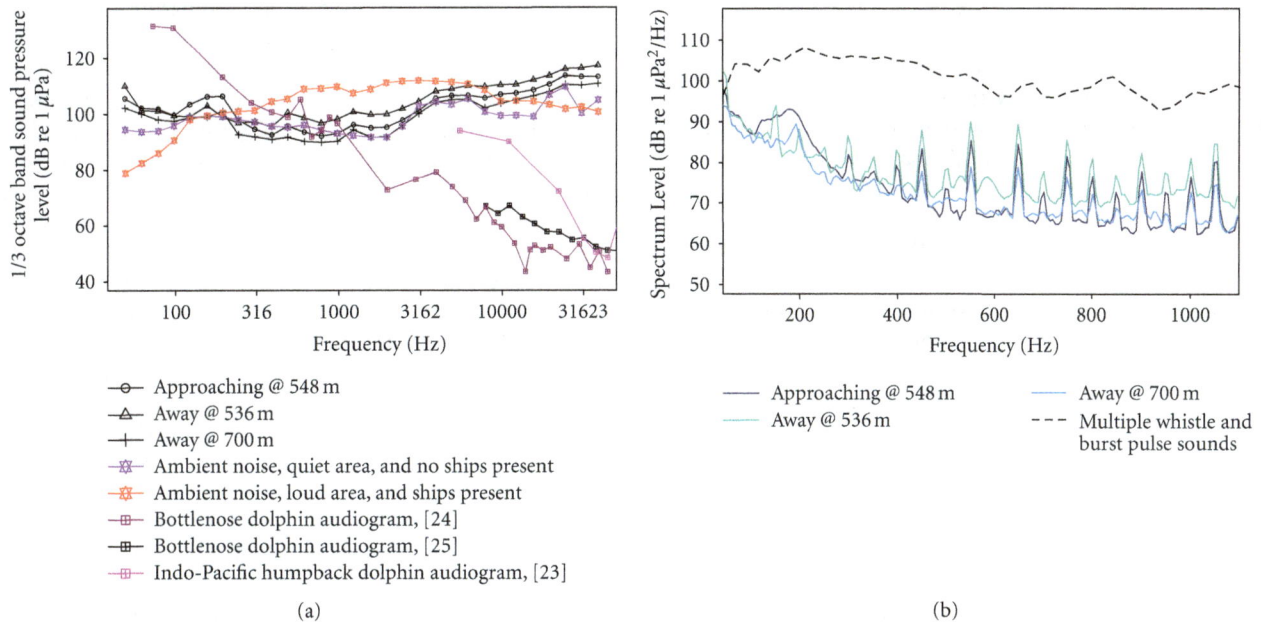

FIGURE 5: (a) The various sound pressure level contributions of a Hi-Speed Ferry (R101), at Northeast Lantau Station number 1 with a
Beaufort sea state of 2. The ambient noise level (represented in purple) was taken from Northwest Lantau number 1, an area with very little
boat traffic, located within the Sha Chau and Lung Kwu Chau Marine Park. The red indicates the South Lantau Vessel Fairway (SLVF), a
busy traffic area, especially for ferries. SLVF was recorded with several ferries and a shrimp trawler present during a Beaufort sea state of 0.
(b) shows selected sound spectra of the above high-speed ferry in comparison to the spectra of humpback dolphin communication sounds
at a distance of ≤100 m.

Sound spectra for the HSFs and small tour boat were
highly variable among vessels. Sound spectra for which
spectral components from all distances converged on each
other indicated that these sounds were likely part of the
ambient background sounds and not due to vessel inputs
(Figure 5(a), from 0 to >1,000 Hz). Some spectra were clearly
distinct in the upper frequencies (e.g., Figure 3(b), from
around 300–800 Hz and Figure 4(b), from 0 to >1,000 Hz),
suggesting that these spectra may have been individual vessel
tonal elements. Distinctive spectral components were also
present in the lower frequencies that appeared only at closer
distances, indicating they were not ambient sounds (e.g.,
Figure 3(b), gray line from 100 to 300 Hz).

The selected humpback dolphin section contained multiple burst pulse and whistle sounds, which are associated with
social behavior and communication [22]. Spectral analysis
revealed high spectrum levels for humpback dolphin sounds,
which were maintained between 95–104 dB re 1 μPa^2/Hz. In
most vessel spectrum plots, the humpback dolphin sounds
were much higher than the vessel sounds present throughout
much of the frequency range (but see Figures 4(b) and 4(d));
however, these dolphin sounds were recorded at distance
≤100 m in comparison to vessel distances, which could range
from <100 to >700 m.

4. Discussion

We determined that vessels contribute appreciable sound
levels to the ambient environment. Greater vessel traffic

appeared to be associated with higher sound pressure levels
across most frequencies. South Lantau Vessel Fairway had the
highest sound pressure levels across most of the frequency
range and the greatest number of vessels present, that is, four.
Though other sites also had relatively high sound level peaks,
none were maintained across the majority of the frequency
band. The other site which maintained relatively high sound
pressure levels, though over smaller frequency range, was WL
number 3, in which a HSF was present during recording. It is
unlikely that these differences can be attributed to Beaufort
sea state (Bss), as SLVF was a Bss of 0, and the other sites,
including WL number 3, were Bss of 3. Because recordings
were taken over such short time periods, seasonal differences
are not likely to be responsible for the observed differences in
sound pressure levels. These results suggest that the presence
of vessel traffic contributes to increases in ambient sound
levels. However, it seems unlikely that the presence of one
HSF at 702 m would contribute to such a high sound pressure
level after ambient sounds were averaged. We propose that
the unexpectedly high sound level peak in WL number 3
may be due to other anthropogenic activity outside the
immediate vicinity, for example distant vessel traffic as WL
number 3 is located directly in the path of the Southern
Lantau vessel route. No recordings were available for WL
number 3 in which vessels were absent, so we are unable
to eliminate the possibility that those sites with vessel traffic
are louder due to factors besides vessel traffic. Nevertheless,
it seems likely that vessels are important contributars to the
current ambient sound environment in both SLVF and WL

number 3, particularly when coupled with the individual sound pressure level data from the HSFs and small tour boat.

While SLVF maintained the highest ambient sound levels across the largest frequency range, NWL number 1 noticeably peaked in the ambient sound to a level at or above that of SLVF at the same frequency (125 and 25,000 Hz resp.). This result is not consistent with the expectation that ambient sounds from quiet areas (NWL number 1) would be lower than those of busy areas (SLVF). We doubt that Bss is the cause of these peaks as they do not appear in the other quiet ambient site having the same Bss of 3 as NWL number 1. An examination of Knudsen's predictions for Bss of 3 (converted to 1/3 octave sound pressure levels in [27]) shows sound levels decreasing at a constant rate from ~94 decibels as frequencies increase from the beginning at 100 Hz. While Bss may account for sounds in the lower frequencies ($\leq$2,500 Hz) for NWL number 1 and SWL number 2, the Knudsen curves do not indicate a similar correspondence for sounds in the upper frequencies (>2,500 Hz) for either NWL number 1 or SWL number 2. Additionally, Knudsen curves predict sound levels beginning around 92 dB at 100 Hz, which SWL number 2 sound levels do not reach until 400 Hz. At 100 Hz, SWL number 2 sound levels are 86 decibels, about 6 decibels below the expected sound pressure level. Thus, Knudsen curves for Bss of 3 may only explain the sound levels present in the lower frequencies of NWL number 1. We hypothesize that nearby traffic accounts for the unexpected sound peaks in NWL number 1, whereas local geography may be reducing sound propagation in SWL number 2. As summarized by Malme et al. [28], sound transmission can vary greatly in shallow environments due to acoustic effects of the bottom and surface.

We used both bottlenose and a humpback dolphin audiograms to compare vessel sound outputs to dolphin hearing. Only one audiogram exists for Indo-Pacific humpback dolphins [23], and recent research indicates that they share similar communication frequencies and repertoire as bottlenose dolphins [12, 22], thus we used the Johnson [24] bottlenose dolphin audiogram as a proxy for humpback hearing sensitivity in frequencies below those in the humpback audiogram. The sound pressure levels for SLVF peak around 110 decibels from around 800–10,000 Hz, at the hydrophone (with unknown levels at a standard 1 m distance from the sound source), well within the lower audible range of bottlenose dolphins [24, 25] and partially so for humpback dolphins [23]. Extrapolation of the bottlenose dolphin audiogram to humpback dolphins in lower frequencies may be questionable based on some of the observed differences where humpback hearing threshold data overlap with those of bottlenose dolphins. However, individual variation in audiograms exists in bottlenose dolphins [24, 25] and humpback dolphins likely exhibit similar differences as well. Thus, any conclusions of species differences or similarities in hearing thresholds should be taken cautiously until more data are available on variability in humpback dolphin hearing thresholds.

It is unknown if ambient noise of the level that we observed may cause physiological damage, increased stress, or behavioral changes since long-term data are not available

for these traits in the Hong Kong population of humpback dolphins. However, humpback dolphins exhibit behavioral changes in response to high levels of traffic, with greater occurrences of longer dives associated with the presence of some oncoming vessels, particularly those at high speeds [8]. It is assumed that increasing diving duration in response to oncoming and high levels of vessel traffic results in elevated stress levels. Furthermore, humpback dolphins increase their whistling rates after a vessel (<1.5 km) has passed, which is hypothesized to function as reestablishing group cohesion [7]. Thus, humpback dolphins may experience increased stress and both physical and communicative behavioral changes in busy traffic environments such as SLVF.

The ambient noise level of SLVF may be a conservatively low estimate since these data were collected during the presence of multiple ships, all of which changed in proximity to the hydrophone throughout the recording. No ships were present during the recordings for SWL number 2 and NWL number 1, so this issue does not pertain to them. Due to the random nature of our selections, it is likely that the represented noise levels are a mixture of both near and far ship distances. Ships closer in proximity will generate higher sound pressure levels, thus our estimated ambient noise level is likely more representative of the average sound levels recorded from the average distance of ships during our recording. This is potentially problematic in determining the effects of noise on the local dolphins, since it is presently unknown what distances dolphins maintain (or attempt to maintain) from ships. Ng and Leung [8] documented differences in humpback dolphin responses to vessel type and distance; however, they did not describe dolphin responses to specific vessel types at varying distances. They report higher rates of vessel avoidance by humpback dolphins in response to high-speed vessels, but it is unknown at what distances these behavioral changes were documented. However, Piwetz et al. [29] found behavioral changes, such as mean leg speed and reorientation rate, in response to small tour boats and trawlers within 1 km. Additionally, many of the vessels present in SLVF are HSFs, which are fast moving vessels, known to make abrupt entrances and departures at high speeds [29]. These HSFs could quickly increase their proximity to dolphins, and sound pressure levels can elevate rapidly, potentially causing startle or other reactions. Indeed, some research indicates increased unpredictability in vessel movement can have stronger effects on dolphin behavior [30, 31]. The potential magnitude of ambient noise levels for SLVF is dependent upon the assumption that the local dolphins maintain distances similar to the average distances between the hydrophone and ships recorded in our analyzed selections. This highlights a need for further research on humpback dolphin proximity and behavior in the presence of various ships to determine potential differences in behavior at varying distances and vessel types.

The difference in most sound pressure levels between vessels and ambient sound recordings highlights a potentially disruptive contribution to the local noise levels, particularly when compared to the quiet ambient background sounds. Most of the vessel sounds exceeded the quiet ambient background at the majority of distances; however, this

High-Speed Vessel Noises in West Hong Kong Waters and Their Contributions Relative to Indo-Pacific
Humpback Dolphins (Sousa chinensis)

35

difference was generally constricted to distances of 100–500 m of apparent sound sources when compared to the busy ambient recordings. Thus, the impact of these increased levels depends on the proximity of dolphins to the vessels. Many of the spectra present at short distances were not present at farther distances, indicating that vessel sounds generally did not propagate to distances ≥600 m. The overlap of spectra above the dolphin sounds suggests that some vessel sounds at distances ≥100 m may disrupt humpback dolphin communication sounds. Considering the fast speeds that these vessels can undertake, dolphins may not have adequate time to distance themselves and may suffer physiological impairment or stress in addition to masking effects on communication. On a long-term scale, this could result in chronic damage, stress, and communication disruption. Indeed, differences in whistles among populations of bottlenose dolphins have been attributed to site differences in vessel traffic and ambient noise and also have been found to vary with the number of vessels present [11, 32]. Thus, it is possible that the differences in the local noise environment may account for differences in communication sounds between the Hong Kong and Australian humpback dolphin populations [12].

One HSF displayed sound pressure level was consistently lower throughout the HSF frequency range for SLVF and around the ambient sound levels for NWL number 1 (Figure 5). This result may be due to individual differences in vessel structure or speed a likely explanation as the other four HSF recordings displayed opposite sound pressure levels, with the majority of their sounds being at or above ambient noise levels. Additionally, the wide variation in HSF sounds suggests individual differences in sound output among all ferries; however, it is unclear whether these discrepancies are from unique ship structures, differences in vessel speeds, or local habitat characteristics [1, 28].

In sum, it appears that the HSFs and small tour boats make important contributions to the local sound environment, although the influence of factors such as local topography and vessel sound propagation and attenuation have yet to be studied. As these vessels are numerous in West Hong Kong waters, management of their speeds and distribution are important in mitigating potential effects on the local dolphin population. Future research should focus on understanding how dolphins distribute themselves spatially relative to these vessels, and how this may vary with differing speeds and distances. The uncertainty in interspecific differences and/or similarities in audiogram hearing thresholds highlights a need for more Indo-Pacific humpback dolphin audiograms to help determine the extent that local and global delphinids may be affected by small high-speed vessels. Additionally, population differences in sound repertoire between Hong Kong and Australian humpback dolphins have yet to be resolved, but provide an opportunity to investigate the potential role of noise pollution in these differences. As an ultimate goal, determination of both the acute and chronic effects of different sound pressure levels on delphinid physiology, behavior, and communication will help to assess and manage anthropogenic ship disturbances of cetacean populations.

Acknowledgments

The authors thank Joe Olson of Cetacean Research Technology for advice on sound analyses. They also thank Diana Raper for preliminary comments on the paper, and an anonymous reviewer for constructive comments on the submitted paper. Thanks are due to Vincent Ho, Keira Yau, Perry Chan, Cherry Yeung, and Mr. and Mrs. Ng for their support during acoustic recordings. The Agriculture, Fisheries and Conservation Department of the Hong Kong SAR Government gave funding support and data collection assistance. P. Q. Sims thanks the National Science Foundation for providing initial funding for the data analyses.

References

[1] C. R. Greene Jr. and S. E. Moore, "Man made noise," in *Marine Mammals and Noise*, W. J. Richardson, C. R. Greene Jr., C. I. Malme, and D. H. Thomson, Eds., chapter 6, Academic Press, San Diego, Calif, USA, 1995.

[2] National Research Council, *Ocean Noise and Marine Mammals*, National Academy Press, Washington, DC, USA, 2003.

[3] J. A. Hildebrand, "Anthropogenic and natural sources of ambient noise in the ocean," *Marine Ecology Progress Series*, vol. 395, pp. 5–20, 2009.

[4] M. M. Holt, D. P. Noren, V. Veirs, C. K. Emmons, and S. Veirs, "Speaking up: killer whales (*Orcinus orca*) increase their call amplitude in response to vessel noise," *Journal of the Acoustical Society of America*, vol. 125, no. 1, pp. EL27–EL32, 2009.

[5] M. L. Melcón, A. J. Cummins, S. M. Kerosky, L. K. Roche, S. M. Wiggins, and J. A. Hildebrand, "Blue whales respond to anthropogenic noise," *PLoS One*, vol. 7, no. 2, article e32681, 2012.

[6] K. C. Buckstaff, "Effects of watercraft noise on the acoustic behavior of bottlenose dolphins, *Tursiops truncatus*, in Sarasota bay, Florida," *Marine Mammal Science*, vol. 20, no. 4, pp. 709–725, 2004.

[7] S. M. van Parijs and P. J. Corkeron, "Boat traffic affects the acoustic behaviour of Pacific humpback dolphins, *Sousa chinensis*," *Journal of the Marine Biological Association of the United Kingdom*, vol. 81, no. 3, pp. 533–538, 2001.

[8] S. L. Ng and S. Leung, "Behavioral response of Indo-Pacific humpback dolphin (*Sousa chinensis*) to vessel traffic," *Marine Environmental Research*, vol. 56, no. 5, pp. 555–567, 2003.

[9] P. J. O. Miller, N. Biassoni, A. Samuels, and P. L. Tyack, "Whale songs lengthen in response to sonar," *Nature*, vol. 405, no. 6789, p. 903, 2000.

[10] S. M. Nowacek, R. S. Wells, and A. R. Solow, "Short-term effects of boat traffic on bottlenose dolphins, Tursiops truncatus, in Sarasota Bay, Florida," *Marine Mammal Science*, vol. 17, no. 4, pp. 673–688, 2001.

[11] T. Morisaka, M. Shinohara, F. Nakahara, and T. Akamatsu, "Effects of ambient noise on the whistles of Indo-Pacific bottlenose dolphin populations," *Journal of Mammalogy*, vol. 86, no. 3, pp. 541–546, 2005.

[12] P. Q. Sims, R. Vaughn, S. K. Hung, and B. Würsig, "Sounds of Indo-Pacific humpback dolphins (*Sousa chinensis*) in West Hong Kong: a preliminary description," *Journal of the Acoustical Society of America*, vol. 131, no. 1, pp. EL48–EL53, 2012.

[13] C. H. Frère, J. Seddon, C. Palmer, L. Porter, and G. J. Parra, "Multiple lines of evidence for an Australasian geographic

boundary in the Indo-Pacific humpback dolphin (*Sousa chinensis*): population or species divergence?" *Conservation Genetics*, vol. 12, no. 6, pp. 1633–1638, 2011.

[14] T. A. Jefferson, S. K. Hung, and B. Würsig, "Protecting small cetaceans from coastal development: impact assessment and mitigation experience in Hong Kong," *Marine Policy*, vol. 33, no. 2, pp. 305–311, 2009.

[15] B. Morton, "Protecting Hong Kong's marine biodiversity: present proposals, future challenges," *Environmental Conservation*, vol. 23, no. 1, pp. 55–65, 1996.

[16] T. A. Jefferson and S. K. Hung, "A review of the status of the Indo-Pacific humpback dolphin (*Sousa chinensis*) in Chinese waters," *Aquatic Mammals*, vol. 30, no. 1, pp. 149–158, 2004.

[17] C. L. H. Hung, R. K. F. Lau, J. C. W. Lam et al., "Risk assessment of trace elements in the stomach contents of Indo-Pacific Humpback Dolphins and Finless Porpoises in Hong Kong waters," *Chemosphere*, vol. 66, no. 7, pp. 1175–1182, 2007.

[18] B. Würsig, C. R. Greene Jr., and T. A. Jefferson, "Development of an air bubble curtain to reduce underwater noise of percussive piling," *Marine Environmental Research*, vol. 49, no. 1, pp. 79–93, 2000.

[19] B. Würsig and C. R. Greene, "Underwater sounds near a fuel receiving facility in western Hong Kong: relevance to dolphins," *Marine Environmental Research*, vol. 54, no. 2, pp. 129–145, 2002.

[20] S. K. Hung, "Monitoring of marine mammals in Hong Kong waters—data collection: final report (2011-12)," Tech. Rep., Agriculture, Fisheries and Conservation Department of Hong Kong SAR Government, 2012.

[21] C. R. Greene Jr., "Acoustic concepts and terminology," in *Marine Mammals and Noise*, W. J. Richardson, C. R. Greene Jr., C. I. Malme, and D. H. Thomson, Eds., chapter 2, pp. 15–32, Academic Press, San Diego, Calif, USA, 1995.

[22] S. M. van Parijs and P. J. Corkeron, "Vocalizations and behaviour of Pacific humpback dolphins Sousa chinensis," *Ethology*, vol. 107, no. 8, pp. 701–716, 2001.

[23] S. Li, D. Wang, K. Wang et al., "Evoked-potential audiogram of an Indo-Pacific humpback dolphin (*Sousa chinensis*)," *The Journal of Experimental Biology*, vol. 215, no. 17, pp. 3055–3063, 2012.

[24] C. S. Johnson, "Sound detection thresholds in marine mammals," in *Marine Bio-Acoustics*, W. N. Tavolga, Ed., pp. 247–260, Pergamon Press, Oxford, UK, 1967.

[25] V. V. Popov, A. Ya. Supin, M. G. Pletenko et al., "Audiogram variability in normal bottlenose dolphins (*Tursiops truncatus*)," *Aquatic Mammals*, vol. 33, no. 1, pp. 24–33, 2007.

[26] R Development Core Team, *R: A Language and Environment For Statistical Computing*, R Foundation for Statistical Computing, Vienna, Austria, 2011, http://www.R-project.org/.

[27] C. R. Greene Jr., "Ambient noise," in *Marine Mammals and Noise*, W. J. Richardson, C. R. Greene Jr., C. I. Malme, and D. H. Thomson, Eds., chapter 5, pp. 87–100, Academic Press, San Diego, Calif, USA, 1995.

[28] C. I. Malme, B. Beranek, and Newman, "Sound propagation," in *Marine Mammals and Noise*, W. J. Richardson, C. R. Greene Jr., C. I. Malme, and D. H. Thomson, Eds., chapter 4, pp. 59–86, Academic Press, San Diego, Calif, USA, 1995.

[29] S. Piwetz, S. Hung, J. Wang, D. Lundquist, and B. Würsig, "Influence of vessel traffic on movements of Indo-Pacific humpback dolphins (Sousa chinensis) off Lantau Island, Hong Kong," *Aquatic Mammals*, vol. 38, no. 3, pp. 325–331, 2012.

[30] R. Constantine, D. H. Brunton, and T. Dennis, "Dolphin-watching tour boats change bottlenose dolphin (*Tursiops truncatus*) behaviour," *Biological Conservation*, vol. 117, no. 3, pp. 299–307, 2004.

[31] D. Lusseau, "Male and female bottlenose dolphins *Tursiops* spp. have different strategies to avoid interactions with tour boats in Doubtful Sound, New Zealand," *Marine Ecology Progress Series*, vol. 257, pp. 267–274, 2003.

[32] L. J. May-Collado and D. Wartzok, "A comparison of bottlenose dolphin whistles in the atlantic ocean: factors promoting whistle variation," *Journal of Mammalogy*, vol. 89, no. 5, pp. 1229–1240, 2008.

4

Fish Larvae Response to Biophysical Changes in the Gulf of California, Mexico (Winter-Summer)

Raymundo Avendaño-Ibarra,[1,2] Enrique Godínez-Domínguez,[1]
Gerardo Aceves-Medina,[2] Eduardo González-Rodríguez,[3] and Armando Trasviña[3]

[1] Departamento de Estudios para el Desarrollo Sustentable de la Zona Costera, Centro Universitario de la Costa Sur (CUCSUR),
Universidad de Guadalajara, Avendia V. Gómez Farías 82, 48980 San Patricio Melaque, JAL, Mexico
[2] Instituto Politécnico Nacional, CICIMAR-IPN, Departamento de Plancton y Ecología Marina, 23096 La Paz, BCS, Mexico
[3] Centro de Investigación Científica y de Educación Superior de Ensenada, Unidad La Paz, 23050 La Paz, BCS, Mexico

Correspondence should be addressed to Raymundo Avendaño-Ibarra; ravendan@ipn.mx

Academic Editor: E. A. Pakhomov

We analyzed the response of fish larvae assemblages to environmental variables and to physical macro- and mesoscale processes in the Gulf of California, during four oceanographic cruises (winter and summer 2005 and 2007). Physical data of the water column obtained through CTD casts, sea surface temperature, and chlorophyll a satellite imagery were used to detect mesoscale structures. Zooplankton samples were collected with standard Bongo net tows. Fish larvae assemblages responded to latitudinal and coastal-ocean gradients, related to inflow of water to the gulf, and to biological production. The 19°C and 21°C isotherms during winter, and 29°C and 31°C during summer, limited the distribution of fish larvae at the macroscale. Between types of eddy, the cyclonic (January) registered high abundance, species richness, and zooplankton volume compared to the other anticyclonic (March) and cyclonic (September). Thermal fronts (Big Islands) of January and July affected the species distribution establishing strong differences between sides. At the mesoscale, eddy and fronts coincided with the isotherms mentioned previously, playing an important role in emphasizing the differences among species assemblages. The multivariate analysis indicated that larvae abundance was highly correlated with temperature and salinity and with chlorophyll a and zooplankton volume during winter and summer, respectively.

1. Introduction

The biological-physical interactions in the oceans play an important role in determining patterns of horizontal distributions of the plankton communities [1], and these interactions occur at a wide range of temporal and spatial scales [2], being the mesoscale processes such as fronts, eddy, and upwelling, the most determinant factors in the spatial distribution and structure of the zooplankton communities on basin and local scales [3]. Mesoscale oceanographic structures such as eddy and fronts can work as mechanisms of retention and concentration of fish larvae [4–11], and upwelling filaments, including eddy, may work as mechanisms of dispersion [12–17].

The Gulf of California is a semienclosed dynamic sea where strong changes in temperature, salinity, and currents [18] are related to the seasonal flux of the Gulf of California

and to tropical surface water masses which provide a unique environment where the southern tropical, subtropical, and northern temperate marine biota develops [19, 20]. The northern region has an anticyclonic circulation most of the year, while in June and September it reverses to a cyclonic eddy [21]. Strong winter upwelling is present in the continental coast, while during the summer it is weak at the peninsular coast [18]. Three to five alternated eddy and jet streams [22, 23] have been registered from south of the Big Islands to the south o the gulf with a markedly seasonal component. All these dynamic features may promote a wide diversity of responses of the fish larvae community to the environment.

It has been shown that the ichthyofauna distribution in the Gulf of California responds to a latitudinal gradient: (a) temperate species are more abundant to the north, (b) tropical affinity species are more abundant to the south, and

(c) an apparent mixture of temperate-tropical fauna has been found in the central zone. Besides these, the abundance is directly related to the thermal gradient [24] in a portion of the Gulf of California. In the same area, several spatially limited studies had shown that oceanographic processes such as fronts, filaments, jets, upwelling, and cyclonic eddy determine local differential ontogenetic distribution patterns in small pelagic fishes [25, 26], maintain differentiated horizontal patterns of the fish larvae assemblages [27–30], and promote retention or eggs and larval drift [24].

In spite of the useful information generated in these previous works, gaps still remain in the knowledge of the effects of oceanographic mesoscale structures on larval fish communities in a broad spatial and seasonal scale. These gaps lead to several questions that remain unsolved such as (a) are the mesoscale structures relevant in the formation of regional patterns of the fish larvae assemblages? Or (b) are the 18°C and 21°C isotherms [24] the most influential factors to explain the spatial and seasonal patterns of the fish larvae assemblages? And (c) what other large spatial gradients, besides latitudinal gradient, are evidenced in the gulf?

In this paper we study the relationship between the macro- and mesoscale oceanographic processes and fish larvae assemblages in a semi-enclosed sea with strong seasonal variability using remotely sensed data sets, field oceanographic measurements, and fish larvae species abundance in the entire area of the Gulf of California, under the two most contrasting seasonal conditions (winter and summer).

2. Methods

Environmental and zooplankton data were collected during four oceanographic cruises made in the winter and summer seasons in the Gulf of California, Mexico. Winter cruises were conducted from February 25 to March 12, 2005 (CGC0503) and from January 13 to 27, 2007 (GOLCA0701), while summer cruises were conducted from September 7 to 19, 2005 (CGC0509) and from July 20 to August 2, 2007 (GOLCA0707) (Figure 1).

Daily high resolution satellite images of sea surface temperature (SST) and chlorophyll a were obtained from MODIS on Aqua (http://oceancolor.gsfc.nasa.gov/), level 2 (1 Km resolution). The level 2 satellite imagery was only georeferenced but not orthorectified. To transform all the images into a Cartesian plane projection, the GRI program (http://gri.sourceforge.net/) was used. Once transformed, the resulting images were averaged for the corresponding days of each survey and used to detect mesoscale processes. To obtain a representative value, the chlorophyll a concentration for each sampling station was calculated from the satellite images using the average value of the following pixels: one corresponding to the location of each station, plus left, right, up, and down respect to the former station. Weekly satellite images composites prior to and after the sampling dates (not showed here) were also generated and analyzed to register the time lag of the eddy observed.

A total of 160 CTD casts (77 in CGC0503, 17 in GOLCA0701, 49 in CGC0509, and 17 in GOLCA0707) were made using calibrated SBE 19, Sea-Bird Electronics and a Mark III, General Oceanics. These data were used to construct temperature profiles from longitudinal transects of the cruises, to calculate the maximum stability depth [31] and to identify water types according to Torres-Orozco [32].

A total of 143 zooplankton samples were collected (39 samples in CGC0503, 26 in GOLCA0701, 43 in CGC0509, and 35 in GOLCA0707). Except for those from CGC0509, all zooplankton samples were obtained with oblique tows using Bongo nets with 505 μm mesh (0.6 m diameter) equipped with a digital flow meter. Oblique tows were made to a maximum depth of 200 m [33]. During CGC0509, samples were collected with a simple conical CalCOFI net at surface (0.6 m diameter, 505 μm mesh). All samples were fixed in 96% ethyl alcohol. Zooplankton volumes were obtained using the displacement volume method [34]. The ichthyoplankton fraction was sorted, and fish larvae were counted and identified to species level [35–40]. Abundance was standardized to number of larvae 10 m^{-2} [33], and species richness was estimated as number of species. Since zooplankton samples during the September survey were taken using surface tows, we only used these data to analyze the relationship of the locations of mesoscale structure. Comparative zooplankton and fish larvae abundance analyses were done only among those surveys in which zooplankton samples were taken with oblique tows.

Canonical ordination methods were used to explore the relationship between the distribution and abundance of fish larvae and the environmental variables for each survey. The larval fish abundance was transformed to the fourth root, and seven environmental variables were used as independent variables: temperature (T10), salinity (S10), density (D10), water type (WM10), maximum stability depth (MSD), sea surface chlorophyll a concentration (CHL), and zooplankton volumes (ZV); the first four were obtained from 10 m depth. To select the most robust canonical analysis, a detrended canonical correspondence analysis (DCA) was performed to obtain the length of the environmental gradients [41]. In DCA, D10 had high orthogonal values so it was not used in further analysis. The species-environment relationship was explored using a redundancy analysis (RDA) [42]. In both statistical analyses, scaling was focused on the interspecies correlations. Species scores were divided by their corresponding standard deviation and centered by species. Further selection of environmental variables was performed automatically and statistical significance was calculated using unrestricted Monte Carlo permutation tests. Only those variables that significantly explained the species variability were included in the final analysis; therefore, water mass variable was also eliminated from the results. Biplots were used for the representation of the biological variables ordination in the environmental multidimensional space [42]. All the statistical multivariate analyses were done with the Canonical Community Ordination (CANOCO) for Windows software ver. 4.56. The naming convention for the groups obtained was to use the first two letters of the month and the number of the group analyzed, for example, Group1 of January = JaG1. A one way analysis of variance (ANOVA) test for significant differences between ZV during each cruise was performed using the STATISTICA V.8 software. CHL and ZV were used

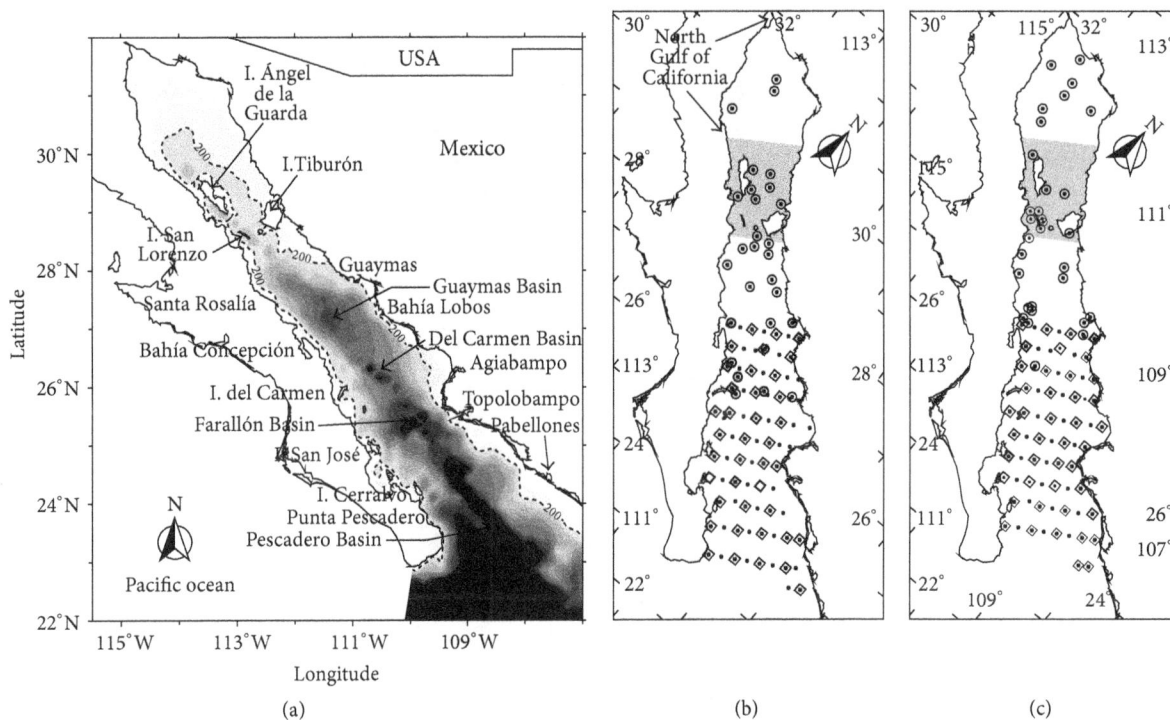

(a)

(b)

(c)

FIGURE 1: Study area and sampling grid stations in the Gulf of California, Mexico. (a) Localities, some basins, and 200 m isobath. Grid stations during (b) winter and (c) summer. Shaded area indicates the Big Islands Archipelago region. Open circles: GOLCA0701 and GOLCA0707 cruises. Open squares: CGC0503 and CGC0509 cruises. Solid circles: CTD casts.

as gross indicators for the biological production and the amount of available food in the environment.

3. Results

3.1. Environmental Conditions. In both winter months (January 2007, Figure 2(a), and March 2005, Figure 3(a)), a north-south gradient of SST observed in satellite images was found in the Gulf of California (mean SST = 16.5°C in January and 20.6°C in March). The 21°C isotherm, used as the limit of recurrent groups on ichthyoplankton [24], and related to the entrance of warm water to the gulf, was registered south of Isla Cerralvo in January, while in March, its position was far north from Bahía Concepción to Topolobampo. During the summer months, an east-west gradient of SST with low values was recorded at the peninsular coast and high values at the continental coast. During July, mean SST was 27.2°C, while during September it was 30.4°C (Figures 4(a) and 5(a)).

The MSD was deeper during the winter months than in the summer when it was very shallow along the gulf (Table 1). During January, a shallow MSD at the continental coast progressively deepened toward the peninsular coast reaching 180 m depth north of Santa Rosalía (Figure 2(b)). During March, MSD was relatively shallow in both coasts and deeper at the most oceanic area. The maximum depth (150 m) was recorded northeast of Isla San José (Figure 3(b)). During July, MSD was shallow at the peninsular coast (Figure 4(b)), and deeper northeast of Isla Tiburón (70 m). In September, MSD was the shallowest of the analyzed cruises reaching 42 m depth south of Bahía Lobos and Isla San José (Figure 5(b)).

3.2. Sea Surface Chlorophyll a. During winter cruises, CHL showed a latitudinal gradient with higher concentrations to the north and lower to the south (Figures 2(d) and 3(d)). In all cruises, high CHL values (>1.0 mg m^{-3}) were recorded in the coastal zone of the gulf, particularly at the continental coast and at the Big Islands region. CHL concentration was higher in the winter than in the summer (mean = 1.147 mg m^{-3} and 0.786 mg m^{-3}, Table 1). During January, high CHL values were distributed from Isla Tiburón to Pabellones in the continental coast (Figure 2(d)), while during March high CHL values were observed covering a wider area along the coasts as well as practically in all the northern gulf's area (Figure 3(d)). During July and September, CHL concentrations were low (<0.2 mg m^{-3}) in almost all the gulf (Figures 4(c) and 5(d)).

3.3. Zooplankton Volumes. Low values of ZV (<200 mL 1000 m^{-3}) were observed in a large area of the gulf in all the oceanographic cruises, with some isolated cores of high ZV (>500 mL 1000 m^{-3}). For those samples taken with oblique trawls, no significant differences in ZV were recorded in both seasons ($F_{7,240} = 10.3$, $P < 0.001$), being slightly higher in the winter (197 mL 1000 m^{-3}) than in the summer (167 mL 1000 m^{-3}). In January, high ZV were found south of Isla Tiburón (841 mL 1000 m^{-3}) and medium ZV (200–500 mL 1000 m^{-3}) in two cores southwest of Guaymas and north of Santa Rosalía (Figure 2(e)). During March, high ZV (729 mL 1000 m^{-3}) were registered southeast of Isla del Carmen and at the continental side of the gulf (Figure 3(e)). In July, high ZV (527 mL 1000 m^{-3}) were found north of Bahía Concepción,

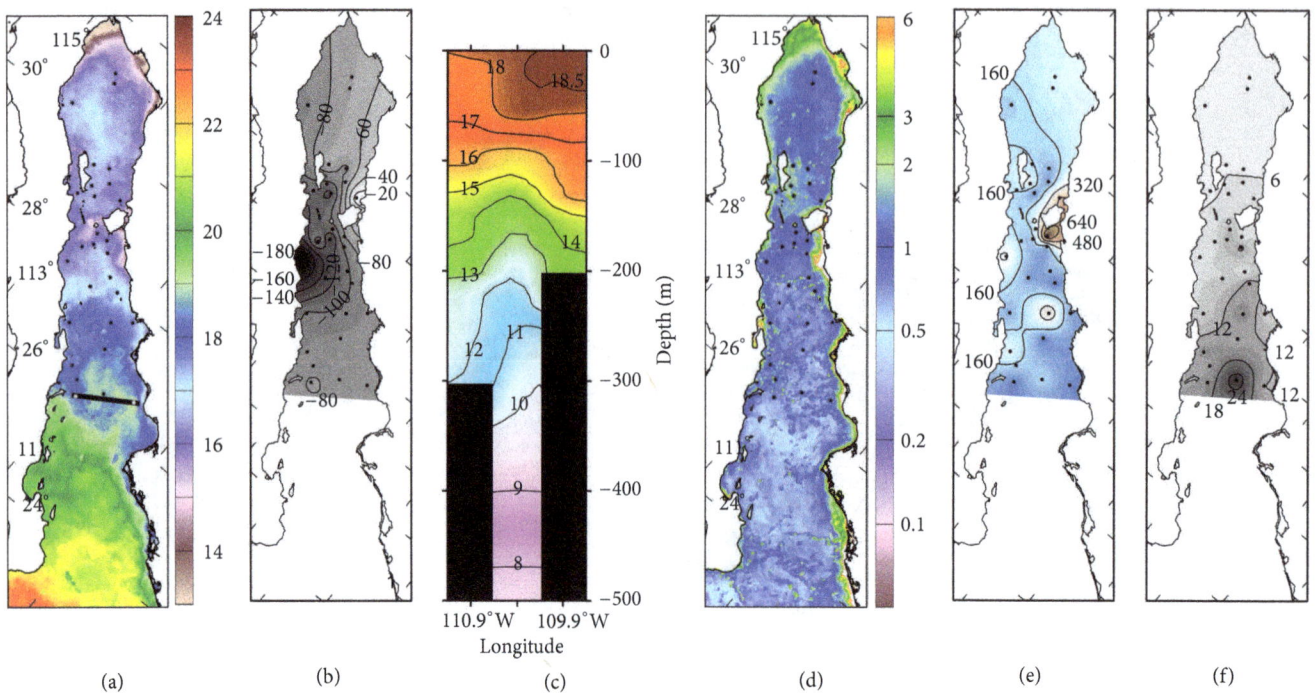

FIGURE 2: Biophysical characteristics during January. Composite of (a) SST (°C) and (d) CHL (mg m^{-3}) satellite images of the Gulf of California. Black line in (a) indicates transect analyzed for water column profile. (c) Temperature profile (°C). Distribution of (b) maximum stability depth (MSD), (e) zooplankton volumes, and (f) species richness.

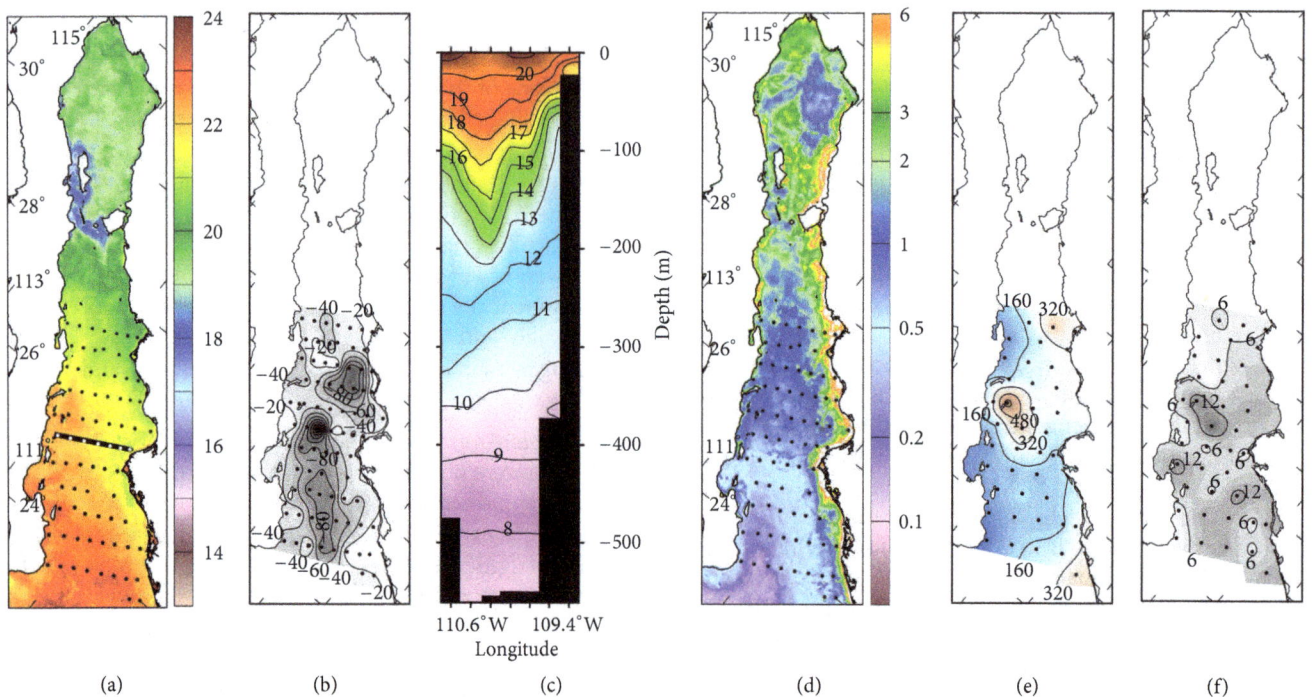

FIGURE 3: Biophysical characteristics during March. Composite of (a) SST (°C) and (d) CHL (mg m^{-3}) satellite images of the Gulf of California. Black line in (a) indicates transect analyzed for water column profile. (c) Temperature profile (°C). Distribution of (b) maximum stability depth (MSD), (e) zooplankton volumes, and (f) species richness.

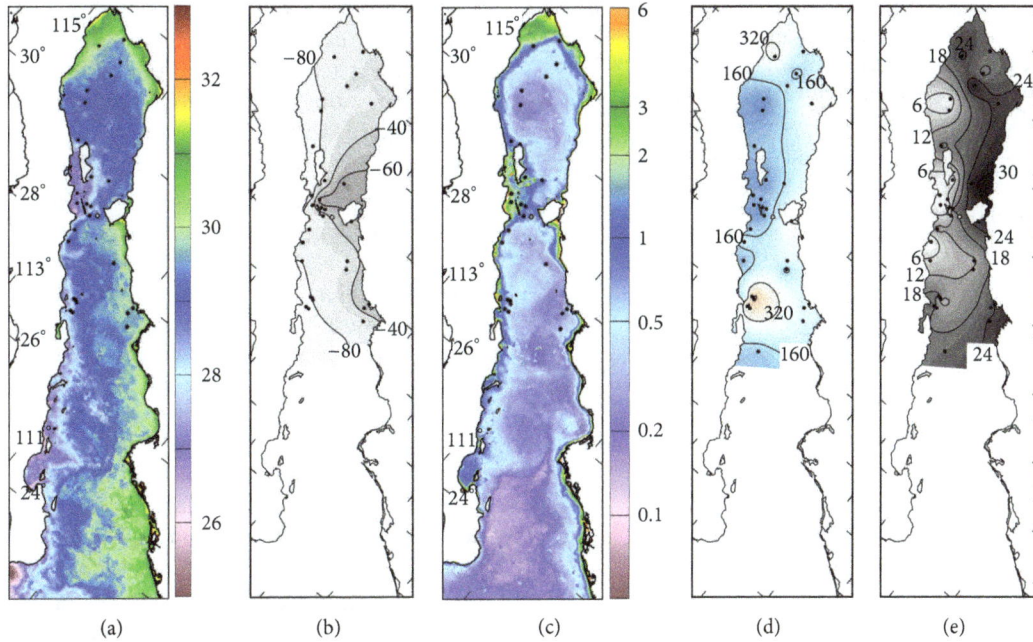

FIGURE 4: Biophysical characteristics during July. Composite of (a) SST (°C) and (c) CHL (mg m^{-3}) satellite images of the Gulf of California. Distribution of (b) maximum stability depth (MSD), (d) zooplankton volumes, and (e) species richness.

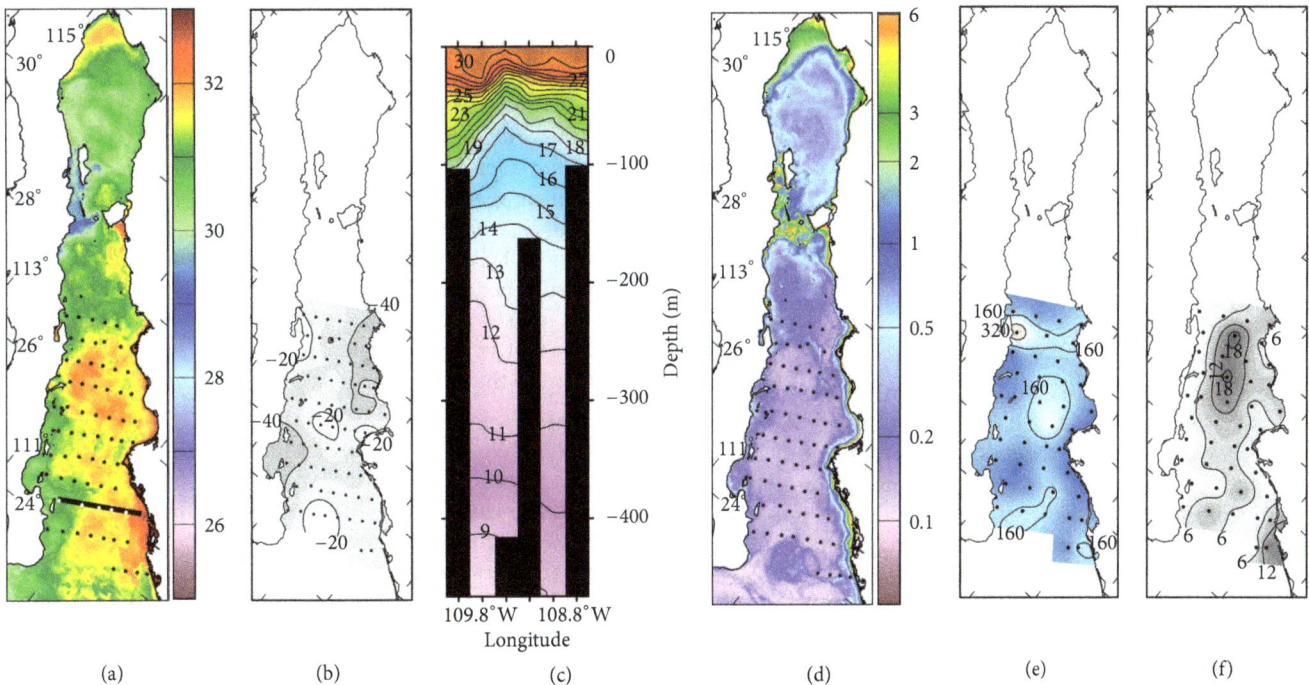

FIGURE 5: Biophysical characteristics during September. Composite of (a) SST (°C) and (d) CHL (mg m^{-3}) satellite images of the Gulf of California. Black line in (a) indicates transect analyzed for water column profile. (c) Temperature profile (°C). Distribution of (b) maximum stability depth (MSD), (e) zooplankton volumes, and (f) species richness.

with medium ZV associated to the continental coast southeast of Isla Tiburón and at the north gulf (Figure 4(d)). During September, only two small cores of medium ZV were registered, one in front of Bahía Concepción and the other southwest of Agiabampo (Figure 5(e)).

3.4. Oceanographic Processes. During January, a frontal thermal zone was observed south of San Lorenzo and Tiburón Islands with 15°C to 16°C, also the curling of a warm water filament located in the oceanic area off Isla del Carmen suggested the presence of a cyclonic eddy about 127 km in

TABLE 1: Larval abundance by faunistic affinity and habitat registered in the Gulf of California, Mexico (winter and summer of 2005 and 2007). Values are given as the average of winter (January 2007 and March 2005) and summer (July 2007 and September 2005) cruises.

	Winter	Summer
Larval abundance		
Average per sample (larvae $10\,m^{-2}$)	1064	4273
Faunistic affinity (%)		
Temperate	7.5	1.8
Tropical	41.8	55.5
Subtropical	49.2	40
Wide distribution	1.5	2.7
Habitat (%)		
Mesopelagic	17.1	6.1
Coastal-pelagic	12.2	20.9
Oceanic-pelagic	2.4	6.8
Shallow-demersal	56.1	60.1
Deep-demersal	6.1	2.7
Bathypelagic	6.1	3.4

diameter (Figure 2(a)). The temperature profile along the transect in Figure 2(a) showed presence of water >18°C in the first 50 m in the continental side and at the center, with lower values in the peninsular side (Figure 2(c)), followed by a relatively homogeneous layer of cold water (17°C to 15°C) down to about 125 m. The isotherms at the center of the transect showed a dome shape with the 16°C isotherm at 90 m, while it was at 100 m in the west and at 140 m in the east. The rising of the isotherms was observed from 90 m to 300 m in depth (Figure 2(c)). In March, the SST satellite image showed also a front from northwest of Isla Ángel de la Guarda to south of Isla Tiburón (Figure 3(a)), and an anticyclonic eddy northeast Isla San José was observed in the CHL satellite image of this month (Figure 3(d)). The temperature profile along the transect in Figure 3(a) showed a warmer water column than in January, with 21.2°C at surface, and 18°C at 90 m depth. A shallow 16°C isotherm (90 m depth in the west and 50 m in the east) was found in this month, and isotherms at the center of the transect were strongly valley shaped between 70 m to 215 m depth (Figure 3(c)). Although the sampling grid station during July did not allow us to generate longitudinal transects profiles, both satellite images for this month showed a front close to the peninsular coast from northwest of Isla Ángel de la Guarda to the south of Isla San Lorenzo. The presence of three eddies, one located in front of Santa Rosalía, the second southeast of Bahía Concepción, and the third northeast of Isla San José were also observed in both images (Figures 4(a) and 4(c)). During September, satellite images did not show mesoscale features in the region where the zooplankton sampling was done (Bahía Concepción to Punta Pescadero). However, the temperature profile (Figure 5(c)) along the transect in Figure 5(a) showed a strongly stratified water column with the highest temperatures (30.7°C) and shallower thermocline at the continental side, dome shaped isotherms at the center of the transect (cyclonic eddy), and a more spread out isotherm

at the peninsular side. The dome shaped isotherms showed that the 16°C isotherm deepened at about 120 m in both coasts, while it reached 95 m at the center of the dome. The rising of the isotherms can be observed at 150 m depth to the surface (Figure 5(c)).

3.5. Fish Larvae Composition and Abundance. A total of 73 families were identified from the four cruises analyzed, with the highest species richness during the summer. During the winter, 41 families, 70 genera, and 98 species were registered, while in the summer, 60 families, 96 genera, and 183 species were identified. See also Supplementary Material available online at http://dx.doi.org/10.1155/2013/176760 for details of species abundance by cruise, by season, and faunistic affinity and habitat.

The main seasonal differences in the larval fish community besides species richness (Table 1) were related to (a) lower total fish larvae abundance during the winter (average per sample 1064 larvae $10\,m^{-2}$) than in the summer (average per sample 4273 larvae $10\,m^{-2}$) in samples taken with oblique trawls, (b) about 4.2 times more temperate species in the winter, (c) lower relative abundance of shallow-demersal, coastal-pelagic, and oceanic-pelagic species (56.1%, 12.2%, and 2.4%, resp.) and more abundant mesopelagic species (17.1%) in the winter than in the summer (60.1%, 20.9%, 6.8%, and 6.1% resp.) and (d) a change in the community structure which included eight species: *E. mordax*, *Vinciguerria lucetia*, *Diogenichthys laternatus*, *Leuroglossus stilbius*, *S. sagax*, *Benthosema panamense*, *Citharichthys fragilis*, and *Scomber japonicus* during the winter and 13 species: *Cetengraulis mysticetus*, *Opisthonema libertate*, *Benthosema panamense*, *Triphoturus mexicanus*, Gobiidae sp. 3, *Oligoplites saurus*, *Auxis* sp. 2, *Scomberomorus sierra*, *V. lucetia*, Sciaenidae sp. 1, *Thunnus* sp. 1, *Eucinostomus dowii*, and *Balistes polylepis* during the summer, accounting for 90% of the relative abundance in each season. The species *V. lucetia* and *B. panamense* were present in both seasons; however, *V. lucetia* was more abundant in the winter while *B. panamense* in the summer.

3.6. Species Richness. Distribution patterns of the species richness showed a southward gradient during January, with low values north of the Big Islands and high richness in the south (Figure 2(f)) related to the cyclonic eddy, while species richness had no pattern during March (Figure 3(f)). During July, sampling stations near the continental coast had the highest species richness of the study with a maximum (30 species) registered in one sampling station north of Isla Tiburón (Figure 4(e)), and during September, high values were recorded only at the oceanic central and southeast gulf (Figure 5(f)). In all months, high values of species richness were coincident with high SST values.

3.7. Data Analyses. During January, the first two RDA canonical axes explained 87.9% of the total variance of the species-environment relationships (Table 2). The first RDA axis was explained by T10 and CHL with negative and positive correlation, respectively, while the second axis was associated with positive correlation values for S10 and negative for MSD

TABLE 2: Summary of the redundancy analysis (RDA) applied to fish larvae abundance and environmental variables in each cruise.

	January		March		July		September	
	Axis 1	Axis 2	Axis 1	Axis 2	Axis 1	Axis 2	Axis 1	Axis 2
Var S-E	67.4	87.9	69.1	91.2	44.3	78.3	53.8	85.9
T10	**-0.689**	0.6294	**0.914**	-0.2303	0.5118	**-0.5589**	0.7056	0.178
S10	0.5016	**0.7568**	-0.4443	0.0594	-0.0723	-0.2364	0.5531	0.0403
CHL	**0.5506**	0.4591	**-0.4807**	**-0.3429**	-0.2869	-0.1223	**0.7882**	-0.0744
ZV	0.0871	0.0607	-0.1282	**0.867**	0.6975	**0.7093**	-0.2496	**0.9036**
MSD	-0.1975	**-0.7277**	0.3368	0.5255	-0.084	-0.2908	**-0.7622**	-0.0677

Bold numbers are the highest correlation values for each axis. Var S-E: cumulative percentage variance of species-environment relation; T10: temperature 10 m depth; S10: salinity 10 m depth; CHL: sea surface chlorophyll a concentration; ZV: zooplankton volumes; MSD: maximum stability depth.

(Table 2). The RDA biplot showed three sampling stations groups in three different environments. The first group (JaG1) was related to high CHL and S10 values (Figure 6(a)) and was located from the Big Islands to the north (Figure 6(b)). Shallow and deep-demersal species of temperate and tropical affinity such as *Prionotus ruscarius*, *Hippoglossina stomata*, and *Merlucius productus* were included in this group (Figure 6(c)). A second group of stations (JaG2) shared an environment where the deepest values of MSD were found (Figure 6(a)), located between the southern part of Isla Tiburón to Bahía Concepción (Figure 6(b)). In this group, we found contrasting distribution among species. Coastal-pelagic (*E. mordax*) and shallow-demersal species (*Citharichthys fragilis* and *Sebastes* sp.) were distributed at the frontal zone near Ángel de la Guarda and San Lorenzo Islands, while species of deep habitats such as mesopelagic (*Triphoturus mexicanus* and *Diogenichtys laternatus*), deep-demersal tropical (*Argentina sialis*), and bathypelagic temperate species (*Leuroglossus stilbius*), as well as oceanic-pelagic (*Trachurus symmetricus*), were registered south, out of the frontal area (Figure 6(c)). In these two groups, fish larvae abundance values were from medium to high (100–999 larvae $10\,\text{m}^{-2}$ and 1000–9999 larvae $10\,\text{m}^{-2}$, resp.), particularly off Isla Ángel de la Guarda and south of Isla Tiburón. The third group (JaG3) was related to the highest T10 values in this cruise (Figure 6(a)), located from Guaymas to the south (Figure 6(b)). Almost all the coastal-pelagic species (*Bregmaceros bathymaster*, *Etrumeus teres*, *Sardinops sagax*, and *Scomber japonicus*) except temperate *E. mordax*, the most abundant mesopelagic (*Benthosema panamense* and *Vinciguerria lucetia*), and bathypelagic (*Stomias atriventer*) species of tropical and subtropical affinity were found in this group (Figure 6(c)). The maximum abundance in this southern group found in station 72 was related to the maximum species richness (Figure 2(f)) and to the center of the mesoscale eddy observed in the SST satellite image for this month (Figure 2(a)).

In March, the first two axes of the RDA explained 91.2% of the variance being T10 and CHL in the first axis, and ZV in the second, the best correlated variables (Table 2). In this month, RDA biplot showed a first group (MaG1) which included sampling stations where T10 had lower values but higher CHL, and higher S10 than in the second group, located from Bahía Concepción to the northern part of Topolobampo (Figures 7(a) and 7(b)). Fish larvae abundance was medium

and homogeneous along the area, with a maximum of 842 larvae $10\,\text{m}^{-2}$ by station. The second group (MaG2) distributed from Isla San José to the south at the highest T10. In this MaG2, station T3 was located at the center of the cyclonic eddy where highest MSD, ZV, fish larvae abundance (1854 larvae $10\,\text{m}^{-2}$), and species richness were registered in this month. The RDA biplot of species-environment showed three species groups, the first related with high S10 and low T10 values, and relatively high ZV, and included two temperate (*E. mordax* and *L. stilbius*) and one subtropical species (*S. japonicus*). The second group gathered most of the species with different affinity and habitat (Gobiidae spp., *Bregmaceros bathymaster*, *Chilara taylori*, and *Diplophos taenia*) related with the highest T10, while the third group included the most abundant tropical mesopelagic (*B. panamense*, *V. lucetia*, *D. laternatus*, and *Hygophum atratum*) and coastal-pelagic species (*S. sagax*) at high T10, and MSD values (Figure 7(c)). This last species were also found at the center of the eddy.

During the summer, the first two RDA canonical axes explained 78.3% in July and 85.9% in September of the total variance of the species-environment relationships (Table 2). In July, ZV and T10 were the best correlated variables in both axes (Table 2). RDA biplot of stations environment showed three groups of stations, JuG1 sharing the coldest and saltiest environment with high CHL values but low ZV (Figure 8(a)), located at the peninsular coast from the northwestern part of Isla Ángel de la Guarda to the southern part of Isla San Lorenzo (Figure 8(b)) in the frontal area observed at the satellite images for this month. In this group, tropical mesopelagic species *B. panamense* and tropical shallow-demersal species *Scorpaenodes xyris*, *Lepophidium negropinna*, and *Abudefduf troschelii* were found (Figure 8(c)). Five stations formed the second group (JuG2) and were related to high ZV (Figure 8(a)) north of Bahía Concepción (Figure 8(b)) with tropical and subtropical mesopelagic (*T. mexicanus*, *V. lucetia*), coastal-pelagic (*O. libertate*, *Caranx caballus*), oceanic-pelagic (*Thunnus* sp. 1), and shallow-demersal species (*Synodus lucioceps*, *Stegastes rectifraenum*) (Figure 8(c)). The third group was related primarily to the highest T10 (Figure 8(a)) observed at the continental coast and northern gulf (Figure 8(b)) and included most of the tropical, subtropical, and temperate affinity species, with coastal-pelagic (*Cetengraulis mysticetus*, *Scomberomorus sierra*, *O. saurus*, and *B. bathymaster*), oceanic-pelagic (*Auxis* sp. 2), and shallow-demersal

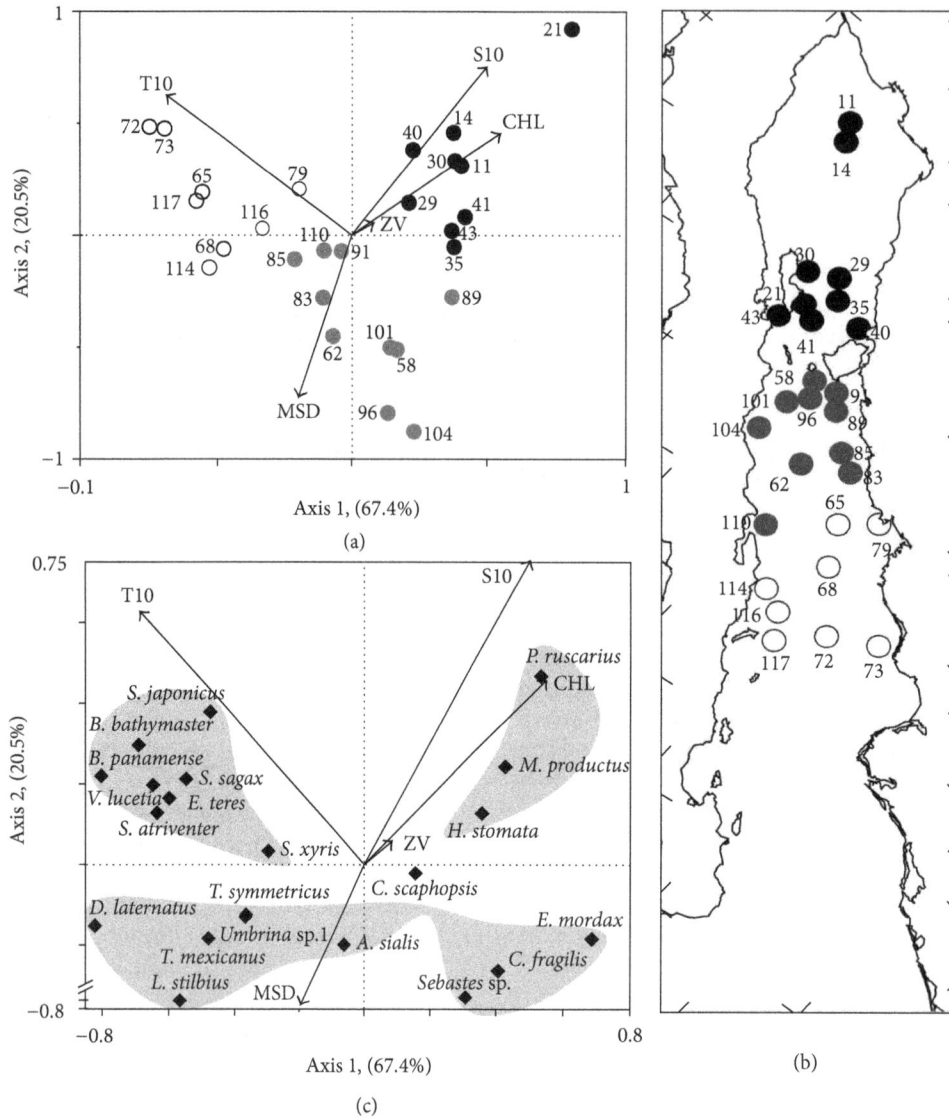

FIGURE 6: Classification and ordination analysis: January. (a) RDA biplot of sampling stations environment, black circles = JaG1; grey circles = JaG2; open circles = JaG3. (b) Distribution of stations groups resulting from RDA. (c) RDA biplot species environment.

species (*Eucinostomus dowii*, *Balistes polylepis*, *Symphurus williamsi*, and *Scorpaena guttata*) (Figure 8(c)). Fish larvae abundance in the three groups was high, but at the northernmost gulf in the third group, we registered the highest fish larvae abundance in this study (59751 larvae $10 \, m^{-2}$ and 16818 larvae $10 \, m^{-2}$).

The canonical axes for September's RDA had the highest positive correlation to CHL and negative to MSD, while the second axis had a high correlation to ZV (Table 2). In the RDA stations-environment biplot, the first group (SeG1) represented mainly coastal stations of both sides of the gulf (Figure 9(b)) with very low fish larvae abundance (maximum of 46 larvae $10 \, m^{-2}$ by station) related to areas with high CHL values and shallow MSD, in a warm environment (Figure 9(a)), in which tropical species of coastal-pelagic (*Opisthonema* spp., *Anchoa* spp., *Oligoplites saurus*, and

Hirundichthys spp.) and shallow-demersal habitat (Gerreidae spp. and Mullidae spp.) were found (Figure 9(c)). In this group, station W3 located at the center of the gulf and of the cyclonic eddy registered only 0.14 larvae $10 \, m^{-2}$ contrasting with stations at the edge of the eddy that reached from 5 to 50 times more abundance. The second group was distributed at many oceanic stations of the center of the gulf, where high ZV and medium to high MSD values were registered during this month (Figures 9(a) and 9(b)). Abundance was also low in this group (maximum of 59 larvae $10 \, m^{-2}$ by station). Here, tropical and subtropical species of oceanic-pelagic (*Auxis* spp., *Katsuwonus pelamis*, and *Cheilopogon heterurus*) and shallow-demersal habitat (*S. ovale*, *Lutjanus* spp., and *Pristigenys serrula*) and the only mesopelagic species *B. panamense* were registered (Figure 9(c)). No temperate species were found in this month.

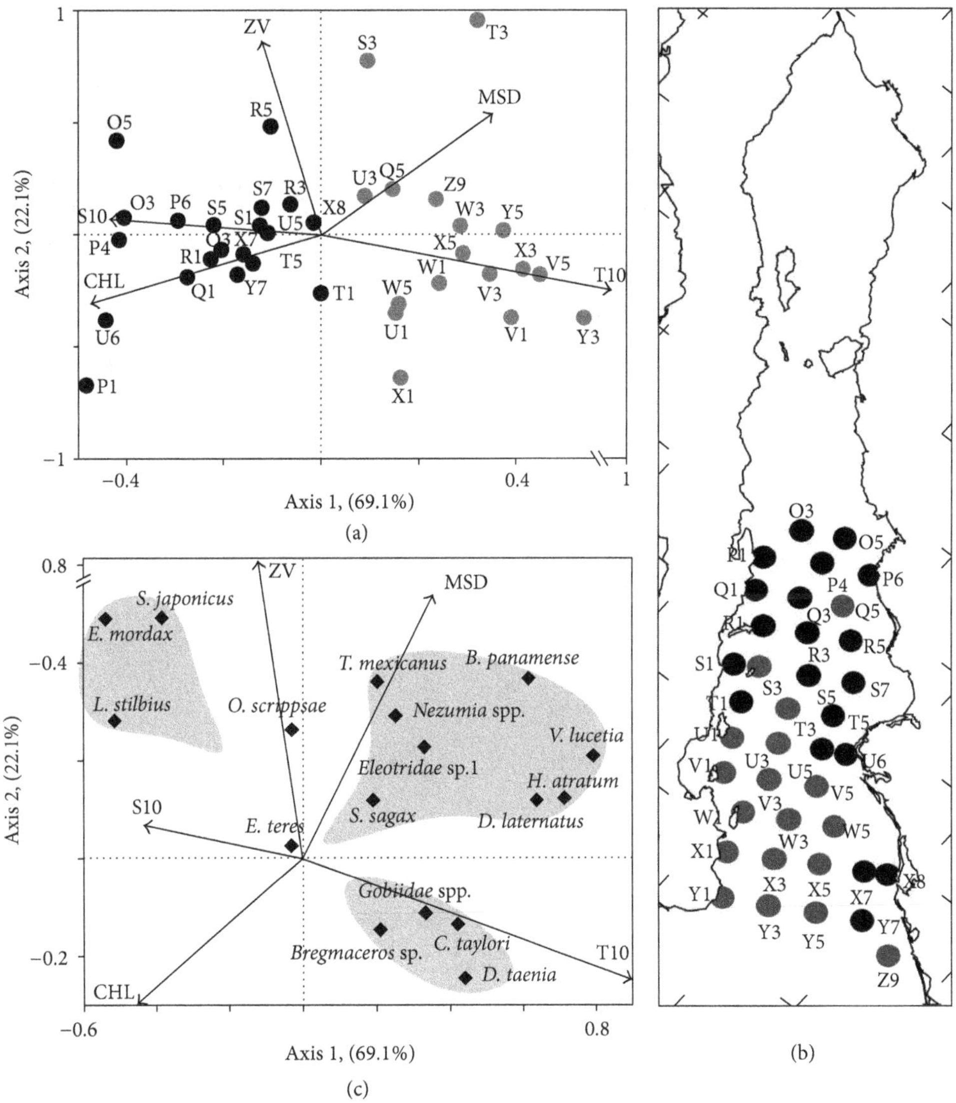

FIGURE 7: Classification and ordination analysis: March. (a) RDA biplot of stations environment, black circles = MaG1; grey circles = MaG2. (b) Distribution of stations groups resulting from RDA. (c) RDA biplot species environment.

4. Discussion

This is the first study that focused on the understanding of the composition, distribution, and abundance of fish larvae assemblages and their relationships with the physical environment in a macroscale analysis, which includes mesoscale processes, such as fronts and eddy structures in the Gulf of California, Mexico, covering the two most important seasons of the year.

The results of this study showed a close relationship between fish larvae abundance and environmental gradients present along the Gulf of California, particularly those of T10, S10, CHL, and ZV present during each season of the year. These relationships were also influenced by the local biophysical gradients related to the mesoscale processes such as fronts and eddy found in this area. In general terms, the responses of the ichthyoplanktonic community were observed as changes in the fish larvae assemblages promoted by north-south and coastal-ocean gradients and accumulation or dispersion effects of the eddy according to their type and stage of formation, and by strong abundance and composition differences among fish larvae assemblages separated by thermal frontal barriers.

In seasonal terms, we found differences in abundance, composition, distribution, and biogeographic affinity between winter and summer fish larvae assemblages observed in the Gulf of California. These differences were related to major seasonal changes and to interannual variability in the oceanographic conditions of the Gulf of California.

4.1. Winter Season. The winter (January and March) mean environmental conditions in the Gulf of California included the presence of a north-south SST gradient, low temperature, high S10 related to the presence of Gulf of California Water (GCW), and a deep MSD in practically all the sampling

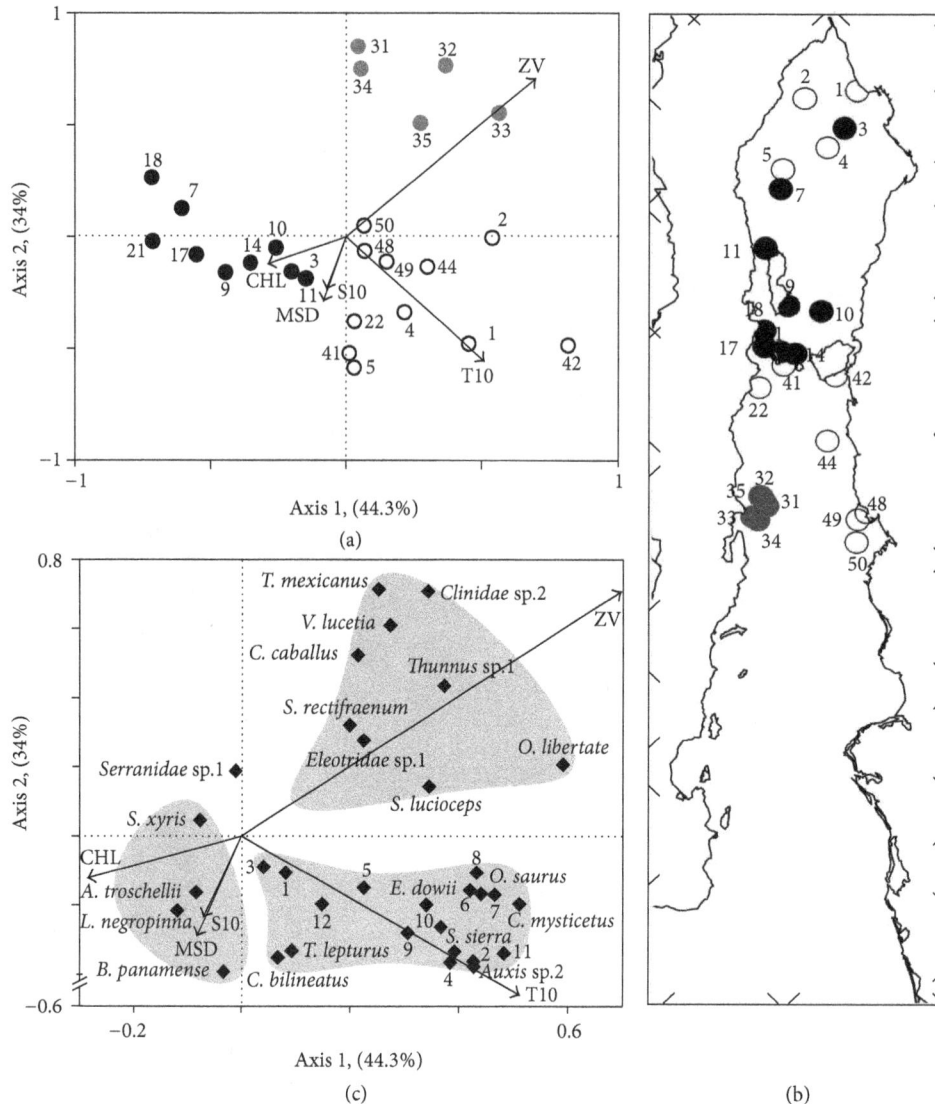

FIGURE 8: Classification and ordination analysis: July. (a) RDA biplot of sampling stations environment, black circles = JuG1; grey circles = JuG2; open circles = JuG3. (b) Distribution of stations groups resulting from RDA. (c) RDA biplot species environment. 1 = *Bregmaceros bathymaster*, 2 = *Balistes polylepis*, 3 = *Epinephelus* sp. 1, 4 = Gobiidae sp. 2, 5 = Gobiidae sp. 3, 6 = *Ophichthus zophochir*; 7 = Sciaenidae sp. 1, 8 = Sciaenidae sp. 2, 9 = *Scorpaena guttata*, 10 = *Syacium ovale*, 11 = *Symphurus williamsi*, 12 = Angiliforme sp. 1.

stations in agreement with winter conditions [18, 43]. In terms of biomass, CHL mean was 1.147 mg m^{-3} and high values found at the Big Islands region and at the peninsular coastal zone were both mostly related to the strong tidal mixing [44, 45] and to coastal upwelling events [46] present in this season, respectively. Also, medium to high ZV (181 mL 1000 m^{-3} and 218 mL 1000 m^{-3}) indicated a productive environment. Fish larvae assemblages in these conditions were primarily correlated to T10 and S10, but also to ZV (Table 2).

Tropical-subtropical and shallow-demersal species dominated the fish larvae composition of the Gulf of California during winter (91% and 56.1%, resp.). This dominance of tropical-subtropical species could be explained by an increased sea surface temperature related to a weak El Niño registered as positive anomalies in the multivariate

ENSO index (http://www.esrl.noaa.gov/psd/enso/mei/table .html), and could be the result of more sampling efforts and a larger sampling area at the southern gulf region if compared with previous studies. A very similar species composition by biogeographic affinity was found during March 2005 by Avendaño-Ibarra et al. [47], but the mesopelagic component dominated. In contrast, Aceves-Medina et al. [20] registered 96% of total larval abundance of temperate affinity during winter season and coastal-pelagic species (63%) as the most abundant.

High relative abundance of fish larvae of small pelagic fishes (*S. sagax*, *E. mordax*, *T. symmetricus*, *S. japonicus*, and *E. teres*) registered in January could be related to the high productivity in terms of CHL and ZV that may have enhanced the reproductive activity of the adults present in the gulf

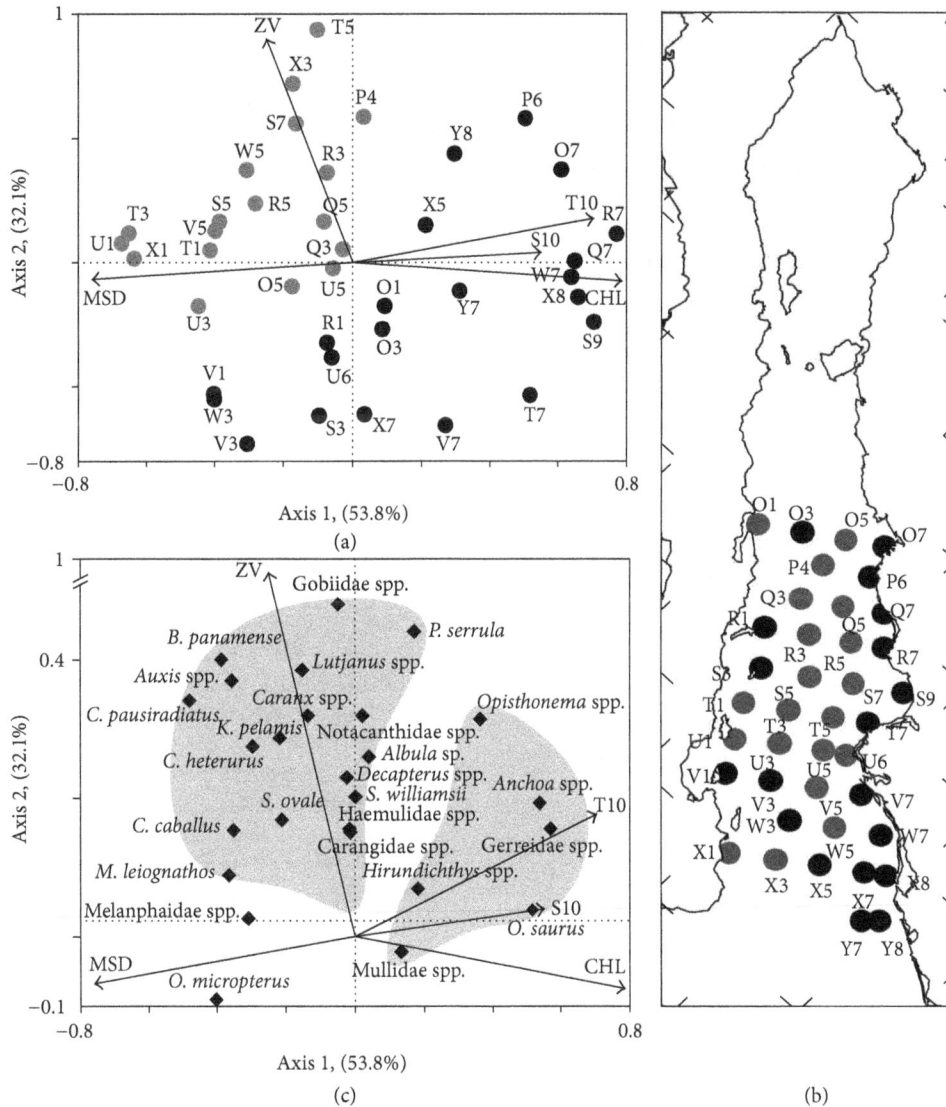

FIGURE 9: Classification and ordination analysis: September. (a) RDA biplot of sampling stations-environment, black circles = SeG1; grey circles = SeG2. (b) Distribution of stations groups resulting from RDA. (c) RDA biplot species-environment.

during the winter. Tidal mixing is important in the north of the Gulf of California and in the archipelago and the shelf south of Isla Tiburón that show conditions for strong internal mixing and upwelling over the sills and in the surrounding area [18, 48, 49]. These features promote the presence of lower temperatures in this area compared to those in the rest of the gulf during almost all the year. These conditions allowed us to detect fronts in this area in all the SST satellite images during our study.

In January, fish larvae of JaG1 were located to the north of the front where an intense water column mix due to strong tidal currents promoted the incorporation and availability of nutrients to the phytoplankton [48] resulting in the highest CHL concentrations of the cruise in a cold environment. Survival of Pacific hake larvae (*Merluccius productus*), an abundant component of this group, is strongly influenced by the environmental conditions (such as upwelling, advection, and water temperature) experienced during the first few

months after spawning at the California Current System [50] and fed upon a wide range of prey particularly on copepod eggs, filter feeding nauplii and copepodites [51]. The enhanced productivity present in the JaG1 area is likely a base for the development of a favorable feeding area for this species inside the Gulf of California.

The JaG2 was distributed over and south of the front in this month. Thermal frontal zones have been recognized as high CHL and ZV areas [52–54]. The high abundance of filter feeding *E. mordax* larvae that characterized JaG2, suggests the presence of an important food supply in the area and also that adults of this species may be spawning at or near this frontal zone. Highest abundance of *E. mordax* and *T. symmetricus* in this region contrasts with *S. sagax* low abundance coinciding with previous reports of spatial segregation between these species [24, 25], and also differences in abundance and composition between the sides of the front line [27, 28]. Surface connectivity in the Gulf of California studied by Marinone

[55] indicates that the export of particles during January is notably large from the north to the south and that particles in the Tiburón and Angel de la Guarda Islands are retained there only 2-3 months because of the strong currents produced by bathymetric restrictions. This southward connectivity and the strong gradient found in this month may limit the degree of connectivity between the north and south sides of the front, affecting the JaG1 and JaG2 species composition (only 44% of the species were present in both regions), in spite of the large connectivity mentioned before.

In January, the cyclonic eddy had low intensity since the isotherms did not reach the surface of the water column probably indicating an early stage of eddy formation [56, 57], it was almost as wide as the gulf extension at the center of the gulf, and was closely related to the Del Carmen Basin [58]. Cyclonic eddy are related to divergence and upwelling at the centers. In these types of eddy, high concentrations of phytoplankton as well as low abundance of zooplankton [59, 60], and fish larvae [11], have been registered [61]. However, the ichthyoplankton sampling made in January allowed us to register the shape of a cold core eddy and the warm jet surrounding it, also observed in the Gulf of Alaska [62]. In this characteristic eddy, coastal surface water is segregated while the deeper water portion of eddy core maintains its physical and biological characteristics resulting in a mix of communities. The response of fish larvae to this oceanographic process resulted in JaG3 being the group with highest species richness [24] and the highest values of ZV (154 mL 1000 m^{-3}) at the center of the eddy at ~18.4°C. Also, the lowest ZV (~45 mL 1000 m^{-3}) was associated with water coming from the southern gulf (~19°C), and with the second highest abundance of fish larvae of the cruise at the center of the eddy. This group had a fish larvae community formed by coastal pelagic species and high abundance of the most important mesopelagic species (*Benthosema panamense* and *Vinciguerria lucetia*). The 19°C isotherm coincided with the northern edge of the cyclonic eddy establishing the boundary between the JaG2 and JaG3 group.

The 21°C isotherm observed in the SST satellite image from March crossing from Bahía Concepción at the peninsular coast to Topolobampo at the continental coast of the gulf showed an apparent inverse circulation of the water with inflow by the peninsular side and outflow at the continental side. This hydrodynamic feature has been observed particularly in this month when a weakening of the poleward eastern coastal current is found, favoring the outflow of water very close to the continental coast of the gulf [23]. This isotherm indicates the boundary in the distribution of larvae fish assemblages MaG1 to the northern portion of the study area (low T10, high CHL and ZV) from MaG2 (high T10) at the south associated to the inflow of warmer and low productive water.

The anticyclonic eddy registered in March was at least 79 km wide and was located over the Farallón Basin, very close to the boundary between the MaG1 and MaG2 groups. It seems to be a recurrent oceanographic process [58]. In spite of the low productivity associated to MaG2, the center of the eddy registered a ZV of 460 mL 1000 m^{-3}, more than twice

the ZV found at the center of the cyclonic eddy of January. Anticyclonic eddy are related to convergence and downwelling at the centers and are also known as warm core eddy. Biophysical conditions registered at the center of the eddy observed in this month indicated a suitable environment for mesopelagic species dominated by *V. lucetia*, *D. laternatus*, *H. atratum*, *T. mexicanus*, and *B. panamense* but not for the coastal pelagic species *E. mordax* and *S. sagax* that distributed at lower T10 than mesopelagic species in the eddy. High concentrations of zooplankton and fish larvae abundance in the March anticyclonic eddy coincide with the same type of eddy [11, 60]. The inflow of warm and low productive water delimited by the 21°C isotherm in a large extension of the southern study area contributed also to a more complex community in MaG2 (high species richness) and was the main factor delimiting larvae fish assemblages in the Gulf of California.

The fish larvae assemblages showed a latitudinal gradient related to SST. Also, the distribution of the 18°C and 21°C isotherms, in January and March, respectively, proposed by Aceves-Medina et al. [24] as the limit of the fish larvae recurrent groups matched the eddy structures edges in these months. The warmer environment registered during our winter cruises, particularly during March, displaced the 18°C isotherm to the north determining the presence of the frontal zone of the Big Islands. Thus, interannual variability may also be playing an important role in the distribution of the fish larvae assemblages at the macroscale.

4.2. Summer Season. July and September conditions included a coastal-ocean gradient, with high T10 and lower S10 resulting from the presence of tropical surface water (TSW) entering the gulf along the continental coast [63]. Also, a stratified water column, lower CHL and ZV, but fish larvae abundance higher than in winter season was present. Assemblages were primarily correlated to the ZV and CHL gradients (Table 2) and the tropical-subtropical species dominated the community (95.5%) coinciding with Aceves-Medina et al. [20], although shallow-demersal species dominated over the mesopelagic component registered by these authors. Epipelagic species (*O. libertate*, *Opisthonema* spp., *Anchoa* spp., *C. mysticetus*, *C. caballus*, *O. saurus*, *S. sierra*, *Thunnus* sp. 1, *Kajikia audax*, and *Katsuwonus pelamis*) replaced the species registered during the winter. A very similar well-mixed water column, highly productive, and cold environment related to the tidal mixing and upwelling process was occupied by mesopelagic and shallow-demersal species present in January and in July (JuG1) in the Big Islands area.

The grouping of species in JuG2 located outside of Bahía Concepción was related to high ZV. Bays, and coastal lagoons are recognized as organic carbon [64] and particulate organic matter [65] exportation zones particularly in summer conditions. These exported organic compounds can be used by mesozooplankton, increasing phytoplankton productivity of the coastal and adjacent oceanic systems [66], providing a suitable environment for the feeding and development of fish larvae.

A very disperse group of sampling station formed JuG3. It was found in shallow and warmer sampling stations located at the continental coast, south of the Big Islands, and at the continental shelf of the northern area of the Gulf of California. In this last zone, extraordinary abundance of larvae of *O. libertate* (24,252 larvae 10 m^{-2}) and *C. mysticetus* (40,149 larvae 10 m^{-2}) filter-feeding coastal-pelagic species evidenced a spawning event of their adults. These larvae were distributed in the northern boundary of the eddy generated by the cyclonic circulation present during the summer in the northern gulf [18, 21, 67, 68], while at the center of the eddy the larvae were scarce. The adults of *O. libertate* and *C. mysticetus* may be spawning at the edge of the cyclonic eddy, as observed in other species (*E. mordax* and *S. sagax*) [26] selecting a very productive area in terms of CHL. Such selection may increase survival of fish larvae at early stages. The thermal front of this month did not influence the formation of groups.

During September, most of the gulf had high temperatures, but the coastal-ocean gradient was lower than in July. The limit between the coastal and oceanic groups (SeG1 and SeG2) formed in this month was independent of T10 and was related to high CHL. The SeG1 was found at the coastal productive environment in both continental and peninsular areas, while the oceanic group (SeG2) at the center of the gulf was correlated with the highest ZV in this cruise. We found a cyclonic low intensity eddy of only ~48 km, located at the northern portion of the Pescadero Basin. The biological component response to the presence of this oceanographic feature was contrary to that observed in the cyclonic and anticyclonic eddy structures observed in the winter. At the center of the September cyclonic eddy, we recorded the lowest ZV, species richness, and fish larvae abundance of the cruise according to eddy records in other areas [11, 59, 60]. The few larvae present at the center of the eddy belonged to one coastal-pelagic (*O. micropterus*) and one shallow-demersal (Gerreidae spp.) species.

Besides the effect of dispersion of this cyclonic eddy, surface tows made in this month influenced the composition and low abundance recorded in this eddy. Methodology of surface tows versus Bongo tows resulted in lower filtering efficiency and collections of mostly surface fish larvae assemblages. Also, few shallow-demersal and meso-bathy pelagic species were found. However, tropical-subtropical biogeographic affinity of the fish larvae registered during surface tows in September was representative of the summer season. Filter efficiency and similar ZV collections in other zooplankton groups have been found [69, 70]. These differences (surface versus oblique tows) may also be important to explain mesoscale effects. However, the analysis of the distribution patterns of fish larvae from surface and oblique tows demonstrates that even with the differences associated to the methodology, the relationships between abundance and environmental variables are similar in both cases. Also, when sampling is not extensively made in time and space over eddy and/or front structures, the oceanographic and biological data may limit some inferences related to its evolution through time, giving only a quick look of the processes present at that moment.

During July and September, the 29°C and 31°C isotherms observed in the corresponding SST satellite images ran across the gulf from the south of the Big Islands to the mouth of the gulf, dividing the gulf in two. The edges of the eddy structures in July coincide with the 29°C isotherm, while the 31°C isotherm registered in September coincided with the west edge of the cyclonic eddy of this month. The similarity in the fish larvae assemblages of both coasts during this last month may be related to transport mechanisms through eddy movement from one coast to the other.

The use of physical and biological variables to understand fish larvae abundance and distribution in the Gulf of California provided a wider overview of their relationships. Satellite imagery not always showed mesoscale processes, as we observed during March and September cruises. This may be explained by the fact that the signal of the dome or valley shape of the water column did not reach the surface. However, water column profiles did show the eddy signals allowing us to confirm their presence and therefore to relate these oceanographic processes to the biological information.

Although most of the previous records in this area established that temperature and physical mesoscale processes are responsible for the presence and distribution of fish larvae assemblages at different scales [24, 30, 71], we found that besides T10 and S10, CHL and ZV gradients used as gross indicators of the biological production and of the amount of available food in the environment were also highly correlated with the fish larvae abundance (Table 2) in the Gulf of California, particularly during summer months.

It is known that spawning activity in fishes is stimulated by environmental gradients, and among them, temperature ones seem to be the most important [26, 72]. Therefore distribution and abundance of fish larvae are highly related to those environmental gradients in large regions because of the physiological requirements of each species, as observed along the Gulf of California. However, in a local scale, mesoscale processes seem to be the main force driving not only the distribution of species but also the biophysical gradients. Eddy structures that retain fish larvae, also affect the distribution of physical variables increasing, for example, the temperature in the cores. In this way, the relationships observed between fish larvae abundance and T10, S10, CHL, and ZV could be the effect of two possibilities. The first is that the coincidence of the distribution of all variables is the result of their dependence on the mesoscale process; the second is that the mesoscale process promotes suitable physical environments that stimulate adults to spawn.

The SST difference between winter and summer in our study was about 10°C and was wide enough to establish low fish larvae abundance and the presence of more temperate (7.5%) species in colder winter conditions than in the summer. The change of T10 and S10 as the main factors affecting fish larvae assemblages during the winter to CHL and ZV during the summer reflects different biophysical conditions for fish larvae inside the gulf related to the seasonal environmental changes. In a not limited food supply environment, because of the presence of a high production originated from strong upwelling during winter [18], the T10 north-south gradients observed in the gulf strongly influenced the formation of fish larvae groups that also distributed in a north-south pattern. In contrast, high T10 along the gulf, but strong food limitation during the summer, related to weak upwelling at the peninsular coast and inflow of low productive TSW with a high-coastal low-ocean gradient, promoted fish larvae groups to distribute according to this pattern.

It is known that the formation and composition of fish larvae assemblages are first determined by the location of the spawning areas of the adults, by timing in spawning, and subsequently by interspecific differences in mortality and development rates associated to larvae behavior, among others [73–75]. In terms of reproductive strategies, Aceves-Medina et al. [26] propose the hypothesis that *E. mordax* fish larvae are abundant when those of *S. sagax* are scarce indicating that the species occupies different positions in the water column, with *S. sagax* mostly above the thermocline while *E. mordax* distributes deeper, and that the adults spawn at different places, not only at the mainland coast.

The cyclonic and anticyclonic eddy registered in our study during winter retain pacific sardine larvae at the center of the gulf, while presence of *E. mordax* in the frontal zone may indicate that the species use these highly productive but different oceanographic processes as nursery areas.

5. Conclusions

The results obtained in this study provide a more comprehensive explanation of the response of the fish larvae community to the environmental complexity of macro, and mesoscale processes in the Gulf of California.

Strong differences between winter (latitudinal gradient) and summer (coast-ocean gradient) environmental conditions as well as in the composition, abundance, biogeographic affinity, and habitat of the fish larvae assemblages were found in the area. In the macroscale, fish larvae abundance was highly related to environmental variables with seasonal changes in the main factors, T10 and S10 during the winter, and CHL and ZV during the summer, affecting the formation of groups. Besides S10, CHL, and ZV, 19°C and 21°C isotherms during the winter, and 29°C and 31°C isotherms during the summer, seemed to be related to the distributional limits between fish larvae assemblages at the macroscale. Changes in location of these limits were related to interannual variability associated to the El Niño, which also represents a major source of variability in the formation of the fish larvae assemblages.

The center of the winter eddy had more abundance than the summer eddy. Among types of eddy, the cyclonic of January recorded the highest abundances of mesopelagic (*V. lucetia*, *D. laternatus*, and *B. panamense*) and coastal-pelagic species (*S. sagax*, *S. japonicus*, and *E. teres*) than in the anticyclonic of March and cyclonic eddy of September, indicating that the evolution of the eddy over the time scale may be important to understand the complexity of the fish larvae community. Eddy structures have an important impact on larval fish assemblages, but these impacts can change depending on the type, size, age, source water mass, dynamics within the eddy and interactions with the surrounding waters, generation time of the organisms, and even by the time and type of sampling. At the mesoscale, eddy and fronts coincided with these isotherms, playing an important role in emphasizing the differences among species assemblages.

Future research on fish larvae ecology on a similar or more complete data base set basis and high resolution sampling of both eddy types and frontal zones, following them over time should be encouraged in order to provide a better understanding of the species-environment relationships. Also, interannual analysis should be attempted in order to understand the macroscale changes in fish larvae assemblages.

Acknowledgments

The authors would like to thank Secretaría de Marina, Armada de México, and Jaime Gómez-Gutiérrez (CICIMAR-IPN, México) for access to CGC and GOLCA samples and oceanographic data, respectively. This research represents part of the first author's PhD dissertation studies at the Universidad de Guadalajara, Mexico. This work was partially funded by CONACyT (FOSEMARNAT-2004-01-C01-144) and COFAA-IPN (SIP 20070782, SIP 20070784). R. Avendaño-Ibarra was supported by a CONACyT postgraduate fellowship and G. Aceves-Medina and R. Avendaño-Ibarra by COFAA-IPN and EDI-IPN grants.

References

[1] D. L. Mackas, K. L. Denman, and M. R. Abbott, "Plankton patchiness: biology in the physical vernacular," *Bulletin of Marine Science*, vol. 37, no. 2, pp. 652–674, 1985.

[2] S. A. Levin, "The problem of pattern and scale ecology," *Ecology*, vol. 73, no. 6, pp. 1943–1967, 1992.

[3] H.-H. Hinrichsen, "Biological processes and links to the physics," *Deep-Sea Research Part II*, vol. 56, no. 21-22, pp. 1968–1983, 2009.

[4] T. D. Iles and M. Sinclair, "Atlantic herring: stock discreteness and abundance," *Science*, vol. 215, no. 4533, pp. 627–633, 1982.

[5] G. R. Bolz and R. G. Lough, "Retention of ichthyoplankton in the Georges bank region during the autumn-winter seasons 1971–1977," *Journal of Northwest Atlantic Fishery Science*, vol. 5, no. 1, pp. 33–45, 1984.

[6] J. S. Wroblewski and J. Cheney, "Ichthyoplankton associated with a warm core ring off the Scotian Shelf," *Canadian Journal of Fisheries and Aquatic Sciences*, vol. 41, no. 2, pp. 294–303, 1984.

[7] T. Kioerboe, P. Munk, K. Richardson, V. Christensen, and H. Paulsen, "Plankton dynamics and larval herring growth, drift and survival in a frontal area," *Marine Ecology Progress Series*, vol. 44, pp. 205–219, 1988.

[8] P. S. Lobel and A. R. Robinson, "Larval fishes and zooplankton in a cyclonic eddy in Hawaiian waters," *Journal of Plankton Research*, vol. 10, no. 6, pp. 1209–1223, 1988.

[9] A. Sabatés and M. Masó, "Unusual larval fish distribution pattern in a coastal zone of the western Mediterranean," *Limnology and Oceanography*, vol. 37, no. 6, pp. 1252–1260, 1992.

[10] A. Sournia, "Pelagic biogeography and fronts," *Progress in Oceanography*, vol. 34, no. 2-3, pp. 109–120, 1994.

[11] J. M. Rodríguez, E. D. Barton, S. Hernández-León, and J. Arístegui, "The influence of mesoscale physical processes on the larval fish community in the Canaries CTZ, in summer," *Progress in Oceanography*, vol. 62, no. 2-4, pp. 171–188, 2004.

[12] G. R. Flierl and J. S. Wroblewski, "The possible influence of warm core Gulf Stream rings upon shelf water larval fish distribution," *Fishery Bulletin*, vol. 83, no. 3, pp. 313–330, 1985.

[13] P. C. Fiedler, "Offshore entrainment of anchovy spawning habitat, eggs, and larvae by a displaced eddy in 1985," *California Cooperative Oceanic Fisheries Investigations Reports*, vol. 27, pp. 144–152, 1986.

[14] R. A. Myers and K. Drinkwater, "The influence of Gulf Stream warm core rings on recruitment of fish in the northwest Atlantic," *Journal of Marine Research*, vol. 47, no. 3, pp. 635–656, 1989.

[15] C. Roy, "An upwelling-induced retention area off Senegal: a mechanism to link upwelling and retention processes," *South African Journal of Marine Science*, vol. 19, no. 1, pp. 89–98, 1998.

[16] K. A. Smith and I. M. Suthers, "Displacement of diverse ichthyoplankton assemblages by a coastal upwelling event on the Sydney shelf," *Marine Ecology Progress Series*, vol. 176, pp. 49–62, 1999.

[17] P. Bécognée, C. Almeida, J. M. Rodríguez et al., "Mesoscale distribution of clupeoid larvae in an upwelling filament trapped by a quasi-permanent cyclonic eddy off northwest Africa," *Deep-Sea Research I*, vol. 56, no. 3, pp. 330–343, 2009.

[18] M. F. Lavín and S. G. Marinone, "An overview of the physical oceanography of the Gulf California," in *Nonlinear Processes in Geophysical Fluid Dynamics. A Tribute to the Scientific Work of Pedro Ripa*, O. Velasco Fuentes, J. Sheinbaum, and J. Ochoa, Eds., Chapter 11, pp. 173–204, Kluwer Academic, Amsterdam, The Netherlands, 2003.

[19] J. L. Castro-Aguirre, E. F. Balart, and J. Arvizu-Martínez, "Contribución al conocimiento del origen y distribución de la ictiofauna del Golfo de California, México," *Hidrobiológica*, vol. 5, no. 1-2, pp. 57–78, 1995.

[20] G. Aceves-Medina, S. P. A. Jiménez-Rosenberg, A. Hinojosa-Medina et al., "Fish larvae from the gulf of California," *Scientia Marina*, vol. 67, no. 1, pp. 1–11, 2003.

[21] A. Jiménez, S. G. Marinone, and A. Parés, "Efecto de la variabilidad espacial y temporal de viento sobre la circulación en el Golfo de California," *Ciencias Marinas*, vol. 31, no. 2, pp. 357–368, 2005.

[22] W. S. Pegau, E. Boss, and A. Martínez, "Ocean color observations of eddies during the summer in the Gulf of California," *Geophysical Research Letters*, vol. 29, no. 9, pp. 1–3, 2002.

[23] L. Zamudio, P. Hogan, and E. J. Metzger, "Summer generation of the Southern Gulf of Colifornia eddy train," *Journal of Geophysical Research C*, vol. 113, no. 6, Article ID C06020, 2008.

[24] G. Aceves-Medina, S. P. A. Jiménez-Rosenberg, A. Hinojosa-Medina, R. Funes-Rodríguez, R. J. Saldierna-Martínez, and P. E. Smith, "Fish larvae assemblages in the Gulf of California," *Journal of Fish Biology*, vol. 65, no. 3, pp. 832–847, 2004.

[25] E. A. Inda-Díaz, L. Sánchez-Velasco, and M. F. Lavín, "Three-dimensional distribution of small pelagic fish larvae (*Sardinops sagax* and *Engraulis mordax*) in a tidal-mixing front and surrounding waters (Gulf of California)," *Journal of Plankton Research*, vol. 32, no. 9, pp. 1241–1254, 2010.

[26] G. Aceves-Medina, R. Palomares-García, J. Gómez-Gutiérrez, C. J. Robinson, and R. J. Saldierna-Martínez, "Multivariate characterization of spawning and larval environments of small pelagic fishes in the Gulf of California," *Journal of Plankton Research*, vol. 31, no. 10, pp. 1–19, 2009.

[27] M. Peguero-Icaza, L. Sánchez-Velasco, M. F. Lavín, and S. G. Marinone, "Larval fish assemblages, environment and circulation in a semienclosed sea (Gulf of California, México)," *Estuarine, Coastal and Shelf Science*, vol. 79, no. 2, pp. 277–288, 2008.

[28] A. Danell-Jiménez, L. Sánchez-Velasco, M. F. Lavín, and S. G. Marinone, "Three-dimensional distribution of larval fish assemblages across a surface thermal/chlorophyll front in a semienclosed sea," *Estuarine, Coastal and Shelf Science*, vol. 85, no. 3, pp. 487–496, 2009.

[29] L. Sánchez-Velasco, M. F. Lavín, M. Peguero-Icaza et al., "Seasonal changes in larval fish assemblages in a semi-enclosed sea (Gulf of California)," *Continental Shelf Research*, vol. 29, no. 14, pp. 1697–1710, 2009.

[30] F. Contreras-Catala, L. Sánchez-Velasco, M. F. Lavín, and V. M. Godínez, "Three-dimensional distribution of larval fish assemblages in an anticyclonic eddy in a semi-enclosed sea (Gulf of California)," *Journal of Plankton Research*, vol. 34, no. 6, pp. 548–562, 2012.

[31] J. H. Simpson, "The shelf-sea fronts: implications of their existence and behaviour," *Philosophical Transactions of the Royal Society of London A, Mathemathical and Physical Sciences*, vol. 302, no. 1472, pp. 531–546, 1981.

[32] E. Torres-Orozco, *Análisis volumétrico de las masas de agua del Golfo de California [M.S. thesis]*, Centro de Investigación Científica y de Educación Superior, 1993.

[33] P. E. Smith and S. L. Richardson, *Standard Techniques for Pelagic Fish Egg and Larva Surveys*, FAO, 1977.

[34] J. R. Beers, "Determination of zooplankton biomass," in *Zooplankton Fixation and Preservation. Monographs on Oceanographic Methodology*, H. F. Steedman, Ed., Chapter 2, pp. 35–84, The UNESCO Press, Paris, France, 1976.

[35] H. G. Moser, W. J. Richards, D. M. Cohen, A. W. J. Kendall, and S. L. Richardson, *Ontogeny and Systematics of Fishes*, Allen Press, 1984.

[36] H. G. Moser, *The Early Stages of Fishes in the California Current Region*, vol. 1, Atlas 33, Allen Press, 1996.

[37] B. S. Beltrán-León and R. Ríos, *Estadios Tempranos De Peces Del Pacífico Colombiano*, vol. 1, Instituto Nacional de Pesca y Acuicultura, 2000.

[38] B. S. Beltrán-León and R. Ríos, *Estadios Tempranos De Peces Del Pacífico Colombiano*, vol. 2, Instituto Nacional de Pesca y Acuicultura, 2000.

[39] W. J. Richards, *Early Stages of Atlantic Fishes. An Identification Guide for the Western Central North Atlantic*, vol. 1, Taylor and Francis, 2006.

[40] W. J. Richards, *Early Stages of Atlantic Fishes. An Identification Guide For the Western Central NorthAtlantic*, vol. 2, Taylor and Francis, 2006.

[41] C. J. F. Ter Braak and I. C. Prentice, "A theory of gradient analysis," *Advances in Ecological Research*, vol. 34, no. 3, pp. 235–282, 2004.

[42] P. Legendre and L. Legendre, *Numerical Ecology*, vol. 20, Elsevier, 1998.

[43] M. F. Lavín, E. Beier, and A. Badan, "Estructura hidrográfica y circulación del Golfo de California: escalas estacionales e interanuales," in *Contribuciones a la Oceanografía Física En México*, M. F. Lavin, Ed., Monografía No. 3, Chapter 7, pp. 141–172, Unión Geofísica Mexicana, Estado de Zacatecas, México, 1997.

[44] S. Álvarez-Borrego and J. Lara-Lara, "The physical environment and primary productivity of the Gulf of California," in *The Gulf*

and Peninsular Province of the Californias, B. R. T. Simoneit and J. P. Dauphin, Eds., vol. 47 of *American Association of Petroleum Geologists Memoir*, pp. 555–567, 1991.

[45] T. L. Espinosa-Carreón and J. E. Valdez-Holguín, "Variabilidad interanual de clorofila en el Golfo de California," *Ecología Aplicada*, vol. 6, no. 1, pp. 83–92, 2007.

[46] S. Álvarez-Borrego, J. A. Rivera, G. Gaxiola-Castro, M. J. Acosta-Ruiz, and R. A. Schwartzlose, "Nutrientes en el Golfo de California," *Ciencias Marinas*, vol. 5, no. 2, pp. 53–71, 1978.

[47] R. Avendaño-Ibarra, R. De Silva-Dávila, G. Aceves-Medina, H. Urías-Leyva, and G. Vázquez-López, *Distributional Atlas of Fish Larvae of the Southern Region of the Gulf of California (February–March 2005)*, Instituto Politécnico Nacional, 2009.

[48] M. L. Argote, A. Amador, and M. F. Lavin, "Tidal dissipation and stratification in the Gulf of California," *Journal of Geophysical Research*, vol. 100, no. 8, pp. 16103–16118, 1995.

[49] R. M. Hidalgo-González, S. Álvarez-Borrego, and A. Zirino, "Mixing in the region of the midrift islands of the Gulf of California: effect on surface pCO_2 ," *Ciencias Marinas*, vol. 23, no. 3, pp. 317–327, 1997.

[50] P. H. Ressler, J. A. Holmes, G. W. Fleischer, R. E. Thomas, and K. C. Cooke, "Pacific hake, *Merluccius productus*, autecology: a timely review," *Marine Fisheries Review*, vol. 69, no. 1–4, pp. 1–24, 2007.

[51] B. Y. Sumida and H. G. Moser, "Food and feeding of Pacific hake larvae, *Merluccius productus*, off southern California and northern Baja California," *California Cooperative Oceanic Fisheries, Investigations Reports*, vol. 21, pp. 161–166, 1980.

[52] J. Gómez-Gutiérrez, S. Martínez-Gómez, and C. J. Robinson, "Influence of thermo-haline fronts forced by tides on near-surface zooplankton aggregation and community structure in Bahía Magdalena, México," *Marine Ecology Progress Series*, vol. 346, pp. 109–125, 2007.

[53] C. J. Robinson, S. Gómez-Aguirre, and J. Gómez-Gutiérrez, "Pacific sardine behaviour related to tidal current dynamics in Bahía Magdalena, México," *Journal of Fish Biology*, vol. 71, no. 1, pp. 200–218, 2007.

[54] J. R. Taylor and R. Ferrari, "Ocean fronts trigger high latitude phytoplankton blooms," *Geophysical Research Letters*, vol. 38, no. 23, Article ID L23601, 2011.

[55] S. G. Marinone, "Seasonal surface connectivity in the Gulf of California," *Estuarine, Coastal and Shelf Science*, vol. 100, pp. 133–141, 2012.

[56] P. L. Richardson, C. Maillard, and T. B. Stanford, "The physical structure and life history of cyclonic Gulf Stream ring Allen," *Journal of Geophysical Research*, vol. 84, no. 12, pp. 7727–7741, 1979.

[57] A. C. Vastano, J. E. Schmitz, and D. E. Hagan, "The physical oceanography of two rings observed by the cyclonic ring experiment. Part I: physical structure," *Journal of Physical Oceanography*, vol. 10, no. 4, pp. 493–513, 1980.

[58] J. M. Figueroa, S. G. Marinone, and M. F. Lavín, "A description of geostrophic gyres in the southern Gulf of California," in *Nonlinear Processes in Geophysical Fluid Dynamics. A Tribute To the Scientific Work of Pedro*, Ripa, O. Velasco Fuentes, J. Sheinbaum, and J. Ochoa, Eds., Chapter 14, pp. 237–255, Kluwer Academic, Amsterdam, The Netherlands, 2003.

[59] M. E. Huntley, A. González, Y. Zhu, M. Zhou, and X. Irigoien, "Zooplankton dynamics in a mesoscale eddy-jet system off California," *Marine Ecology Progress Series*, vol. 201, pp. 165–178, 2000.

[60] A. Bakun, "Fronts and eddies as key structures in the habitat of marine fish larvae: opportunity, adaptive response and competitive advantage," *Scientia Marina*, vol. 70, no. 2, pp. 105–122, 2006.

[61] T. L. Espinosa-Carreón, G. Gaxiola-Castro, E. Beier, P. T. Strub, and J. A. Kurczyn, "Effects of mesoscale processes on phytoplankton chlorophyll off Baja California," *Journal of Geophysical Research C*, vol. 117, no. 4, Article ID C04005, 2012.

[62] E. Atwood, J. T. Duffy-Anderson, J. K. Horne, and C. Ladd, "Influence of mesoscale eddies on ichthyoplankton assemblages in the Gulf of Alaska," *Fisheries Oceanography*, vol. 19, no. 6, pp. 493–507, 2010.

[63] G. I. Roden, "Oceanographic aspects of Gulf of California," in *Marine Geology of the Gulf of California: A Symposium*, T. H. Van Andel and G. G. Shor Jr., Eds., pp. 30–54, 1964.

[64] D. C. Escobedo-Urías, A. Martínez-López, A. Jiménez-Illescas, A. E. Ulloa-Pérez, and A. Zavala-Norzagaray, "Intercambio de carbono orgánico particulado del sistema lagunar San Ignacio-Navachiste, Sinaloa, con el mar adyacente," in *Carbono En Ecosistemas Acuáticos De México*, B. Hernández de la Torre and G. Gaxiola-Castro, Eds., Chapter 11, pp. 171–185, Secretaría de Medio Ambiente y Recursos Naturales. Instituto Nacional de Ecología; Centro de Investigación Científica y de Educación Superior de Ensenada, Baja California, México, 2007.

[65] A. Martínez-López and I. Gárate-Lizárraga, "Variación diurna de la materia orgánica particulada en una laguna costera del Golfo de California," *Revista De Biología Tropical*, vol. 45, no. 4, pp. 1310–1317, 1997.

[66] J. García-Pamanes, J. R. Lara-Lara, and C. Bazán-Guzmán, "Pastoreo por el mesozooplancton en la región central del Golfo de California: un estudio estacional," in *Carbono En Ecosistemas Acuáticos De México*, B. Hernández de la Torre and G. Gaxiola, Eds., Chapter 9, pp. 141–155, Secretaría de Medio Ambiente y Recursos Naturales. Instituto Nacional de Ecología; Centro de Investigación Científica y de Educación Superior de Ensenada, Baja California, México, 2007.

[67] E. Beier and P. Ripa, "Seasonal gyres in the Northern Gulf of California," *Journal of Physical Oceanography*, vol. 29, no. 2, pp. 305–311, 1999.

[68] V. Makarov and A. Jiménez, "Corrientes básicas barotrópicas en el Golfo de California," *Ciencias Marinas*, vol. 29, no. 2, pp. 141–153, 2003.

[69] K. Saitô, "Distribution of paralarvae of *Ommastrephes bartrami* and *Eucleoteuthis luminosa* in the eastern waters off Ogasawara Islands," *Fisheries Research Institute*, no. 58, pp. 15–23, 1994.

[70] M. D. Ohman and P. E. Smith, "A comparison of zooplankton sampling methods in the CalCOFI time series," *California Cooperative Oceanic Fisheries Investigations Reports*, vol. 36, pp. 153–158, 1995.

[71] H. G. Moser, E. H. Ahlstrom, D. Kramer, and E. G. Stevens, "Distribution and abundance of fish eggs and larvae in the Gulf of California," *California Cooperative Oceanic Fisheries Investigations Reports*, vol. 17, pp. 112–128, 1974.

[72] R. J. Lynn, "Variability in the spawning habitat of Pacific sardine (*Sardinops sagax*) off southern and central California," *Fisheries Oceanography*, vol. 12, no. 6, pp. 541–553, 2003.

[73] G. W. Boehlert and B. C. Mundy, "Ichthyoplankton assemblages at seamounts and oceanic islands," *Bulletin of Marine Science*, vol. 53, no. 2, pp. 336–361, 1993.

[74] H. G. Moser and P. E. Smith, "Larval fish assemblages and ocean boundaries," *Bulletin of Marine Science*, vol. 53, no. 2, pp. 283–289, 1993.

[75] T. J. Miller, "Assemblages, communities, and species interactions," in *Concepts in Fishery Science: The Unique Contributions of Early Life Stages*, L. A. Fuiman and R. G. Werner, Eds., Chapter 8, pp. 183–205, Blackwell Sciences, Malden, Mass, USA, 2002.

5

Reef Flat Community Structure of Atol das Rocas, Northeast Brazil and Southwest Atlantic

Adriana C. Fonseca,[1,2] Roberto Villaça,[3] and Bastiaan Knoppers[2]

[1] Instituto Chico Mendes de Conservação da Biodiversidade (ICMBIO), Rodovia Mauricio S. Sobrinho s/n°, km 2, 88053-700, Florianópolis, SC, Brazil
[2] Programa de Pós-Graduação em Geoquímica, Instituto de Química, Universidade Federal Fluminense, Outeiro de São João Batista, s/n°, 24020-141, Niterói, RJ, Brazil
[3] Programa de Pós-Graduação em Biologia Marinha, Instituto de Biologia, Universidade Federal Fluminense, Outeiro de São João Batista, s/n°, 24001-970, Niterói, RJ, Brazil

Correspondence should be addressed to Adriana C. Fonseca, adricarvalhal@globo.com

Academic Editor: Jakov Dulčić

This study was conducted during 1999 to 2002 and addresses the community structure and some ecological aspects of the benthic reef flat assemblages of Atol das Rocas, located offshore the NE brazilian coast. It corresponds to the sole atoll of the SW Atlantic, which characterized by a shallow topography and is almost completely built by coralline algae. The turf forming red macroalgae *Digenea simplex* and the crustose coralline *Hydrolithon pachydermum* were the dominant species of the reef flat. The crustose green macroalgae *Dictyosphaeria ocellata* and the turf forming red macroalgae *Gelidiella acerosa* were the subdominant species. Biomass values of *D. simplex* were about twice higher than the other species, pointing out to its relevance in the community structure of this reef zone. Biodiversity indices indicated a high equitability within the few species observed and a relative temporal stability of the community structure. Some local spatial variations were found in the community structure of the reef flat zone, enabling the definition of three subhabitats. The patterns of distribution and abundance of the benthic organisms seem to be related to the environmental conditions of the reef flat, such as low water turbulence, lengthy periods of aerial exposure, and low herbivore pressure.

1. Introduction

Reefs of tropical atolls have been described as harboring well-defined geomorphic zones, including a fore-reef, a reef crest, a reef flat, and a lagoonal reef zone [1–3]. Theoretically, each zone is subject to intrinsic environmental conditions which circumscribe unique habitat types and communities. Benthic communities within each zone were thought to exhibit a great similarity between each other. However, local small scale variability of physical and biotic factors in a well-defined geomorphic zone may harbor microcosms, each supporting unique benthic communities [4–6].

The reef flat zone is a complex area with gradients of environmental factors such as temperature, turbidity, and tidal related exposure of communities [3]. These gradients coupled with differences in depth and substrate type provide a great number of habitats that have resulted in subdivisions of this zone [7–10]. For further understanding of the similarities and differences within this geomorphic zone, it becomes essential to obtain more information on the organization of the benthic community assemblages and how they operate in accordance to the physical-geomorphological features of their habitats [11].

The Marine Biological Reserve of Atol das Rocas, NE Brazil, is a pristine reef area which offers an ideal system to study natural variations of benthic community assemblages within geomorphic zones. The reserve was created in 1979 encompassing a region of about $360 \, km^2$ around the reef ring formerly reaching depths down to about 1.000 m. In contrast to the atolls of the Indo-Pacific and Caribbean waters, mainly constructed by hermatypic corals, the reef framework of Atol das Rocas is largely composed of crustose coralline algae, foraminiferans, and shells of mollusk gastropods [12, 13].

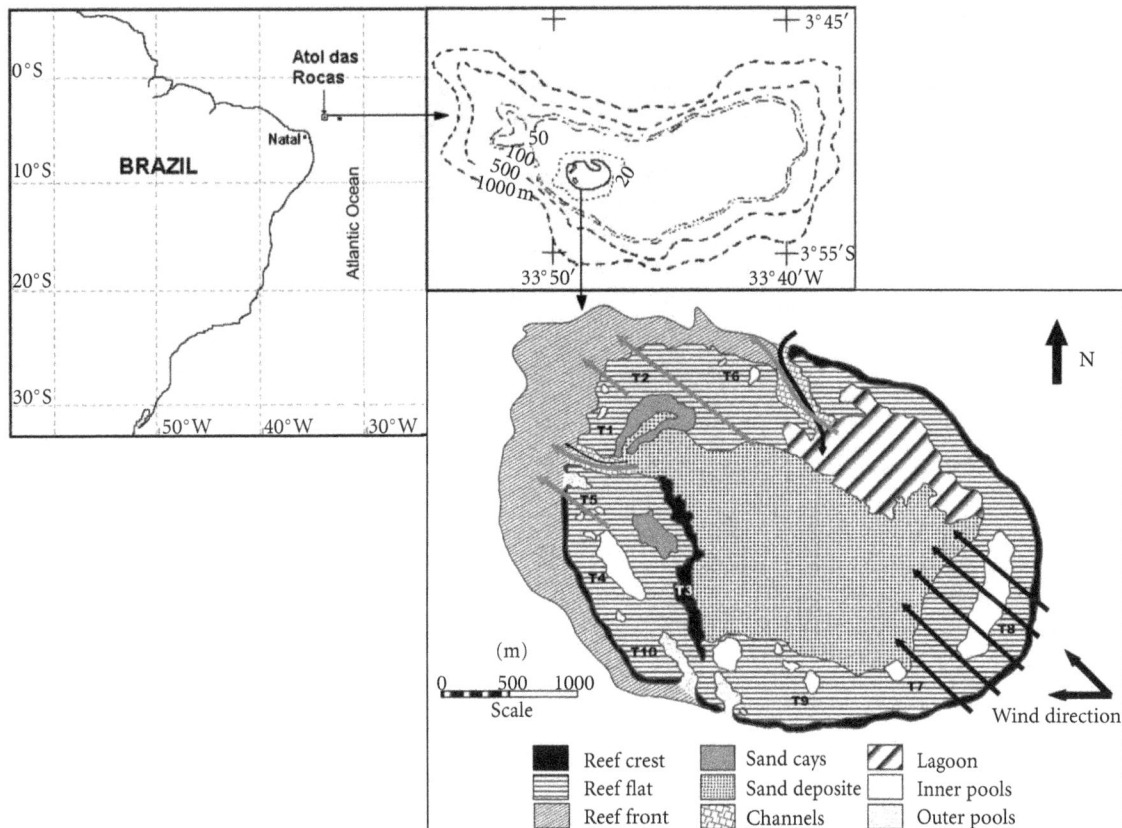

FIGURE 1: Study area showing the sampling sites on the reef flat (T1–T10) and the pattern of internal current circulation: black arrows = flood tide, gray arrows = ebb tide.

In Atol das Rocas, frondose macroalgae, algal turfs, and coralline algae are the major space occupiers in many benthic areas [14, 15]. Until now, however, few studies focused on benthic community assemblages of Atol das Rocas [13, 16–18] and none of these assessments included the algal communities as a whole.

The present paper addresses the benthic communities with emphasis on the algal assemblages of the reef flat zone of Atol das Rocas. The main goals were to determine (1) the distribution and abundance of benthic organisms; (2) the communities' structure and diversity; (3) the spatial and seasonal similarities between the benthic assemblages; (4) also infer about the environmental factors that govern the distribution and abundance of benthic organisms. The results should help to understand the organization of benthic communities in the reef flat giving support to the management of the marine reserve of Atol das Rocas.

2. Material and Methods

2.1. Study Area. Atol das Rocas is located in the Southwest Atlantic Ocean at 3°51′S and 33°49′W, 266 km offshore from the city of Natal, Rio Grande do Norte State, NE Brazil (Figure 1). It corresponds to the sole atoll in the South Atlantic and is one of the smallest atolls in the world. It has an elliptical shape, measuring 3,5 km along its main E-W axis and 2,5 km along its N-S axis. The reef flat proper has a total

area of about 2,62 km^2 with a width varying between 100 and 800 m. The reef flat surface is composed by associations of coralline algae-vermetid gastropods growing as linear ridges with an elevation of around 2 m above the Mean Seawater Level (MSL). Its contour is almost continuous, being interrupted by two tidal channels to the ocean, one set at the western and the other at the northern flank. These channels divide the reef flat into a windward and a leeward arch. An algal dominated reef crest of about 5 m wide and height of up to 0.5 m above the reef flat surface limits the entire windward arch. At the leeward side, this feature is almost inexistent [12, 13, 19] (Figure 1). The windward arch presents higher elevations than the leeward one, since the geological development of the former is older than the latter [19, 20].

The South Equatorial Current (SEC) and SE to E trade winds dominate the Rocas area. SEC has a constant westward drift with a mean speed of 30 cm s^{-1} and the winds attain maximum speeds of 11 m s^{-1} (13 and others cited therein). Waves pound largely upon the eastern windward portion of the atoll at periods of 4 s to 7 s and heights not beyond 2 m. However, in summer months (December to March), when the Intertropical Convergence Zone (ITCZ) crosses the equator and attains its southernmost position, higher waves with longer periods emerge from the North Atlantic on the leeward side of the atoll (19 and others cited therein). The regional climate is tropical and dry, with a minimum air temperature of 26°C and annual average precipitation of 109 mm [21].

Local tides are semidiurnal and attain a maximum range of 3,8 m [13]. At flood tide, water flows into the reef through the northern channel and over the windward side of the reef flat. At ebb tides, water outflow to the sea occurs via both the northern and western channels [16] (Figure 1). The reef flat is entirely emerged during low tide and submerged at high tide. Surface water temperatures at the fore reef vary between 27°C and 29°C and salinities between 36 and 38. In shallow waters of the back reef, the water temperature may vary from 24°C to 36°C and salinities from 36 to 42.

2.2. Sampling. The biological data were obtained from four (three-weekly) field surveys, being two set in "austral winter" (July 1999 and June 2000) and two in "austral summer" (December 2000 and March 2002). The abundance of benthic organisms was estimated as the percentage cover at 10 different sites (T1 to T10) on the reef flat (Figure 1). At each site, 15 quadrat frames, with a size of 25×25 cm each, were set in a distance of five meters between each other, forming a grid. Each grid, corresponded to a total area of $93,7$ m^2 and was set between the reef crest and the inside edge of the reef flat. The quadrats were sampled by a visual estimate method [22]. The site positions were marked by GPS (Garmin Etrex) and manually positioned local markings and revisited during each campaign. The standing stock of the most representative macroalgal species of the reef flat was estimated using percentage cover-biomass (dry weight) conversion factors. To obtain the factors, the total cover of each macroalgae species (wet-weight) was scratched from some quadrats and the constant dry weight asserted in the laboratory. At least five samples were obtained for each species and mean factors were computed. The standing stock of each species on the reef flat was thus estimated using mean percentage cover data.

2.3. Data Analysis. Biological variables such as the Shannon Index of Diversity (H′), Taxonomic Richness, and Evenness (Pielou's J) [23] were calculated for each site of the reef flat, and campaign. The similarity between sites was analyzed by nonmetric multidimensional scaling ordination (MDS), driven by taxa percentage cover data (by square-root transformation) and the Bray-Curtis index. Formal significance tests for differences between the groups were performed using the 1-way ANOSIM test. The taxa contributing for the similarity of the groups and the dissimilarities within them were investigated using the similarities percentages procedure (SIMPER). The statistical package PRIMER 6.0 (Primer-E) was used for these analyses.

Seasonal differences in taxa cover in the different habitats were tested using 1-way ANOVA. These biological data were determined by averaging site data. Bartlett's test was used to test normality and where necessary, data were $\log(x + 1)$ transformed or the nonparametric Kruskal-Wallis test was applied. The post hoc multiple comparisons tests

Tukey-Kramer, for parametric data, and Dunn's, for nonparametric data, were used when significant differences were detected ($p < 0.05$) [24]. The statistical package GraphPad Instat (GraphPad Software) was used.

3. Results

3.1. Benthic Community Structure. The reef flat benthic community presents a physiognomic monotony, with few taxa replacing each other amongst the different sites. During all surveys only twenty taxa were recorded, being the macroalgae the dominant organisms (Table 1). The overall Shannon Index (H′) ranged between 2,75 and 3,18, Richness between 13 and 16, and Evenness (Pielou's J) between 0,70 and 0,84 (Table 2).

The benthic community stands were, in general, bi- or tri-stratified, composed by a crustose and an erect stratum with less than 10 cm in height. The erect stratum is basically formed by turf-like mat species, mainly the Rodophyceae *Digenea simplex* (17–29%) and *Gelidiella acerosa* (7–14%). The crustose stratum was mainly composed by the red coralline *Hydrolithon pachydermum* (15–30%) and the Chlorophyceae *Dictyosphaeria ocellata* (10–29%). These four species achieved together 68% to 95% of the total benthic cover on the reef flat. The biomass data exhibited values of about 969 g dw m^{-2} for *D. simplex*, 484 g dw m^{-2} for *D. ocellata*, and 274 g dw m^{-2} for *G. acerosa*. Figure 2 shows the variations of biomass estimates for these species in the reef flat as a whole.

Despite the dominance of macroalgae, sessile fauna were also important. The zoanthid *Zoanthus sociatus* dominates great areas of the reef flat (18–20%), showing high competitive abilities, as in certain places all others organisms were absent. Other important components of the reef benthic fauna were the encrusting tube worms (0,2–14%) and the mollusc Vermetidae *Dendropoma* sp. (0,2–6%), which participates in reef construction as secondary framework builders [12, 13]. Though the majority of taxa, either of the flora and the invertebrates, were distributed along almost the entire reef flat, some exceptions, such as *G. acerosa* and *G. setacea*, were more restricted to specific sites, mainly in the leeward reef flat.

3.2. Data Analysis. The MDS ordinations of taxa cover on the reef flat identified four major groups, mainly reflecting spatial differences of the community structure along this reef zone (Figure 3). Group I, composed the sites T1, T2, and T10; group II, mainly the sites T3, T5, and T6, and groups III and IV, the sites T4, T7, T8, and T9. The results of the ANOSIM tests confirmed the differences between the groups ($R = 0,879; p = 0,001$). The pairwise tests showed that the most significant differences were between group I and the remainder (Table 3).

The Similarity Percentage Analysis (SIMPER) showed that variations on the cover of *Digenea simplex*, *Gelidiella acerosa*, *Zoanthus sociatus*, polychaeta tubes, and cyanobacteria (*Lyngbya confervoides* and another nonidentified species) contributed most to the break-down of similarity between

TABLE 1: Mean cover of benthic species recorded in Atol das Rocas reef flat. The values correspond to the mean of the 10 sites sampled. Algae nomenclatures are in accordance with AlgaeBase (2009). Other species in accordance with Catalogue of Life (2009).

Species	Mean cover (%) ± SD			
	Winter 1999	Winter 2000	Summer 2000	Summer 2002
Bacteria				
Cyanophyceae (sp.1)	5.7 ± 7.6	2.9 ± 4.1	8.2 ± 5.3	4.8 ± 7.8
Lyngbya confervoides	2.5 ± 5.8	10.5 ± 15.3	0.2 ± 0.6	7.6 ± 19.4
Plantae				
Chlorophyta				
Cladophoropsis membranacea	0.6 ± 1.9	0.2 ± 0.7	0.1 ± 0.4	1.5 ± 2.9
Phyllodictyon anastomosans	—	—	0.1 ± 0.2	—
Dictyosphaeria ocellata	21.1 ± 19.4	23.1 ± 17.3	29.0 ± 22.1	9.6 ± 11.2
Dictyosphaeria versluysii	0.1 ± 0.1	0.2 ± 0.3	0.3 ± 0.7	3.7 ± 7.0
Rhizoclonium sp.	1.3 ± 2.2	0.1 ± 0.2	—	—
Rhodophyta				
Digenea simplex	29.2 ± 34.0	28.5 ± 31.8	17.3 ± 24.9	29.3 ± 28.7
Herposiphonia secunda	—	1.1 ± 2.7	—	—
Gelidiella acerosa	14.1 ± 24.4	13.0 ± 22.7	9.1 ± 14.7	7.5 ± 20.0
Gelidiella setacea	2.5 ± 4.0	3.0 ± 7.3	1.3 ± 2.3	1.3 ± 2.3
Corallinaceae (sp.1)	0.2 ± 0.8	0.1 ± 0.2	0.1 ± 0.4	—
Hydrolithon pachydermum	15.1 ± 12.6	30.1 ± 21.3	27.7 ± 19.3	22.0 ± 15.9
Jania adhaerens	0.3 ± 0.1	—	—	2.4 ± 5.1
Phaeophyta				
sp. 1 (crustose form)	5.9 ± 18.5	—	—	—
Animalia				
Zoanthus sociatus	18.2 ± 28.3	17.9 ± 22.9	19.9 ± 27.4	18.0 ± 28.5
Siderastrea stellata	0.1 ± 0.1	—	—	—
Dendropoma sp.	2.1 ± 4.6	0.2 ± 0.6	5.8 ± 9.4	—
Plakortis sp	—	—	0.3 ± 0.6	1.2 ± 1.9
Polychaeta (sp.1—encrusting tube worm)	12.1 ± 16.4	14.5 ± 21.3	0.2 ± 0.8	7.9 ± 13.5

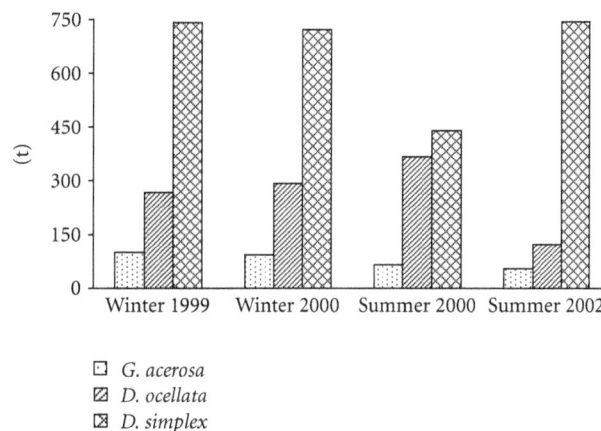

Legend:
- ☐ *G. acerosa*
- ▨ *D. ocellata*
- ☒ *D. simplex*

FIGURE 2: Biomass variation of the most representative macroalgae species, *Gelidiella acerosa*, *Dictyosphaeria ocellata*, and *Digenea simplex*, on the entire reef flat. The values correspond to the mean of the 10 sites sampled.

the groups (Table 4). The most important taxa in characterizing group I were *Z. sociatus* and *G. acerosa*. *D. simplex* was the dominant of group II. *Hydrolithon pachydermum* was well represented in all groups, but more relevant in groups III and IV. *Dictyosphaeria ocellata* occurred in both groups III and IV and *L. confervoides* and polychaeta tubes in group IV.

Although the variance analyses pointed out to some significant differences ($p < 0,05$ and $p < 0,01$) in taxa coverage between the different sampling periods, it could not be attributed to a clear seasonal pattern, being more likely to local differences in the degree of patchiness. MDS ordination also revealed some differences in taxa cover in summer 2000 for sites T4, T7, T8, and T9 (group III of MDS). SIMPER

TABLE 2: Diversity data within the ten sampling sites: Shannon index (H'), Species Richness (S), and Pielous's Evenness (J); W99 = winter 1999, W00 = winter 2000, S00 = summer 2000, S02 = summer 2002.

	Stations of years	(S)	(H')	(J)
Site 1	W99	7.00	1.59	0.57
	W00	8.00	1.50	0.50
	S00	6.00	1.91	0.74
	S02	6.00	1.79	0.69
Site 2	W99	8.00	2.33	0.78
	W00	7.00	2.44	0.87
	S00	7.00	2.44	0.87
Site 3	W99	5.00	0.94	0.40
	W00	7.00	1.31	0.47
	S00	7.00	1.78	0.63
	S02	7.00	1.73	0.62
Site 4	W99	7.00	1.96	0.70
	W00	8.00	2.22	0.74
	S00	5.00	1.67	0.72
Site 5	W99	8.00	1.83	0.61
	W00	7.00	2.02	0.72
	S00	7.00	2.22	0.79
	S02	5.00	1.30	0.56
Site 6	W99	7.00	2.19	0.78
	W00	8.00	2.26	0.75
	S00	11.00	2.22	0.64
	S02	7.00	2.00	0.71
Site 7	W99	5.00	1.82	0.79
	W00	4.00	1.92	0.96
	S00	5.00	1.94	0.83
	S02	5.00	1.64	0.71
Site 8	W99	6.00	2.10	0.81
	W00	5.00	2.16	0.93
	S00	6.00	1.74	0.68
	S02	8.00	2.61	0.87
Site 9	W99	6.00	2.43	0.94
	W00	7.00	2.31	0.82
	S00	5.00	1.44	0.62
	S02	7.00	2.26	0.81
Site 10	W99	10.00	2.29	0.69
	W00	8.00	2.33	0.78
	S00	6.00	1.84	0.71
	S02	8.00	1.80	0.60
Overall	W99	16.00	3.18	0.80
	W00	15.00	2.98	0.76
	S00	15.00	2.75	0.70
	S02	13.00	3.12	0.84

analyses revealed that these were attributed to variations in the cover of the polychaeta tubes and cyanobacteria. Nevertheless, these patterns were not observed in summer 2002. As such, it is assumed that these were not brought about by seasonal variability, but more likely due to more

TABLE 3: R-statistic values for the global ANOSIM and significance of pairwised tests for differences in community structure between reef flat groups of stations.

Global test	R	p
	0.879	0.001
Pairwised tests		
1×2	0.912	0.001
1×3	0.998	0.001
1×4	1	0.001
2×3	0.490	0.003
2×4	0.780	0.001
3×4	0.991	0.002

specific short term local variability of the organisms at the individual sampling sites.

3.3. Benthic Community Habitats. The groups defined by the statistical analysis were considered as distinct habitats. However, groups III and IV were considered as a single habitat, as both included T4, T7, T8, and T9 differing only for the summer 2000 campaign in the cover of polychaeta tubes and cyanobacteria. As such three main habitats could be defined for Atol das Rocas reef flat.

Habitat I was characterized by a large coverage of the zoanthid *Z. sociatus* (41–57%) along with the turf forming red algae *Gelidiella acerosa* (29–47%) (Figure 4(a)). A particular feature of this habitat is the absence of the turf forming red macroalgae *Digenea simplex*. The Shannon Index (H') ranged between 1,50 and 2,44, Richness 6 and 10, and Evenness between 0,50 and 0,87.

Habitat II was dominated by the turf forming *D. simplex* with a coverage of 67–72%. In contrast to Habitat I, *G. acerosa* was absent (Figure 4(b)). The Shannon Index (H') ranged between 0,94 and 2,26, Richness between 5 and 11, and Evenness between 0,40 and 0,79.

Habitat III was composed by a high cover of the crustose coralline *Hygrolithon pachydermum* (23–48%) along with the crustose flesh macroalgae *Dictyosphaeria ocellata* (17–51%). The presence of polychaeta (1–29%) and the cyanobacteria *Lyngbya confervoides* (0,5–26%) was also relevant (Figure 4(c)). The Shannon Index (H') ranged between 1,44 and 2,61, Richness between 4 and 8, and Evenness between 0,62 and 0,96.

4. Discussion

4.1. Benthic Community Structure. Atol das Rocas is the sole atoll of the South Atlantic Ocean. It corresponds to one of the world's smallest atolls with a reef flat set about 2 m above MSL and an enclosed lagoon type system. The reef flat exhibits a great scarcity of hermatypic coral species and cover and is near to solely dominated by turf forming and crustose macroalgae species. Kikuchi and Leão [19] considered the lack of hermatypic corals as the most striking feature of Atol das Rocas, when compared to other Atlantic and Pacific

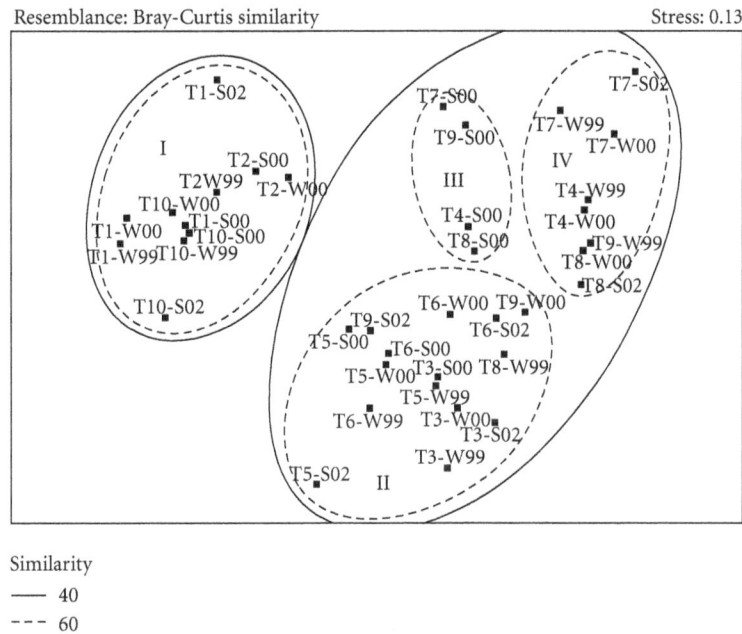

FIGURE 3: MDS ordination of the reef flat stations based on species cover; T1–T10 = sites one to ten; W99 = winter 1999, W00 = winter 2000, S00 = summer 2000, S02 = summer 2002.

atolls, where corals dominate most of the reef surface [7, 25–27] or at least appear as an important component of the benthic community [11].

The turf forming red macroalgae *Digenea simplex* and the crustose coralline *Hydrolithon pachydermum* were the dominant species on most of the Rocas reef flat. The crustose green macroalgae *Dictyosphaeria ocellata* and the turf forming red macroalgae *Gelidiella acerosa* were the subdominant species. *D. simplex* exhibited biomass values of about $969\,g\,dw\,m^{-2}$, more than twice higher than the subdominant species, corroborating its relevance to the community structure.

In general, the Shannon Index (H') showed a relative similarity with other reef flat communities of unpolluted reef areas in Pacific and Caribbean waters, but Richness values were below than those observed in the same areas [5, 8, 9]. On the other hand, the high values of Evenness reflected a relative equitability between the few taxa of the Atol das Rocas reef flat. The small variance of these biological variables between the summer and winter surveys spread over two years suggests that the community structure is subject to stable conditions in time.

As a whole, the main factors controlling the distribution and zonation of benthic organisms within reef systems are wave energy (turbulence of water and abrasion), high light penetration (as a function of depth and low water turbidity), bottom topography, and sedimentation [2, 3, 28, 29]. In open atolls the community structure is strongly influenced by wave action, a product of the prevailing winds [3], and the exposure to oceanic conditions. High physical energy provides an adequate environment for dominance of corals and encrusting calcareous algae [30]. In contrast, the enclosed Atol das Rocas with a reef flat set 2 m above MSL exhibits a scarcity of coral species and coverage which

may be primarily attributed to its high degree of protection from the sea by wave action. According to Kench and Brander [28], reef flats with high elevations above to MSL are efficient in filtering and dissipating wave energy, resulting in a lower potential of geomorphic work, less intense sediment transport, and a higher sediment accumulation within the reef flat. Henceforth, it is expected that the elevated reef flat of Atol das Rocas is also governed by these features. Another striking feature of Atol das Rocas reef flat is the lengthy periods of aerial exposure of the benthic communities, attaining over more than 8 hrs per semidiurnal tidal cycle. The detrimental effects of sedimentation and desiccation on coral diversity and abundance have also been reported elsewhere [3, 7, 30–32].

The high percentage coverage of turf and crustose macroalgae as compared to fleshy macroalgae on the reef flat of Atol das Rocas reflects their ability to successfully colonize and persist under the environmentally unfavorable conditions encountered (i.e., desiccation). The dominance of filamentous and turf forming algae may be attributed to their short generation time, rapid growth rates, opportunistic life histories, and wider tolerance to changing conditions, allowing them to colonize available space rapidly and reach their full growth potential in a shorter time period [30]. Macroalgae that form dense entangled turf may hold water among their branches during exposure, thus, avoiding desiccation [33, 34]. Also, the few fleshy macroalgae species encountered at the shallow reef flat may have been strongly affected by intense solar radiation, as also reported elsewhere by Morrison [35]. Marques et al. [36] conducted experiments at Atol das Rocas and observed that individuals of some macroalgal species transplanted from the fore reef to the reef flat developed a whitish color, indicative of light damage.

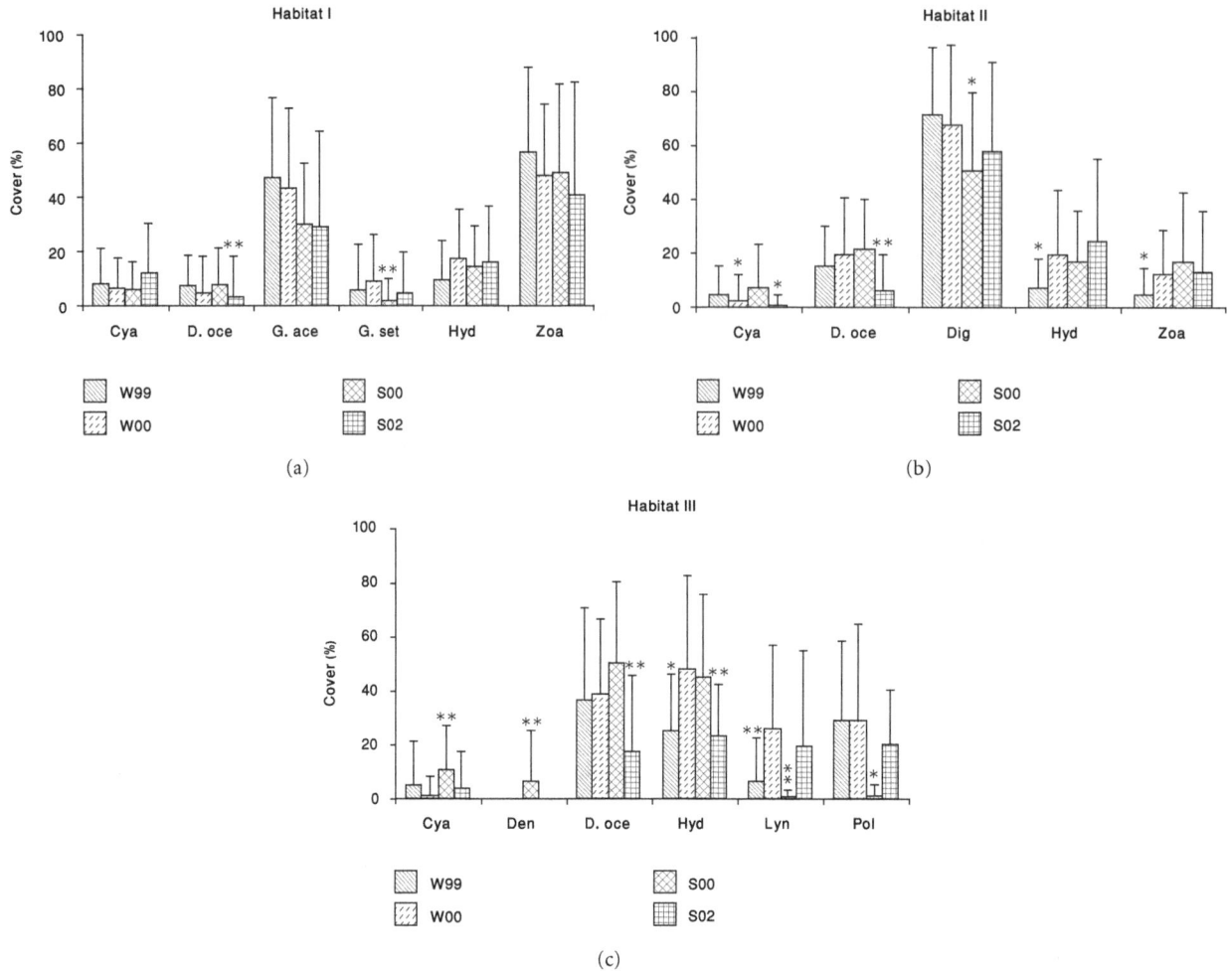

FIGURE 4: Mean percentage cover and SD for the more representative benthic organisms in the three habitats of the reef flat, pointing the significant post hoc multiple comparisons tests results (* $p < 0.05$; ** $p < 0.01$); Cya: Cyanobacteria, Den: *Dendropoma* sp., Dig: *Digenea simplex*, D.oce: *Dictyosphaeria ocellata*, G.ace: *Gelidiella acerosa*, G.set: *Gelidiella setacea*, Hyd: *Hydrolithon pachydermum*, Lyn: *Lyngbya confervoides*, Pol: Polychaeta tube, Zoa: *Zoanthus sociatus*; W99: winter 1999, W00: winter 2000, S00: summer 2000, S02: summer 2002.

Another causal agent that may influence the structure of the reef flat benthic community in Atol das Rocas is the low pressure from herbivory, which is considered to play a primary role in determining the distribution and abundance of algae and corals in reef habitats [37–41]. Although the total number of fish species and biomass at Atol das Rocas is similar to other Atlantic islands and small atolls in, for example, French Polynesia [42], several factors argue in favor for a low herbivory upon the benthic habitats within the reef flat, these include (a) the high elevation of the reef flat induces a shallow and bare environment which prevents access of large herbivorous fish and the vulnerability to predators [8]; (b) the low abundance and scarcity of sea urchins represented by only two species *Diadema antillarum* and *Tripneustes variegatus* which are restricted to small areas in the lagoon and some pools [36]; (c) only one genus of the coralline algae parrot fish grazer (*Sparisoma*) seems to play a role within Atol das Rocas [42]. Also, the absence of the more common and powerful grazer of the genus *Scarus* may play an important role in the enhancement of the coralline algae in the atoll [19], as the grazing activity of parrot fish is one of

the most relevant ecological constraints to the development of coralline algae [43].

4.2. Comparisons between Reef Habitats. The statistical analysis revealed the presence of at least three distinct habitats set within the reef flat. The habitats almost harbor the same taxa composition but differ with respect to their abundance. Although environmental conditions were not measured for each specific habitat, the differences observed are likely the result of small scale variations of sedimentation, herbivory, and hydrodynamic conditions.

Observations during samplings of Habitat I, prevailing the leeward arch of the atoll, indicated the presence of high sediment deposition. The leeward side is considered as the youngest geological sector of the atoll and exhibits lower elevation than the windward side [19, 20]. Consequently, it also might be conditioned to a relatively higher grazing pressure, since the depths during high tide are higher than on the windward side. These conditions promote the development of sediment-resistant species and forms which might be more resistant to grazing pressure. Turf forming

TABLE 4: Summary results of SIMPER show the percentage contribution of species for the similarity into the groups of station and the dissimilarities within them. Bold type values correspond to species with more than 15% contribution.

Average similarity (%)	Group 1	Group 2	Group 3	Group 4		
	76.30	77.34	70.29	74.00		
Contribution (%)						
Cyanobacteria	11.93	8.90	21.88			
Dendropoma sp.			7.66			
Dictyosphaeria ocellata	11.09	23.44	34.13	21.27		
Digenea simplex		33.75				
Gelidiella acerosa	22.67					
Gelidiella setacea	9.92					
Hydrolithon pachydermum	17.53	21.09	32.46	27.81		
Lyngbya confervoides				17.72		
Polychaeta tube				25.08		
Zoanthus sociatus	23.98	7.22				
Cumulative (%)	97.11	94.39	96.13	91.88		
Average dissimilarity (%)	1 × 2	1 × 3	1 × 4	2 × 3	2 × 4	3 × 4
	50.70	55.01	67.89	39.39	47.00	45.11
Contribution (%)						
Cyanobacteria	6.31	4.17	9.06	9.14	10.84	18.85
Dendropoma sp.	5.52	8.51		12.01	3.67	11.85
Dictyosphaeria ocellata	5.78	10.52	6.35	7.97	5.95	6.05
Dictyosphaeria versluysii	3.86			5.82	5.88	6.26
Digenea simplex	22.48	7.41	7.31	22.32		11.08
Gelidiella acerosa	18.65	19.50	15.04		15.24	
Gelidiella setacea	9.41	10.37	7.44			
Hydrolithon pachydermum		5.74	4.07	8.45	6.20	
Lyngbya confervoides			11.92	5.89	15.69	17.18
Polychaeta tube	5.75	4.02	13.09	8.00	17.20	21.57
Zoanthus sociatus	12.54	20.70	15.97	13.81	11.00	
Cumulative (%)	90.27	90.94	90.26	93.40	91.68	92.85

algae are known to have a substantial resistance to sediment deposition [44, 45]. Herbivory experiments in Atol das Rocas [36] revealed that *Gelidiella acerosa* were hardly consumed due to the morphological configuration of their thalli. The same experiments also revealed that *Digenea simplex* was intensively grazed by fishes, which may explain the absence of the species in this habitat. Furthermore, the zoanthid *Zoanthus sociatus* also with a high tolerance to environmental variability [46, 47] achieves its highest cover in this habitat.

Habitat II prevailed on the central sheltered areas of the reef flat. In contrast to Habitat I, the slightly higher elevations of the flat results in very shallow waters even at high tide, which prevent the access of large herbivorous fishes enabling the better development of *D. simplex*. The absence of *G. acerosa* is probably related to higher competitive abilities of *D. simplex* under these environmental conditions. The lower Shannon Index and Evenness reflected the dominance of *D. simplex* over other species.

Habitat III set on the windward arch of the atoll and was thus exposed to a more intense hydrodynamic scenario than the others habitats. During the highest tidal stages, the benthic community was entirely washed over by waves. These

conditions improve the development of species relatively resistant to higher turbulent waters. The dominance of crustose corallines in reef zones submitted to higher energy conditions were described elsewhere [8, 11, 30]. These species are physically very resistant and are almost the single organisms at windward reef crests [2, 3]. The prostate shape of the crustose macroalgae *Dictyosphaeria ocellata* also enhances its abilities to support higher hydrodynamic conditions [48]. The lower Richness and higher Evenness values reflected a great equitability between the few species observed in this habitat.

In general, the study indicates that the reef flat of Atol das Rocas harbors a benthic community adapted to low water turbulence, lengthy periods of aerial exposure, and low herbivore pressure. There were no data about significant human impacts until now on this reef, affecting the general stability of the community studied. These conditions should be maintained due to the strict environmental Brazilian laws implemented for the preservation of this Marine Reserve, unless other climatic and oceanographic factors controlling the atoll change over time.

Acknowledgments

The authors would like to thank the Brazilian Environmental Agency "Instituto Chico Mendes de Conservação da Biodiversidade" (ICMBio) for the research facilities at the Marine Biological Reserve of Atol das Rocas, and M. Brito, the reserve manager; The foundation "Fundação O Boticário de Proteção à Natureza" for the financial support; CAPES which provided the first author with two years of a Ph.D. scholarship; P. S. Figueiredo, L. Santi, A. Almeida, C. Pezzela (in memoriam), A. B. Villas Bôas, and F. Fonseca for essential help in fieldwork. B. Knoppers. is a senior research fellow of CNPq (Grant no. 306157/2007-1).

References

[1] C. M. Lalli and T. R. Parsons, *Biological Oceanography: An Introduction*, Elsevier, Oxford, UK, 2nd edition, 1997.

[2] J. S. Levinton, *Marine Biology: Function, Biodiversity, Ecology*, Oxford University Press, Oxford, UK, 2nd edition, 2001.

[3] J. W. Nybakken and M. D. Bertness, *Marine Biology: An Ecological Approach*, Pearson/Benjamin Cummings, San Francisco, Calif, USA, 6th edition, 2004.

[4] M. M. Littler and D. S. Littler, "Models of tropical reef biogenesis: the contribution of algae," *Progress in Phycological Research*, vol. 3, pp. 323–364, 1984.

[5] M. A. Huston, "Patterns of species diversity on coral reefs," *Annual Review of Ecology and Systematics*, vol. 17, pp. 149–177, 1985.

[6] J. H. Connell, T. P. Hughes, and C. C. Wallace, "A 30-year study of coral abundance, recruitment, and disturbance at several scales in space and time," *Ecological Monographs*, vol. 67, no. 4, pp. 461–488, 1997.

[7] H. T. Odum and E. P. Odum, "Trophic structure and productivity of a windward coral reef community on eniwetok Atol," *Ecological Monographs*, vol. 25, pp. 291–320, 1955.

[8] M. M. Littler, P. R. Taylor, D. S. Littler, R. H. Sims, and J. N. Norris, "Dominant macrophyte standing stocks, productivity and community structure on a Belizean barrier reef," *Atoll Research Bulletin*, vol. 302, pp. 1–24, 1987.

[9] C. L. Rodrigues, S. Caeiro, and S. V. Raikar, "Marine macrophyte communities on the reef flat at Agatti atoll (Lakshadweep, India)," *Botanica Marina*, vol. 40, no. 6, pp. 557–568, 1997.

[10] T. Nakamura and T. Nakamori, "Estimation of photosynthesis and calcification rates at a fringing reef by accounting for diurnal variations and the zonation of coral reef communities on reef flat and slope: a case study for the Shiraho reef, Ishigaki Island, southwest Japan," *Coral Reefs*, vol. 28, no. 1, pp. 229–250, 2009.

[11] P. S. Vroom, K. N. Page, K. A. Peyton, and J. K. Kukea-Shultz, "Spatial heterogeneity of benthic community assemblages with an emphasis on reef algae at French Frigate Shoals, Northwestern Hawai'ian Islands," *Coral Reefs*, vol. 24, no. 4, pp. 574–581, 2005.

[12] D. F. M. Gherardi and D. W. J. Bosence, "Modeling of the ecological succession of encrusting organisms in recent coralline-algal frameworks from Atoll das Rocas, Brazil," *Palaios*, vol. 14, no. 2, pp. 145–158, 1999.

[13] D. F. M. Gherardi and D. W. J. Bosence, "Composition and community structure of the coralline algal reefs from Atol das Rocas, South Atlantic, Brazil," *Coral Reefs*, vol. 19, no. 3, pp. 205–219, 2001.

[14] R. C. Villaça, A. C. Fonseca, C. A. C. Pezzella, and V. K. Jensen, "Ecology of macroalgae from atol das rocas reef," *Phycologia*, vol. 40, supplement 4, p. 113, 2001.

[15] R. C. Villaça, A. G. Pedrini, S. M. B. Pereira, and M. A. O. Figueiredo, "Flora bentônica das ilhas oceânicas brasileiras," in *Ilhas Oceânicas Brasileiras: Da Pesquisa ao Manejo*, R. J. V. Alves and J. W. A. Castro, Eds., pp. 105–146, Ministério do Meio Ambiente, Brasília, Brazil, 2006.

[16] C. A. Echeverria, D. O. Pires, M. S. Medeiros, and C. B. Castro, "Cnidarians of the atol das rocas," in *Proceedings of the 8th International Coral Reef Symposium*, vol. 1, pp. 443–446, 1997.

[17] S. A. Netto, M. J. Attrill, and R. M. Warwick, "Sublittoral meiofauna and macrofauna of Rocas Atoll (NE Brazil): indirect evidence of a topographically controlled front," *Marine Ecology Progress Series*, vol. 179, pp. 175–186, 1999.

[18] S. A. Netto, R. M. Warwick, and M. J. Attrill, "Meiobenthic and macrobenthic community structure in carbonate sediments of Rocas atoll (north-east, Brazil)," *Estuarine, Coastal and Shelf Science*, vol. 48, no. 1, pp. 39–50, 1999.

[19] R. K. P. Kikuchi and Z. M. A. N. Leão, "Rocas (Southwestern Equatorial Atlantic, Brazil): an atoll built primarily by coralline algae," in *Proceedings of the 8th International Coral Reef Symposium*, vol. 1, pp. 731–736, 1997.

[20] D. F. M. Gherardi and D. W. J. Bosence, "Late Holocene reef growth and relative sea-level changes in Atol das Rocas, equatorial South Atlantic," *Coral Reefs*, vol. 24, no. 2, pp. 264–272, 2005.

[21] O. Höflich, "Climate of the South Atlantic Ocean," in *Climates of the Oceans*, H. Van Loon, Ed., pp. 1–192, Elsevier, Amsterdam, The Netherlands, 1984.

[22] C. M. Sabino and R. Villaça, "Estudo comparativo de métodos de amostragem de comunidades de costão," *Revista Brasileira de Biologia*, vol. 59, no. 3, pp. 407–419, 1999.

[23] P. Legendre and L. Legendre, *Numerical Ecology. Developments in Environmental Modelling*, Elsevier, Amsterdam, The Netherlands, 2nd edition, 1998.

[24] J. H. Zar, *Biostatistical Analysis*, Prentice Hall, London, UK, 5th edition, 2009.

[25] D. R. Stoddart, "Three caribbean atolls: turneffe Islands, Lighthouse Reef and Glover's Reef, British Honduras," *Atoll Research Bulletin*, vol. 87, pp. 1–151, 1962.

[26] J. D. Milliman, "The geomorphology and history of hogsty reef, a bahamian atoll," *Bulletin of Marine Science*, vol. 17, no. 3, pp. 519–543, 1967.

[27] J. D. Milliman, "Four southwestern carribbean atolls: courtown cays, albuquerque cays roncador bank and serrana bank," *Atoll Research Bulletin*, vol. 129, pp. 1–41, 1969.

[28] P. S. Kench and R. W. Brander, "Wave processes on coral reef flats: implications for reef geomorphology using Australian case studies," *Journal of Coastal Research*, vol. 22, no. 1, pp. 209–223, 2006.

[29] J. M. Díaz, G. Díaz-Pulido, and J. A. Sánchez, "Distribution and structure of the southernmost Caribbean coral reefs: golfo de Uraba, Colombia," *Scientia Marina*, vol. 64, no. 3, pp. 327–336, 2000.

[30] J. Morrissey, "Community structure and zonation of microalgae and hermatypic corals on a fringing reef flat of magnetic island (Queensland, Australia)," *Aquatic Botany*, vol. 8, pp. 91–139, 1980.

[31] K. J. Roy and S. V. Smith, "Sedimentation and coral reef development in turbid waters: fanning lagoon," *Pacific Science*, vol. 25, no. 2, pp. 234–248, 1971.

[32] Y. Loya, "Effects of water turbidity and sedimentation on the community structure of Puerto Rico corals," *Bulletin of Marine Science*, vol. 26, no. 4, pp. 450–456, 1976.

[33] T. A. Norton, "Conflicting constraints on the form of intertidal algae," *British Phycological Journal*, vol. 26, no. 3, pp. 203–218, 1991.

[34] L. Airoldi, M. Fabiano, and F. Cinelli, "Sediment deposition and movement over a turf assemblage in a shallow rocky coastal area of the Ligurian Sea," *Marine Ecology Progress Series*, vol. 133, no. 1–3, pp. 241–251, 1996.

[35] D. Morrison, "Comparing fish and urchin grazing in shallow and deeper coral reef algal communities," *Ecology*, vol. 69, no. 5, pp. 1367–1382, 1988.

[36] L. V. Marques, R. Villaça, and R. C. Pereira, "Susceptibility of macroalgae to herbivorous fishes at Rocas Atoll, Brazil," *Botanica Marina*, vol. 49, no. 5-6, pp. 379–385, 2006.

[37] R. C. Carpenter, "Partitioning herbivory and its effects on coral reef algal communities," *Ecological Monographs*, vol. 56, no. 4, pp. 345–363, 1986.

[38] M. E. Hay, "Fish-seaweed interactions on coral reefs: effects of herbivorous fishes and adaptations of their prey," in *The Ecology of Fishes on Coral Reefs*, P. F. Sale, Ed., pp. 96–119, Academic Press, San Diego, Calif, USA, 1991.

[39] T. R. McClanahan, "Primary succession of coral-reef algae: differing patterns on fished versus unfished reefs," *Journal of Experimental Marine Biology and Ecology*, vol. 218, no. 1, pp. 77–102, 1997.

[40] T. R. McClanahan, "Predation and the control of the sea urchin Echinometra viridis and fleshy algae in the patch reefs of Glovers Reef, Belize," *Ecosystems*, vol. 2, no. 6, pp. 511–523, 1999.

[41] B. E. Lapointe, P. J. Barile, C. S. Yentsch, M. M. Littler, D. S. Littler, and B. Kakuk, "The relative importance of nutrient enrichment and herbivory on macroalgal communities near Norman's Pond Cay, Exumas Cays, Bahamas: a "natural" enrichment experiment," *Journal of Experimental Marine Biology and Ecology*, vol. 298, no. 2, pp. 275–301, 2004.

[42] R. S. Rosa and R. L. Moura, "Visual assessment of reef fish community structure in the Atol Das Rocas Biological Reserve, off northeastern Brazil," in *Proceedings of the 8th International Coral Reef Symposium*, vol. 1, pp. 983–986, 1997.

[43] R. S. Steneck, "The ecology of coralline algal crusts: convergent patterns and adaptative strategies," *Annaul Review of Ecology and Systematics*, vol. 17, pp. 273–303, 1986.

[44] L. Airoldi and F. Cinelli, "Effects of sedimentation on sub-tidal macroalgal assemblages: an experimental study from a mediterranean rocky shore," *Journal of Experimental Marine Biology and Ecology*, vol. 215, no. 2, pp. 269–288, 1997.

[45] L. Airoldi and M. Virgilio, "Responses of turf-forming algae to spatial variations in the deposition of sediments," *Marine Ecology Progress Series*, vol. 165, pp. 271–282, 1998.

[46] K. P. Sebens, "Intertidal distribution of zoanthids on the caribbean coast of Panama: effects of predation and desiccation," *Bulletin of Marine Science*, vol. 32, no. 1, pp. 316–335, 1982.

[47] Y. I. Sorokin, "Biomass, metabolic rates and feeding of some common reef zoantharians and octocorals," *Australian Journal of Marine & Freshwater Research*, vol. 42, no. 6, pp. 729–741, 1991.

[48] S. T. Larned and M. J. Atkinson, "Effects of water velocity on NH_4 and PO_4 uptake and nutrient-limited growth in the macroalga Dictyosphaeria cavernosa," *Marine Ecology Progress Series*, vol. 157, pp. 295–302, 1997.

6

Histology and Mucous Histochemistry of the Integument and Body Wall of a Marine Polychaete Worm, *Ophryotrocha* n. sp. (Annelida: Dorvilleidae) Associated with Steelhead Trout Cage Sites on the South Coast of Newfoundland

H. M. Murray,[1] D. Gallardi,[1,2] Y. S. Gidge,[1,3] and G. L. Sheppard[1]

[1] *Fisheries and Oceans Canada, 80 White Hills Road, P.O. Box 5667, St. John's, NL, Canada A1C 5X1*
[2] *School of Fisheries, Marine Institute of Memorial University of Newfoundland, St. John's, NL, Canada A1C 5R3*
[3] *Department of Environmental Science, Memorial University of Newfoundland, St. John's, NL, Canada A1C 5S7*

Correspondence should be addressed to H. M. Murray, harry.murray@dfo-mpo.gc.ca

Academic Editor: Garth L. Fletcher

Histology and mucous histochemistry of the integument and body wall of a marine polychaete worm, *Ophryotrocha* n. sp. (Annelida: Dorvilleidae) associated with Steelhead trout cage sites on the south coast of Newfoundland. A new species of polychaete (*Ophryotrocha* n. sp. (Annelida: Dorvilleidae)) was identified from sediment below Steelhead trout cages on the south coast of Newfoundland, Canada. The organisms were observed to produce a network of mucus in which groups of individuals would reside. Questions regarding the nature and cellular source of the mucus were addressed in this study. Samples of worms were taken from below cages and transported to the laboratory where individuals were fixed for histological study of the cuticle and associated mucus histochemistry. The body wall was organized into segments with an outer cuticle that stained strongly for acid mucopolysaccharides. The epidermis was thin and supported by loose fibrous connective tissue layers. Channels separating individual segments were lined with cells staining positive for Alcian blue. Mucoid cellular secretions appeared thick and viscous, strongly staining with Alcian blue and Periodic Acid Schiff Reagent. It was noted that lateral channels were connected via a second channel running through the anterior/posterior axis. The role of mucus secretion is discussed.

1. Introduction

The impact of aquaculture on the diversity of benthic fauna below sea cages has been well studied [1–4]. It has been observed that as the microenvironment beneath cages changes over time so does the benthic community [4]. The majority of these effects are due to excessive sedimentation resulting from feed pellets and fecal material [1, 3]. The build-up of organic sediment can result in conditions of high sulphur content and low oxygen due to increased microbial activity within the top layers of sediment [5]. Only fauna able to tolerate low oxygen can survive under these conditions [3, 6–8]. It has been shown that some benthic fauna are sulfide tolerant (e.g., some polychaete species) [9–12]. Opportunistic polychaete complexes (OPCs) are frequently found beneath salmon and Steelhead trout cage sites and are considered an

indicator of benthic impact related to aquaculture activities. The presence/absence of these "indicator" species or faunal groups under and around finfish aquaculture sites may show transitions from low (background) levels of organic matter to high deposition rates caused by unconsumed feed pellets and fish feces in areas subject to low transport [8, 10, 12, 13].

Intensive salmonid aquaculture on the south coast of Newfoundland and Labrador has been occurring for approximately 30 years [14]. Characteristically, aquaculture sites in this area are quite deep (>100 m) with hard bottoms. Recent surveys of species diversity beneath cage sites in this region have shown the occurrence of large assemblages of polychaete worms (unpublished data). Samples taken from these assemblages have indicated that the worms present are a new species, *Ophryotrocha* n. sp. of the Annelid family

Dorvilleidae (G. Pohle and H. Wiklund, personal communication). Generally, *Ophryotrocha sp.* are small opportunistic polychaete worms of the Annelid family Dorvilleidae. Thornhill et al. [15] provide an excellent description of the basic life history traits of this genus. They typically exhibit a rounded or blunt prostomium and distinct setigers. The jaws have been noted as one of the most interesting characteristics of this group and are most commonly described at *P* type. They are commonly found in soft sediments associated with polluted and nutrient rich habitats such as harbours and appear to graze on a variety of food types found in the substrate (e.g., bacteria, eukaryotic microbes, and detritus) [15, 16]. While relatively low population densities have been noted in some commonly studied species, others are opportunistic or stress tolerant and can reach high abundance in environments that can be inhibitory to other organisms [15]. Some of these habitats include whale-falls and the organic rich environments beneath salmonid aquaculture cage sites [17].

While little is currently known of the basic biology of this new species of *Ophryotrocha* and its role in the ecology of the environment below aquaculture cages, some preliminary field and laboratory observations have shown that they produce and frequently reside in mucus complexes or networks. The function of these complexes in this species and this environment is not yet clear. However, the importance of mucus production and functionality has been explored in a number of other worm species including some of the genus *Ophryotrocha* [15, 18]. For example, some polychaete species (i.e., *Paralvinella palmiformis*) resident near deep sea hydrothermal vents have been noted to produce a continual secretion of mucus [19]. It was suggested that, in this species, the purpose might be to clear its body wall of particulate debris or in other related species to eliminate accumulated toxins, that is, elemental sulphur or metallothioneins [19]. In some *Ophryotrocha* sp. mucous has been noted to be associated with reproductive behaviour including the production of mucous-lined tubes and trails [15]. The excessive production of mucus in the Newfoundland species also suggests some physiological and/or ecological importance although that importance is not yet clear. The present study is an investigation designed to characterize the histology of the integument and body wall, identify sources of mucus production, and describe the distribution of mucus secreting cells in *Ophryotrocha* n. sp. and provide some preliminary data on the chemistry of the mucus secretions through histochemical staining. This data will help to develop our understanding of the role of this species in the microenvironment beneath salmonid aquaculture cages.

2. Materials and Methods

2.1. Field Sampling. Polychaetes were collected at an active steelhead trout aquaculture site (Margery Cove; 47.047°N, −055.411°W) located near St. Alban's, Newfoundland during November 26, 2010. The polychaetes were sampled by dragging a small modified egg net (45 cm by 30 cm oval opening reinforced with metal tubing weighted with two 1.0 kg lead bullet weights; mesh size of 500 μm) over the bottom at a depth of 50 m. The net was deployed by hand tossing it at horizontal distances of 10–15 m in various directions from the vessel, predominantly in areas located between aquaculture cages. Once the net settled to the expected depth it was slowly dragged along the bottom and subsequently hauled aboard the vessel. The net was then inverted inside a 20 L bucket filled with seawater collected at a depth of 20 m using a small Niskin bottle. For transport back to the laboratory (Northwest Atlantic Fisheries Centre, St. John's, NL), worms were stored in coolers containing freshly collected seawater from an appropriate depth. On average, roughly 100 polychaetes were collected during each net deployment. Samples were largely free of debris or mud which suggested the net sampled the top layer of the bottom as intended. After arrival at the laboratory polychaetes were transferred to 4 L plastic containers partially filled with sand and seawater (water temperature = 3-4°C). The containers holding worms were subsequently put in a second holding tank filled with sand filtered chilled water (water temperature = 3-4°C) and air stones were added to provide aeration. The water flow into each tank allowed for approximately one turnover per hour. The oxygen saturation on average was between 90 and 95%. Aluminum foil was placed over each individual container to reduce light exposure. Crushed trout feed was added to each container as a potential food source. Worms were kept under the above conditions for one month prior to tissue sampling for histology and mucus histochemistry.

2.2. Histology. Groups of five to ten worms were fixed in 10% neutral buffered formalin for 24 to 48 hrs at 4°C. Following fixation the samples were dehydrated through ethanol series, cleared in two changes of xylene, and infiltrated and embedded in paraffin for both longitudinal and transverse sections. Six to eight micron sections were cut using a rotary microtome (Leica, RM2265), placed on uncoated glass slides, dried overnight at 37°C, and stored at room temperature. Sections were stained with either haematoxylin and eosin (H&E) or Alcian blue (AB) pH 2.5 and Periodic Acid Schiff's reagent (PAS) [20]. Slides were stained in batches using a Leica, Auto Stainer XL and were examined using a Zeiss Axio Imager-A1 compound microscope with attached AxioCam HRc camera and associated software. Image plates were created using Photoshop Elements 7.0.

3. Results

Within hours of transfer to the holding containers the worms were noted to produce copious amounts of a clear mucus-like substance. Frequently, groups of individuals were observed to be suspended in networks of mucus situated just off the tank bottom normally in areas associated with added intact food pellets (Figure 1(a)).

Individuals were observed to be pink to red in colour with a mean length of 10.25 mm ($n = 50$) and a mean mass of 12.8 mg ($n = 50$) (Figures 1(a) and 1(b)). Morphologically, they were divided into approximately 36 segments with defined parapodia and chaetae (Figure 1(b)). The anterior

FIGURE 1: General appearance and size of polychaetes (*Ophyrotrocha* sp.) from salmonid aquaculture sites located on the south coast of Newfoundland. (a) Congregation of worms in mucus complex around food pellet in the laboratory tanks (scale bar = 8 mm). (b) Worm showing upper size range of individuals (approx. 16 mm) (scale bar = 2 mm).

region exhibited two peristomal achaetous segments with jaw (Figure 1(b)). Longitudinal histological sections through the body indicated that each segment was defined by a single pair of parapodia supported internally with a club-shaped muscular structure surrounding one to two chaetae (Figure 2(a)). Each segment adjacent to the parapodia was defined by a narrow channel lined with a single layer of cells. Generally the secretory cells were cuboidal with a basal nucleus and eosinophilic apical cytoplasm (Figure 2(d)). This epithelium seemed to be continuous with the integumental surface of the parapodia (Figures 2(a) and 2(b)). The luminal surface adjacent to parapodia was coated by a relatively thin layer of mucoid secretion (Figures 2(a) and 2(b)). Near the anterior region of the worm, other tubular structures appeared to produce a thicker or more viscous layer of mucus (Figures 2(c) and 2(d)). It is unclear as to whether these regions are associated with the previously mentioned channels or represent another type of secretory structure.

Histochemical staining of longitudinal sections through the body wall with Alcian Blue (pH 2.5) and PAS showed that the outer portion of the integument was distinctly Alcian blue positive as was a layer of connective tissue immediately below (Figures 3(a), 3(b), and 3(c)). The total thickness of the outer integumental layer was approximately 0.5 to 1 μm. Each body segment was defined by large PAS positive muscle blocks lying just adjacent to the integument (Figures 3(a) and 3(b)). Interestingly, no distinct cellular integumental epidermis was discernable following histochemical staining. The outer cuticular region also stained strongly with Alcian blue but showed no evidence of distinct secretory cells (Figure 3(e)). Channels separating individual segments were lined with cells staining slightly positive for Alcian blue

(Figures 3(c) and 3(d)). Mucoid cellular secretions appeared thick and viscous, strongly staining with Alcian blue and PAS (Figures 3(c), 3(d), and 3(f)). It was also noted that lateral channels were connected via a second channel running through the anterior/posterior axis (Figure 3(b)). The point of connection appeared to form a small collecting sinus (Figure 3(b)).

4. Discussion

Samples for the present study were collected near the end of the annual production cycle for steelhead trout on the south coast of Newfoundland so the introduction of organic material would have gone through maximum input. Visual sampling revealed that the worms appear to reside in mucus complexes associated with this organic layer. This behaviour was subsequently verified through tank observations. The production of mucus in this species appears significant to its basic ecology and warranted further investigation. Individuals from the present study appeared most similar morphologically to *Ophryotrocha craigsmithi*, originally described from both a Minke whale carcass and sediment collected from below a fish farm in Norway [17]. Histology was not discussed in the original description of *O. craigsmithi* and no mention was made of mucus secretion. The information provided in the present study is a first description of the basic histology and mucous histochemistry of the integument in a new species of *Ophyrotrocha* found to congregate below salmonid aquaculture sites on the south coast of Newfoundland.

Histochemical staining using Alcian blue pH 2.5 and Period Acid/Schiff reagent provided a comprehensive view of tissue diversity in histological sections in the current study.

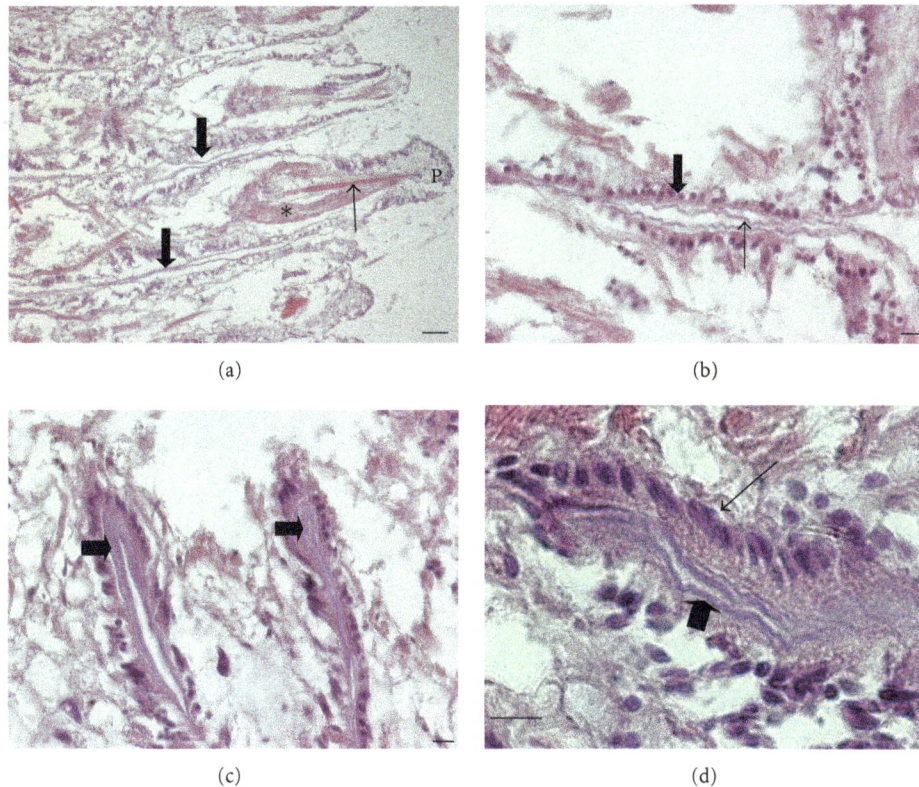

FIGURE 2: Histology of the integument of Newfoundland *Ophryotrocha* sp. (a) Overview of the epidermis and cuticle showing detail of parapodia (note muscular support structure (asterisk) and chaeta (thin arrow)) and orientation of mucus channels. Thick arrows indicate a thin layer of mucus (scale bar = 5 μm). (b) Detail of mucus channel showing secretory cells with basal nuclei and eosinophilic apical cytoplasm (thick arrow). Note thin mucoid secretion on the luminal surface of cells lining channel (thin arrow) (scale bare = 1 μm). (c) Detail of mucus channel from anterior portion of worm. Note the presence of a more viscous mucoid secretion (thick arrows) in this region compared to the previous (Scale bar = 1 μm). (d) Detail of cells lining anterior glandular structure. Note basal nucleus and granular apical cytoplasm (large arrow) with associated viscous secretion (arrowhead) (scale bar = 1 μm).

The outer most region of the integument (cuticle) stained strongly with Alcian blue suggesting a high content of acid mucopolysaccaride but did not show identifiable secretory cells. Surface mucus secretion was not noted in this region and may suggest that the fixation was not effective enough to preserve mucus on the outer integumental surface. Interestingly, mucus secretion was localized to defined channels that appeared to run perpendicular to each other forming what seemed to be collection sinuses at points of intersection in longitudinal orientation. Hausen [21] indicated that secretory cells release their contents via pores in the cuticle. This was not directly evident in this investigation and suggests that the observation of cell lined secretory channels in this species may be a unique character and a first description.

Histologically, within the species from the current study, cells lining secretion channels would sometimes appear continuous with the cuticular/epidermal region or alternatively not appear to gain access to the cuticle at all. Any mucus production would then potentially be secreted directly to the outer integument through the channels. This secretion could then be used as a low friction surface for locomotion or as observed in the laboratory, as a mesh or net structure. It is not clear as to the function of the mucous mesh but one could hypothesize that it may be used as a way of providing colonial cohesiveness, particle trapping, and/or a reproductive role. It is interesting to note that in histological section occasionally the mucoid secretions would appear to vary in volume and apparent viscosity. This may be an indicator of variable function. Further work will be necessary to elucidate the respective roles of this secretion in this species.

Mucus production has been investigated in other species of polychaete and has been noted to have a number of functions related to physiology and ecology [21]. In some species mucous secretions are used for maintaining a film on the body surface or in the production of mucus feeding traps and transport of food particles to the mouth [22, 23]. In other species it has been noted to be utilized in the production of brood chambers or egg cases and in the lining of burrows or tubes [21].

The cellular source of mucus secretion is variable and can be species-specific. Generally, cells localized to the epidermis and the cuticles are found to be associated with the majority of the mucus production in polychaetes [21]. Numerous studies have investigated the relationship between structure and function in this tissue using a variety of techniques including ultrastructural, histological, and histochemical

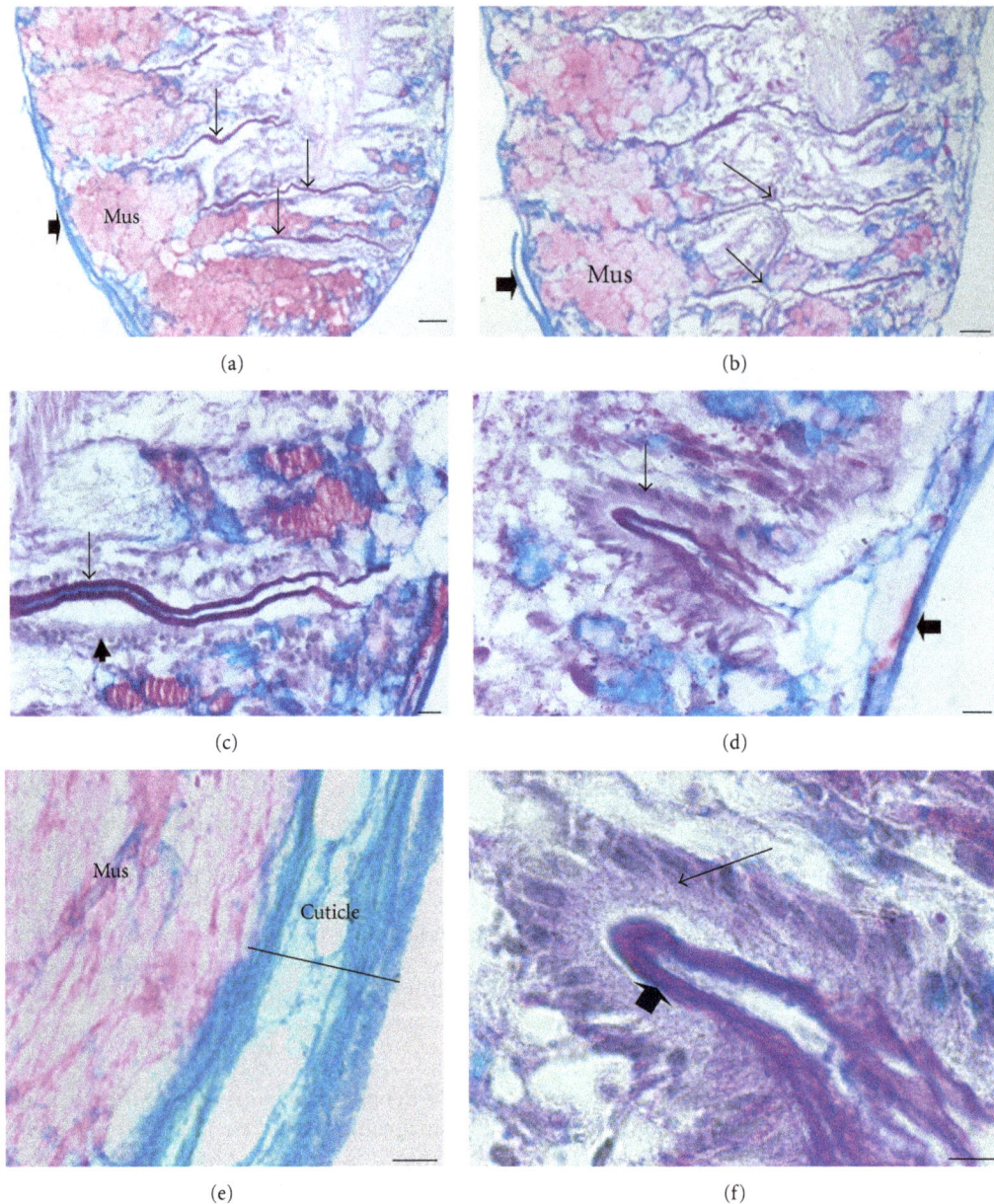

FIGURE 3: Alcian blue (pH 2.5)/PAS staining of longitudinal sections cut through the body wall showing segmental muscle blocks (mus) with detail of adjacent mucus channels. (a) Low magnification of a histological section through the worm showing latero-lateral orientation of mucus channels (thin arrows). Note also the strong Alcian blue (pH 2.5) reaction in the cuticle (arrowhead) (Scale bar = 5 μm). (b) Low magnification of a histological section through the worm showing anterior/posterior orientation of channels perpendicular to laterals. Note presence of collection sinuses at channel intersections (thin arrows). The cuticle again stained strongly with Alcian blue (arrowhead) (Scale bar = 5 μm). (c) High magnification of lateral mucus channel showing intense staining of secretion with Alcian blue (pH 2.5)/PAS (large arrow). Note lightly stained secretory cells with basal nuclei lining the channel (arrowhead) (Scale bar = 1 μm). (d) High magnification of lateral mucus channel opening toward the integumental surface. Note secretory cells with basal nuclei (small arrow) and deeply stained mucoid secretion (large arrow). The cuticle is indicated by the arrowhead (Scale bar = 1 μm). (e) High magnification of outer cuticle adjacent to a muscle block (Mus) showing strong Alcian blue reactivity in the region (Scale bar = 1 μm). (f) Detail of glandular structure showing secretory cells exhibiting basal nucleus and granular apical cytoplasm (large arrow). Note viscous secretion associated with the luminal surface (arrowhead). (Scale bar = 1 μm).

analysis (reviewed in [21]). The structure of the cuticle in polychaetes is generally dependent on its life history and basic ecology. For example, in nontube dwelling worms the cuticle is thicker and consists of collagen fibers arranged in layers whereas tube dwellers have a thinner cuticle or none at all [24–27]. Anton-Erxleben [24] noted that the polychaete cuticle was composed of two classes of organic material, a carbohydrate component and a protein component (collagen). Various histochemical investigations have shown that epidermal cells can secrete many different substances, for example, glycosaminoglycans, different mucopolysaccharides, and mucoproteins [21, 26, 28].

The orientation and organization of the mucus secreting cells from the species in this study certainly is novel, based on available literature descriptions, and warrants further investigation. In addition, observations on the utilization of the mucus in this species, both in the field and laboratory, raise further questions toward understanding its significance in the ecology and physiology of this species especially with reference to the transitional communities found beneath salmonid aquaculture sites.

Acknowledgments

This work was supported by a grant provided through the Aquaculture Collaborative Research Development Program (ACRDP) and the Department of Fisheries and Oceans Canada. The authors would like to thank Mr. Danny Ings for his many informative discussions during the progress of this project.

References

[1] R. H. Findlay, L. Watling, and L. M. Mayer, "Environmental impact of salmon net-pen culture on marine benthic communities in Maine: a case study," *Estuaries*, vol. 18, no. 1, pp. 145–179, 1995.

[2] P. Tomassetti and S. Porrello, "Polychaetes as indicators of marine fish farm organic enrichment," *Aquaculture International*, vol. 13, no. 1-2, pp. 109–128, 2005.

[3] B. T. Hargrave, G. A. Phillips, L. I. Doucette et al., "Assessing benthic impacts of organic enrichment from marine aquaculture," *Water, Air, and Soil Pollution*, vol. 99, no. 1–4, pp. 641–650, 1997.

[4] H. Yokoyama, K. Abo, and Y. Ishihi, "Quantifying aquaculture-derived organic matter in the sediment in and around a coastal fish farm using stable carbon and nitrogen isotope ratios," *Aquaculture*, vol. 254, no. 1–4, pp. 411–425, 2006.

[5] D. J. Wildish and G. W. Pohle, "Benthic macrofauna changes resulting from finfish mariculture," in *The Handbook of Environmental Chemistry. Part M Environmental Effects of Marine Finfish Aquaculture*, B. T. Hargrave, Ed., vol. 5, pp. 275–304, Springer, Berlin, Germany, 2005.

[6] R. J. Diaz and R. Rosenberg, "Marine benthic hypoxia: a review of its ecological effects and the behavioural responses of benthic macrofauna," *Oceanography and Marine Biology*, vol. 33, pp. 245–303, 1995.

[7] H. C. Nilsson and R. Rosenberg, "Succession in marine benthic habitats and fauna in response to oxygen deficiency: analysed by sediment profile-imaging and by grab samples," *Marine Ecology Progress Series*, vol. 197, pp. 139–149, 2000.

[8] K. M. Brooks, "Evaluation of the relationship between salmon farm biomass, organic inputs to sediments, physicochemical changes associated with the inputs, and the infaunal response—with emphasis on total sediment sulfides, total volatile solids, and oxygen reduction potential as surrogate end-points for biological monitoring," *Report to the Technical Advisory Group, BC Ministry of the Environment*, pp.183, 2080-A Labieux, Road, Nanaimo, Canada V9T 6J9, 2001.

[9] B. T. Hargrave, D. E. Duplisea, E. Pfeiffer, and D. J. Wildish, "Seasonal changes in benthic fluxes of dissolved oxygen and ammonium associated with marine cultured Atlantic salmon," *Marine Ecology Progress Series*, vol. 96, no. 3, pp. 249–257, 1993.

[10] P. Pocklington, D. B. Scott, C. T. Schaffer et al., "Polychaete response to different aquaculture activities," in *Proceedings of the Actes de la erne Conference Internationale des Polychetes Mem Mus Natn 'ist Nat*, J. C. Dauvin, L. Laubier, and D. J. Reish, Eds., pp. 511–520, 1994.

[11] D. E. Duplisea and B. T. Hargrave, "Response of meiobenthic size-structure, biomass and respiration to sediment organic enrichment," *Hydrobiologia*, vol. 339, no. 1–3, pp. 161–170, 1996.

[12] H. K. Dean, "The use of polychaetes (Annelida) as indicator species of marine pollution: a review," *Revista de Biologia Tropical*, vol. 56, supplement 4, pp. 11–38, 2008.

[13] D. P. Weston, "Quantitative examination of macrobenthic community changes along an organic enrichment gradient," *Marine Ecology Progressive Series*, vol. 61, pp. 233–244, 1990.

[14] M. F. Tlusty, V. A. Pepper, and M. R. Anderson, "Realizing the potential of frontier regions in aquaculture-the newfoundland salmonid experience," *World Aquaculture*, vol. 31, pp. 50–54, 2000.

[15] D. J Thornhill, T. G. Dahlgren, K. M. Halanych et al., "Evolution and ecology of Ophryotrocha (Dorvilleidae, Eunicida)," in *Annelids in Modern Biology*, D. H. Shain, Ed., pp. 242–252, John Wiley & Sons, 2009.

[16] T. G. Dahlgren, B. Åkesson, C. Schander, K. M. Halanych, and P. Sundberg, "Molecular phylogeny of the model annelid Ophryotrocha," *Biological Bulletin*, vol. 201, no. 2, pp. 193–203, 2001.

[17] H. Wiklund, A. G. Glover, and T. G. Dahlgren, "Three new species of Ophryotrocha (Annelida: Dorvilleidae) from a whale-fall in the North-East Atlantic," *Zootaxa*, no. 2228, pp. 43–56, 2009.

[18] V. Storch, "The ultrastructure of polychaeta. I. Integument," in *Microfauna Marina*, W. Westheide and C. O. Hermans, Eds., vol. 4, pp. 13–36, Verlag, Stuttgart, Germany, 1988.

[19] K. Alain, M. Olagnon, D. Desbruyères et al., "Phylogenetic characterization of the bacterial assemblage associated with mucous secretions of the hydrothermal vent polychaete Paralvinella palmiformis," *FEMS Microbiology Ecology*, vol. 42, no. 3, pp. 463–476, 2002.

[20] J. D. Bancroft and H. C. Cook, *Manual of Histological Techniques*, Longman Group Limited, New York, NY, USA, 1984.

[21] H. Hausen, "Comparative structure of the epidermis in polychaetes (Annelida)," *Hydrobiologia*, vol. 535, no. 1, pp. 25–35, 2005.

[22] N. McDaniel and K. Banse, "A novel method of suspension feeding by the maldanid polychaete Praxillura maculata," *Marine Biology*, vol. 55, no. 2, pp. 129–132, 1979.

[23] P. R. Flood and A. Fiala-Médioni, "Structure of the mucous feeding filter of Chaetopterus variopedatus (Polychaeta)," *Marine Biology*, vol. 72, no. 1, pp. 27–33, 1982.

[24] F. Anton-Erxleben, "Investigations on the cuticle of the polychaete elytra using energy dispersive X-ray analysis," *Helgoländer Meeresuntersuchungen*, vol. 34, no. 4, pp. 439–450, 1981.

[25] L. M. Gustavsson and C. Erséus, "Cuticular ultrastructure in some marine oligochaetes (Tubificidae)," *Invertebrate Biology*, vol. 119, no. 2, pp. 152–166, 2000.

[26] L. M. Gustavsson, "A Comparative study of the cuticle in some aquatic oligochaetes (Annelida: Clitellata)," *Journal of Morphology*, vol. 248, no. 2, pp. 185–195, 2000.

[27] M. Mastrodonato, E. Lepore, M. Gherardi, S. Zizza, M. Sciscioli, and D. Ferri, "Histochemical and ultrastructural analysis of the epidermal gland cells of *Branchiomma luctuosum* (Polychaeta, Sabellidae)," *Invertebrate Biology*, vol. 124, no. 4, pp. 303–309, 2005.

[28] A. Licata, A. Mauceri, L. Ainis et al., "Lectin histochemistry of epidermal glandular cells in the earthworm *Lumbricus terrestris* (Annelida Oligochaeta)," *European Journal of Histochemistry*, vol. 46, no. 2, pp. 173–178, 2002.

7

Cyanobacteria in Coral Reef Ecosystems: A Review

L. Charpy,[1] B. E. Casareto,[2] M. J. Langlade,[1] and Y. Suzuki[2]

[1] *Mediterranean Institute of Oceanography (MIO), IRD, UR235 Center of Tahiti, BP 529, 98713 Papeete, French Polynesia*
[2] *Graduate School of Science and Technology, Shizuoka University, 836 Ohya, Suruga-ku, Shizuoka 422-8529, Japan*

Correspondence should be addressed to B. E. Casareto, casaretobe@aol.com

Academic Editor: Horst Felbeck

Cyanobacteria have dominated marine environments and have been reef builders on Earth for more than three million years (myr). Cyanobacteria still play an essential role in modern coral reef ecosystems by forming a major component of epiphytic, epilithic, and endolithic communities as well as of microbial mats. Cyanobacteria are grazed by reef organisms and also provide nitrogen to the coral reef ecosystems through nitrogen fixation. Recently, new unicellular cyanobacteria that express nitrogenase were found in the open ocean and in coral reef lagoons. Furthermore, cyanobacteria are important in calcification and decalcification. All limestone surfaces have a layer of boring algae in which cyanobacteria often play a dominant role. Cyanobacterial symbioses are abundant in coral reefs; the most common hosts are sponges and ascidians. Cyanobacteria use tactics beyond space occupation to inhibit coral recruitment. Cyanobacteria can also form pathogenic microbial consortia in association with other microbes on living coral tissues, causing coral tissue lysis and death, and considerable declines in coral reefs. In deep lagoons, coccoid cyanobacteria are abundant and are grazed by ciliates, heteroflagellates, and the benthic coral reef community. Cyanobacteria produce metabolites that act as attractants for some species and deterrents for some grazers of the reef communities.

1. Cyanobacteria

Cyanobacteria are oxy-photosynthetic bacteria. One of the characteristics of cyanobacteria is their thylakoids, the seats of photosynthesis, respiration, and in some species, molecular nitrogen fixation. One of the earliest signs of life on Earth was the formation of stromatolite reefs, which exist now as fossil structures in the oldest rocks known [1]. This cyanobacterial fossil record is among the oldest of any group of organism, possibly reaching back to 3500 million years (myr) ago. Throughout the succeeding 3000 myr, many shallow reefs arose and provided a habitat for cyanobacteria. Modern corals are a relatively recent phenomenon; indeed, scleractinian corals first appeared 230 myr ago in the Triassic [2]. Although cyanobacteria have been supplanted to an extent by eukaryotic algae on modern coral reefs, especially by the dinoflagellate *Symbiodinium sp.* (zooxanthellae) and coralline red and green algae, they play an essential role in the ecology of modern reefs. Nowadays, cyanobacteria are present in the benthos and plankton compartments of coral reef ecosystems. In this paper, we discuss the contribution of cyanobacteria to photosynthetic biomass and their role in coral reef ecosystems.

2. Benthic Cyanobacteria

2.1. Microbialites. Microbialites are organosedimentary deposits of trapped benthic microbes and detrital sediment and/or mineral precipitation [3]. Thus, microbialites may display various degrees of mineral induration. Based on their internal structure, Burne and Moore [4] divided microbialites into stromatolites characterized as sedimentary structures containing lithified laminae [5], thrombolites (clotted texture), cryptic microbialites (vague, mottled or patchy texture), oncolites (concentric lamination), and spherulitic microbialites (spherular aggregates).

Microbialites may represent a major structural component of the reef. Microbialites consist exclusively of millimetre- to centimetre-thick thrombolite crusts. In the barrier reef-edge of Tahiti, they may form 80% of the rock by volume and reflect at least 13,500 years of continuous reef formation. However, the development of microbialites in the cryptic niches of the reef framework ceased about 6000 years ago when the sea level approached its present level [6].

Soft, biscuit-shaped, internally finely laminated stromatolitic structures, with substantial quantities of fine grain (micritic) carbonate, have been discovered in a lagoon on

FIGURE 1: Stromatolitic structures built by filamentous, sheathed, non-heterocystous cyanobacteria recognized as two new species of *Phormidium*. Ahe lagoon, 25 m depth (Tuamotu Archipelago).

Tikehau atoll (Tuamotu Archipelago, French Polynesia) at depths of 15–23 m [7]. These modern stromatolites cover large areas of the lagoon floor and are especially numerous around patch reefs (Figure 1). They consist of filamentous, sheathed, non-heterocystous cyanobacteria recognized as two new species of *Phormidium* [3]. The constructional elements of carbonate precipitates fall into two categories characterized by distinctive forms and size ranges: micrometre-sized (0.5–2.0 μm) mineral fibres, rounded (0.1–0.2 μm) bodies, and grape-like clusters [3]. The growth of modern marine stromatolites represents a dynamic balance between sedimentation and intermittent lithification of cyanobacterial mats [8].

2.2. Endolithic Cyanobacteria.

Carbonate skeletons of hermatypic corals harbour diverse populations of microboring organisms. Skeletons of live colonies are bored from the inside outward by Chlorophyta, while dead and denuded parts of coral skeletons are colonized at the surface and bored inward by a succession of euendoliths, starting with Chlorophyta and followed by cyanobacteria, to establish a stable Chlorophyta-dominated endolith community within 2 years [9].

The distribution of boring cyanobacteria generally depends on light level and depth; however some other factors may also influence their distribution. In Jamaica, in clear water, the boring cyanobacteria community structure changes below 20–30 m [10]. Boring cyanobacteria can also infest shells. In French Polynesia, infestations of cyanobacteria identified as *Hyella*, *Mastigocoleus*, and *Plectonema* destroy the commercially valuable shells of the black oyster *Pinctada margaritifera* [11].

In the carbonate cycle, cyanobacteria play an important and sometimes decisive role. Cycling of carbon and carbonate is linked to biological processes: some build up specific carbonate structures, some destroy carbonate substrates, and others do both simultaneously [12]. The photosynthetic activity of cyanobacteria, their extracellular polymeric substances, and possibly also their adherent heterotrophic bacteria are responsible for the construction of various carbonate structures and the ability to penetrate carbonate

material [13]. The boring activity of euendoliths results in biological corrosion and disintegration of carbonate surfaces. Grazing organisms on carbonate surfaces colonized by epi- and endolithic cyanobacteria produce specific biokarst forms and specific grains that can contribute to near-shore sedimentation [14]. Biological corrosion and abrasion together constitute bioerosion.

Endolithic phototrophs (cyanobacteria and Chlorophytes) are one of the major primary producers in dead coral substrates in a wide range of coral reef environments [15]. In an investigation of the photosynthetic activity and N_2 fixation rates of coral rubble endoliths in fringing reefs at La Reunion Island (France) and Sesoko Island (Okinawa, Japan), the main endolith flora was composed of the cyanobacteria *Hyella* (cf.) *caespitosa*, *Plectonema* (cf.) *terabrans*, *Mastigocoelus testarumin,* and *Scytonema* (cf.) *conchophyllum* (the last two species with heterocysts). Their primary production rate varied seasonally between 1.6 and 4.8 μg C μg chl^{-1} day^{-1} and were comparable to those of scleractinian corals [16].

2.3. Symbiotic Cyanobacteria.

Marine sponges can host a variety of cyanobacterial and bacterial symbionts. For example, the filamentous cyanobacterium *Oscillatoria spongeliae* is found in the sponge *Dysidea* on the Great Barrier Reef (Australia) and also in three species of *Dysidea* found around Guam [17]. In the Western Central Pacific reefs from Taiwan to the Ryukyu Archipelago, the encrusting sponge *Terpios hoshinata* is associated with unicellular cyanobacteria first described as *Aphanocapsa raspaigellae* [18, 19] and later reclassified using molecular tools as closely related to *Prochoron* sp. [20]. In the shallow waters of the Caribbean Sea, the encrusting sponges *Terpios manglaris* and *T. belindae* are associated with the cyanobacterium *Hypheothrix sp.* (Oscillatoriales, Schizotrichaceae) [18, 19]. The sponge *Terpios* sp. aggressively competes for space by killing and overgrowing live corals and is responsible for devastating wide areas of coral reef. Phylogenetic analyses of 16S rRNA sequences of sponge-associated cyanobacteria have shown them to be polyphyletic. Many sequences are affiliated with *Synechococcus* and *Prochlorococcus* species [21, 22]. Cyanobacteria fill the cortical region of the sponge and penetrate inward into the choanosomal region [23]. Microbial symbionts may produce many of the pharmaceutically active compounds isolated from marine sponges [24, 25]. These compounds can serve a variety of ecological functions, from predator and competitor deterrence and resistance to malignant microbial infections. Because cyanobacterial symbionts can also overgrow and kill their host sponge, it is not known whether sponges can actively regulate their symbiont populations [26].

2.4. Epiphytes.

Benthic marine species of *Phormidium* with narrow trichomes and *Plectonema* are common epiphytes on cyanobacteria and algae. These organisms attach externally onto sheaths of other cyanobacteria, while *Spirulina* tend to crawl inside their sheaths. Small coccoid epiphytic cyanobacteria (<0.8 μm diameter), which attach to sheaths of large

Lyngbya majuscula (>80 μm), illustrate the enormous cell size range of marine cyanobacteria [27].

Cyanobacteria are frequently observed as epiphytes of seagrass on the Great Astrolabe Reef, Fiji [28], and on the Great Barrier Reef, Australia [29], as well as epiphytes of algal turf on Virgin Island [30] in French Polynesia [31].

2.5. Microbial Mats. Microbial mats are associations of organisms dominated by cyanobacteria in association with photosynthetic bacteria, sulphur bacteria, and other microorganisms. They generally form flat, extensive mats of several millimetres in thickness on sand or mud. In coral reef ecosystems, microbial mats are found in soft muddy floors of lagoons comprised alternatively by different gliding filamentous cyanobacteria. The diversity of cyanobacterial mats inhabiting different environments has been the focus of several recent studies that applied molecular methods to natural populations. To explore the identity and distribution of natural populations of benthic marine cyanobacteria, polyphasic approaches have been used on Tikehau atoll (French Polynesia) [32], in New Caledonia [33], in the western Indian Ocean in Zanzibar (Tanzania) [34], in La Reunion Island, and in Okinawa [36]. These studies identified three types of organosedimentary structures that regularly occur on the lagoon floor: horizontally spreading mats, cobweb-like soft gelatinous masses, and hemispherical to spherical domes. These structures differ in appearance, species composition, mode of growth, and in their relationship to the substrate.

For example, on Tikehau Atoll, mats were dominated by *Hydrocoleum cantharidosmum*, *H. coccineum*, *Spirulina subsalsa*, *Symploca hydnoides* (Figure 2), and various species of *Phormidium* [32], whereas those in New Caledonia were dominated by heterocystous (*Nodularia harveyana*) and non-heterocystous (*Hydrocoleum cantharidosmum*, *H. lyngbyaceum*) [33]. In Page reef, Zanzibar, mats were dominated by filamentous non-heterocystous genera such as *Lyngbya*, *Microcoleus*, *Spirulina*, and *Oscillatoria* as well as by genera within Pseudanabaenaceae. Unicellular taxa were also represented, while heterocystous taxa were encountered only rarely [34]. In Broward County, Florida, USA, the blooms were dominated by *Lyngbya polychroa* [35]. Finally, on La Reunion Island and Sesoko Island, *Anabaena sp.* among heterocystous (Figure 3) and *Hydrocoleum majus* and *Symploca hydnoides* among non-heterocystous cyanobacteria occurred in microbial mats at both sites, whereas *Oscillatoria bonnemaisonii* and *Leptolyngbya* spp. occurred only on La Reunion Island, and *Hydrocoleum coccineum* and *Phormidium laysanense* dominated on Sesoko Island. Mats dominated by *Hydrocoleum lyngbyaceum* and *Trichocoleus tenerrimus* occurred at lower frequencies [16, 36].

Biological N$_2$ fixation performed by cyanobacteria provides these organisms and microbial mat communities with a particular advantage when growing under N-limited conditions, which are most common in marine environments. Biological N$_2$ fixation by cyanobacteria appears to make a major contribution to N supply in coral reef ecosystems [36]. Not all cyanobacteria can fix atmospheric nitrogen.

FIGURE 2: Tufts dominated by *Symploca hydnoides* in Mayotte lagoon at 10 m depth.

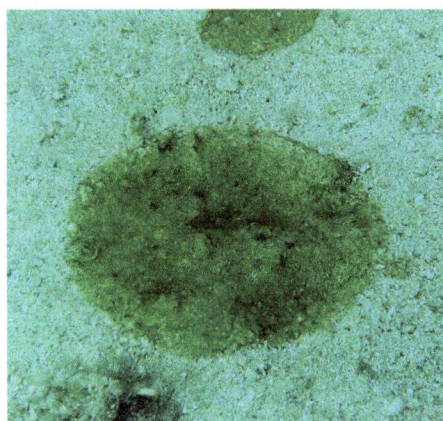

FIGURE 3: Cyanobacteria mats dominated by *Anabaena* sp. (heterocystous) in Mayotte lagoon at 14 m depth.

The process is oxygen-sensitive and energetically expensive, which constrains its implementation in oxygenic cyanobacteria; these bacteria separate the processes of carbon and nitrogen fixation either in space (i.e., heterocyst) or time [37].

The contribution of N$_2$ fixation to that required for primary production is between 2% and 21% in Tikehau atoll lagoon [38] and New Caledonia [33]. Casareto et al. [16] compared N$_2$ fixation rates of three different subenvironments (coral rubbles, microbial mats, and sandy bottoms) on La Reunion and Sesoko Islands. They found that N$_2$ fixation rates of microbial mats are one order of magnitude higher than that of other subenvironments and can contribute up to 95% of their primary production [16].

2.6. Harmful Effects. Cyanobacteria are becoming increasingly prominent on declining reefs, as these microbes can tolerate strong solar radiation [39]. Changes in land use or seabird distribution that lead to alter dissolved organics, iron, and phosphorus input enhance proliferation of noxious blooms of cyanobacteria [40]. The production of deterrent secondary metabolites by benthic cyanobacterial and similar microbial assemblages facilitates the formation of cyanobacterial blooms on coral reefs [41].

Kuffner et al. [42] found evidence that algae and cyanobacteria use tactics beyond space occupation to inhibit coral recruitment. On reefs experiencing phase shifts or temporary algal blooms, the restocking of adult coral populations may be slowed due to recruitment inhibition by cyanobacteria, thereby perpetuating reduced coral cover and limiting coral community recovery. Cyanobacterial mats act as a poison for scleractinian corals and are able to kill live coral tissue [43]. About 30 diseases of corals have been recognised since they were first discovered more than 30 years ago. Little is known of the causes and effects of coral disease, although they can be caused by bacteria, fungi, algae, worms [44], and viruses [45–47].

Black band disease (BBD) of corals is caused by a pathogenic microbial consortium that exists as a horizontally migrating, laminated microbial mat. The consortium is structurally directly analogous to the cyanobacterial mats found in many illuminated, sulfide-rich benthic environments such as hot spring outflows and sediments of hypersaline lagoons, but is unique in that the entire mat community migrates across the surface of coral colonies, completely degrading coral tissue. BBD is one of a number of coral diseases believed to play an important role in the observed decline of coral reefs. The black band microbial consortium is dominated by *Phormidium corallyticum* [48] in the Caribbean, but other cyanobacteria species were described in Palau reefs [49] and in the Red Sea where a new cyanobacterium species, *Pseudoscillatoria coralii* gen. nov., sp. nov., dominates the BBD consortium on *Favia* sp. corals [50]. *P. corallyticum* can perform oxygenic photosynthesis in the presence or absence of sulfide but cannot conduct anoxygenic photosynthesis with sulfide as an electron donor. This species is not capable of fixing N_2 [51]. Recent discoveries [52, 53] indicate that different species and strains of BBD cyanobacteria, which can occur in the same BBD infection, may contribute to BBD pathobiology by producing different types and amounts of toxins at different stages of the disease process. Understanding the interactions between coral larvae and benthic bloom-forming cyanobacteria may be important in managing coral reef ecosystems [54].

2.7. Benthic Cyanobacteria Grazing. Cyanobacteria are generally considered to be a poor food source due to toxicity, low nutritional value, or a morphology that makes ingestion difficult. Despite these factors, there are grazers that are adapted to feeding on cyanobacteria [55].

Lyngbya majuscula constitutes a major portion of the diet of a Pomacentrid damselfish on Davies Reef, Australia [56], and on Orpheus Island, Australia [57]. Many cyanobacteria, including species of *Calothrix*, *Lyngbya*, *Oscillatoria*, and *Phormidium*, have been found in the plate of fish *Hemiglyphi-dodon plagiometopon* [58].

During the life of the coral, the endoliths are protected from grazers, but in dead coral skeletons endolith cyanobacteria are exposed to grazing by molluscs, echinoderms and scarid fish [9]. The importance of epiphytic cyanobacteria as a food source for heterotrophs in coral reef ecosystems was also reported by Yamamuro [28]. Thacker et al. [59] found that coral reef fishes can learn to avoid defensive secondary metabolites, but that this learning does not occur when access to food is limited. This strategy may indicate

that the effectiveness of the chemical defences of an alga or cyanobacteria is dependent on the state of the consumer and the defences of other prey in the environment. Thacker et al. [60] observed selected grazing on the cyanobacteria of Guam coral reefs, stressing the critical role of herbivory in determining coral reef community structure.

Some tropical benthic cyanobacteria are preferred foods for specialized consumers in the size range of mesograzers. Therefore, a diverse fauna may depend on cyanobacterial mats. Tropical mesograzers exploit considerably different food resources, with some species adapted to consume cyanobacterial mats. Benthic cyanobacteria may play important roles as food and shelter for marine consumers and may indirectly influence local biodiversity through their associated fauna [61, 62]. The cyanobacterial genus *Lyngbya* includes free-living, benthic, filamentous cyanobacteria that form periodic nuisance blooms in lagoons, reefs, and estuaries. *Lyngbya* spp. are prolific producers of biologically active compounds (metabolites). *Lyngbya majuscule* produces a wide variety of secondary metabolites, as well as lyngbyatoxin A (LTA). LTA production varies in different locations worldwide [63]. Specific metabolites produced by *Lyngbya majuscula* act as both feeding attractants to the specialist herbivore *Stylocheilus longicauda*, and as effective feeding deterrents to the generalist fishes [64]. One species, identified as *Lyngbya* cf. *confervoides*, produces a diverse array of bioactive peptides and depsipeptides [65].

Opisthobranchs may also play a role in top-down control of toxic cyanobacterial blooms, as was demonstrated for toxic *Lyngbya* by Capper and Paul [66].

Microbial mats can also be ingested by filter feeders. Identification of homoanatoxin-a from benthic marine cyanobacteria (*Hydrocoleum lyngbyaceum*) samples collected in Lifou (Loyalty Islands, New Caledonia) was recently reported [67]. This cyanobacterium was suspected to cause giant clam (*Tridacna maxima*) intoxications.

3. Planktonic Cyanobacteria

Planktonic cyanobacteria found in coral reef plankton are mainly filamentous and unicellular.

3.1. Planktonic Filamentous Cyanobacteria. Large blooms of *Trichodesmium*, a filamentous nitrogen-fixing cyanobacterium, are observed frequently in coral reef ecosystems [68]. They have been documented in the eastern Indian Ocean and western Pacific [69], in the central region of the Great Barrier Reef [70–73], in the Gulf of Thailand [74], and in the southwestern Tropical Pacific [75]. *Trichodesmium* spp. have been described to be nontoxic, sometimes toxic, or always toxic to a range of organisms [76–81]. Recent studies have provided unprecedented evidence of the toxicity of *Trichodesmium* spp. from the New Caledonia lagoon [82], demonstrating the possible role of these cyanobacteria in ciguatera fish poisoning.

Trichodesmium is the most well-studied marine N_2-fixing organism and perhaps one of the most important. The rate of nitrogen fixation by *Trichodesmium* species in surface waters is close to 2 pmol N trichome^{-1} h^{-1} [83]. It is difficult to quantify the importance of *Trichodesmium* diazotrophy because of the stochastic nature of the blooms. However, it is estimated that *Trichodesmium* contributes about 0.03–20% of the total CO_2 fixation in the coastal surface waters of Tanzania [84].

The pelagic harpacticoid copepod *Macrosetella gracilis* is usually found in association with blooms of *Trichodesmium* in tropical and subtropical waters. This copepod is one of the few direct grazers of these often toxic cyanobacteria [79, 85].

The study of Villareal [86] in the Belizean barrier reef showed significant grazing of *Trichodesmium* by the coral reef community.

3.2. Planktonic Unicellular Cyanobacteria. Oligotrophic waters surrounding coral reef ecosystems and lagoons are dominated by the small coccoid unicellular cyanobacteria *Synechococcus* and *Prochlorococcus* [87–95]. In coral reef waters, *Synechococcus* has a size of 1 μm and an abundance ranging from 10×10^3 to 500×10^3 cells mL^{-1}, while *Prochlorococcus* has a size of 0.6 μm and an abundance ranging from 10×10^3 to 400×10^3 cells mL^{-1}.

The contribution of unicellular cyanobacteria to phytoplankton biomass and production varies according to the ecosystem. In Tuamotu lagoon (French Polynesia), *Synechococcus* is the predominant group in terms of abundance and carbon biomass and has the highest planktonic primary production among lagoons. As it is generally scarce in deep water with limited light availability, its biomass contribution is reduced in deep lagoons. In very shallow lagoons, no general trend has been observed, as the dominant group appears to depend on the water residence time within the lagoon [89–96]. In Tuamotu lagoon and Miyako Island (Okinawa) picoplankton primary production represents 65–80% of total phytoplankton production [97, 98].

In the Great Astrolabe Reef lagoon (Fiji), *Synechococcus* is the most abundant group (85–95%), followed by picoeukaryotes (5–10%) and *Prochlorococcus* (<4%) [90]. Picoplankton primary production makes up 53.2% of the total phytoplankton production [90].

Ayukai [99] reported that on the Great Barrier Reef, the average abundance of cyanobacteria (*Synechococcus*) is 0.16–2.41 $\times 10^4$ cells mL^{-1}. Later, Crosbie and Furnas [92], using a flow cytometer, observed that *Synechococcus* was more abundant and had a greater biomass than *Prochlorococcus* at most inshore and mid-shelf sites in central regions (17°S) and at all shelf sites in southern areas (20°S) of Great Barrier Reef. Moreover, *Synechococcus* and *Prochlorococcus* abundance was better correlated with salinity, shelf depth, and chlorophyll *a* concentration than with nutrient concentrations.

At Sesoko Island (Okinawa), Tada et al. [100] found that picoplankton dominated the phytoplankton community with an average contribution to the total chlorophyll-*a* biomass of 52%. At Miyako Island (Okinawa, Japan), the contribution of picophytoplankton to total phytoplankton biomass is 45–100% [101]. In another study, Ferrier-Pagès and Furla [96] found that the picophytoplankton contribution to total chlorophyll was 32–73%. *Prochlorococcus*,

Synechococcus, and picoeukaryote abundance was on average 64 ± 11, 12 ± 2, and 4 ± 0.7 $\times 10^3$ cells mL^{-1}, respectively. Their contribution to picoplankton biomass was 10, 49, and 41%, respectively, and the contribution of picoplankton primary production to total phytoplankton production is 65%. On Miyako Island, Okinawa (Japan), *Synechococcus* spp. represented 65% of the chlorophyll (<3 μm), 53% of autotrophic carbon, and 67% of the nitrogen [101]. In Mayotte (south-western Indian Ocean), particles <10 μm accounted for 74% of the chlorophyll-*a* concentration and for 47% of the total living carbon [102].

In one study in New Caledonia's coral lagoon, unicellular diazotrophic cyanobacteria of 1–1.5 μm were found along a nutrient gradient using whole-cell hybridization with specific Nitro 821 probes [103]. Their abundance ranged from 3 to 140 cells mL^{-1}. These cells may contribute to N_2 fixation (from the <10 μm size fraction) which was estimated to be 4.4–8 nmol N^{-1} d^{-1}.

Very few studies have investigated grazing of unicellular cyanobacteria in coral reef waters [101, 102]. In Tikehau lagoon (Tuamotu), González et al. [104] showed that phagotrophic nanoflagellates were the major grazers of picocyanobacteria. Ciliates and heterotrophic dinoflagellates appeared to be grazing mostly on nanoplankton, both autotrophic and heterotrophic cells, showing the important contribution of coccoid cyanobacteria to the microbial food web.

In Takapoto (Tuamotu), the grazing rates of <200 μm protozoa on cyanobacteria represented 74% of their growth rates [105]. In the lagoonal waters of the two largest atolls of French Polynesia (Rangiroa and Fakarava), 75% of the cyanobacteria production was consumed by <10 μm fractions, equal to 0.05–0.5 $\times 10^4$ cyanobacteria mL^{-1} h^{-1} [96]. In the water over a fringing coral reef at Miyako Island (Japan), 30–50% of picocyanobacteria production was grazed by heterotrophic flagellates and ciliates, which themselves were grazed (50–70% of the production) by higher trophic levels [101].

On Conch Reef, Florida Keys, sponges are a net sink for picocyanobacteria [106]. In the Gulf of Aqaba, Red Sea, measurements of depletion of phytoplankton cells and pigments over coral reefs have revealed that *Synechococcus* contributes >70% of the total depleted carbon in summer. The grazing of cyanobacteria appears to be an important component of benthic-pelagic coupling in coral reefs [102, 107]. Another study by Yahel et al. [108] demonstrated that sponges removed significant amounts of picocyanobacteria but suggested that DOC may play a major role in the trophic dynamics of coral reefs. In Caribbean coral reef communities, gorgonian corals do not appear to graze significantly on picocyanobacteria [109].

4. Conclusions

Cyanobacteria are ubiquitous in coral reef ecosystems:

(i) as a part of the reef (Microbialites),

(ii) inside (endoliths) and above (epiliths and epiphytes) the coral reef,

(iii) as symbionts of sponges,

(iv) covering soft bottoms as microbial mats,

(v) in the water column.

In addition, they have the following.

(i) They help build and erode the reef.

(ii) They are important primary producers.

(iii) They represent an organic source for planktonic and benthic heterotrophic organisms.

(iv) They enrich the ecosystem with nitrogen.

Acknowledgments

This work was supported by the Institute of Research for Development (IRD) and by grants from the Ocean Development Sub-Committee of France-Japan S&T Cooperation, Mitsubishi cooperation, the Ministry of Education, Science, Sport, and Culture of Japan.

References

[1] S. M. Awramik, J. W. Schopf, and M. R. Walter, "Filamentous fossil bacteria from the Archean of Western Australia," *Precambrian Research*, vol. 20, no. 2–4, pp. 357–374, 1983.

[2] J. E. N. Veron, *Corals in Space and Time: The Biogeography and Evolution of the Scleractinia*, UNSW Press, Sydney, Australia, 1995.

[3] S. Sprachta, G. Camoin, S. Golubic, and T. Le Campion, "Microbialites in a modern lagoonal environment: Nature and distribution, Tikehau atoll (French Polynesia)," *Palaeogeography, Palaeoclimatology, Palaeoecology*, vol. 175, no. 1-4, pp. 103–124, 2001.

[4] R. V. Burne and L. S. Moore, "Microbialites: organosedimentary deposits of benthic microbial communities," *Palaios*, vol. 2, no. 3, pp. 241–254, 1987.

[5] T. F. Steppe, J. L. Pinckney, J. Dyble, and H. W. Paerl, "Diazotrophy in modern marine Bahamian stromatolites," *Microbial Ecology*, vol. 41, no. 1, pp. 36–44, 2001.

[6] G. F. Camoin, P. Gautret, L. F. Montaggioni, and G. Cabioch, "Nature and environmental significance of microbialites in Quaternary reefs: the Tahiti paradox," *Sedimentary Geology*, vol. 126, no. 1–4, pp. 271–304, 1999.

[7] C. Charpy-Roubaud, T. Le Campion, S. Golubic, and G. Sarazin, "Recent cyanobacterial stromatolites in the lagoon of Tikehau Atoll (Tuamotu Archipelago, French Polynesia): preliminary observations," in *Marine Cyanobacteria*, L. Charpy and A. W. D. Larkum, Eds., vol. 19, pp. 121–125, Monaco Musée Océanographique, Monaco, 1999.

[8] R. P. Reid, P. T. Visscher, A. W. Decho et al., "The role of microbes in accretion, lamination and early lithification of modern marine stromatolites," *Nature*, vol. 406, no. 6799, pp. 989–992, 2000.

[9] T. Le Campion Alsumard, S. Golubic, and P. Hutchings, "Microbial endoliths in skeletons of live and dead corals: *Porites lobata* (Moorea, French Polynesia)," *Marine Ecology Progress Series*, vol. 117, no. 1–3, pp. 149–158, 1995.

[10] C. T. Perry and I. A. Macdonald, "Impacts of light penetration on the bathymetry of reef microboring communities: implications for the development of microendolithic trace assemblages," *Palaeogeography, Palaeoclimatology, Palaeoecology*, vol. 186, no. 1-2, pp. 101–113, 2002.

[11] L. Mao Che, T. Le Campion-Alsumard, N. Boury-Esnault, C. Payri, S. Golubic, and C. Bézac, "Biodegradation of shells of the black pearl oyster, *Pinctada margaritifera* var. *cumingii*, by microborers and sponges of French Polynesia," *Marine Biology*, vol. 126, no. 3, pp. 509–519, 1996.

[12] J. Schneider and T. Le Campion-Alsumard, "Construction and destruction of carbonates by marine and freshwater cyanobacteria," *European Journal of Phycology*, vol. 34, no. 4, pp. 417–426, 1999.

[13] G. Arp, A. Reimer, and J. Reitner, "Calcification in cyanobacterial biofilms of alkaline salt lakes," *European Journal of Phycology*, vol. 34, no. 4, pp. 393–403, 1999.

[14] J. Schneider and H. Torunski, "Biokarst on limestone coasts, morphogenesis and sediment production," *Marine Ecology*, vol. 4, no. 1, pp. 45–63, 1983.

[15] A. Tribollet, C. Langdon, S. Golubic, and M. Atkinson, "Endolithic microflora are major primary producers in dead carbonate substrates of Hawaiian coral reefs," *Journal of Phycology*, vol. 42, no. 2, pp. 292–303, 2006.

[16] B. E. Casareto, L. Charpy, M. J. Langlade et al., "Nitrogen fixation in coral reef environments," in *Proceedings of the 11th International Coral Reef Symposium*, vol. 2, pp. 896–900, Fort Lauderdale, Fla, USA, 2008.

[17] R. Hinde, F. Pironet, and M. A. Borowitzka, "Isolation of Oscillatoria spongeliae, the filamentous cyanobacterial symbiont of the marine sponge Dysidea herbacea," *Marine Biology*, vol. 119, no. 1, pp. 99–104, 1994.

[18] K. Rutzler and K. Muzik, "Terpios hoshinata, a new cyanobacteriosponge threatening Pacific reefs," *Scientia Marina*, vol. 57, no. 4, pp. 395–403, 1993.

[19] K. Rutzler and K. P. Smith, "The genus Terpios (Suberitidae) and new species in the "Lobiceps" complex," *Scientia Marina*, vol. 57, no. 4, pp. 381–393, 1993.

[20] S.-L. Tang, M.-J. Hong, M.-H. Liao et al., "Bacteria associated with an encrusting sponge (Terpios hoshinota) and the corals partially covered by the sponge," *Environmental Microbiology*, vol. 13, no. 5, pp. 1179–1191, 2011.

[21] L. Steindler, D. Huchon, A. Avni, and M. Ilan, "16S rRNA phylogeny of sponge-associated cyanobacteria," *Applied and Environmental Microbiology*, vol. 71, no. 7, pp. 4127–4131, 2005.

[22] P. M. Erwin and R. W. Thacker, "Cryptic diversity of the symbiotic cyanobacterium Synechococcus spongiarum among sponge hosts," *Molecular Ecology*, vol. 17, no. 12, pp. 2937–2947, 2008.

[23] E. Gaino, M. Sciscioli, E. Lepore, M. Rebora, and G. Corriero, "Association of the sponge Tethya orphei (Porifera, Demospongiae) with filamentous cyanobacteria," *Invertebrate Biology*, vol. 125, no. 4, pp. 281–287, 2006.

[24] G. G. Harrigan and G. Goetz, "Symbiotic and dietary marine microalgae as a source of bioactive molecules-experience from natural products research," *Journal of Applied Phycology*, vol. 14, no. 2, pp. 103–108, 2002.

[25] P. Proksch, R. A. Edrada, and R. Ebel, "Drugs from the seas—current status and microbiological implications," *Applied Microbiology and Biotechnology*, vol. 59, no. 2-3, pp. 125–134, 2002.

[26] R. W. Thacker and S. Starnes, "Host specificity of the symbiotic cyanobacterium Oscillatoria spongeliae in marine sponges, Dysidea spp," *Marine Biology*, vol. 142, no. 4, pp. 643–648, 2003.

[27] S. Golubic, T. Le Campion-Alsumard, and S. E. Campbell, "Diversity of marine cyanobacteria," in *Marine Cyanobacteria*, L. Charpy and A. W. D. Larkum, Eds., vol. 19, pp. 53–76, Monaco Musée Océanographique, Monaco, 1999.

[28] M. Yamamuro, "Importance of epiphytic cyanobacteria as food sources for heterotrophs in a tropical seagrass bed," *Coral Reefs*, vol. 18, no. 3, pp. 263–271, 1999.

[29] H. Iizumi and M. Yamamuro, "Nitrogen fixation activity by periphytic blue-green algae in a seagrass bed on the great barrier reef," *Japan Agricultural Research Quarterly*, vol. 34, no. 1, pp. 69–73, 2000.

[30] J. M. Hackney, R. C. Carpenter, and W. H. Adey, "Characteristic adaptations to grazing among algal turfs on a Caribbean coral reef," *Phycologia*, vol. 28, no. 1, pp. 109–119, 1989.

[31] S. Le Bris, T. Le Campion-Alsumard, and J. C. Romano, "Characteristics of epilithic and endolithic algal turf exposed to different levels of bioerosion in French Polynesian coral reefs," *Oceanologica Acta*, vol. 21, no. 5, pp. 695–708, 1998.

[32] R. M. M. Abed, S. Golubic, F. Garcia-Pichel, G. F. Camoin, and S. Sprachta, "Characterization of microbialite-forming cyanobacteria in a tropical lagoon: Tikehau Atoll, Tuamotu, French Polynesia," *Journal of Phycology*, vol. 39, no. 5, pp. 862–873, 2003.

[33] L. Charpy, R. Alliod, M. Rodier, and S. Golubic, "Benthic nitrogen fixation in the SW New Caledonia lagoon," *Aquatic Microbial Ecology*, vol. 47, no. 1, pp. 73–81, 2007.

[34] K. Bauer, B. Díez, C. Lugomela, S. Seppälä, A. J. Borg, and B. Bergman, "Variability in benthic diazotrophy and cyanobacterial diversity in a tropical intertidal lagoon," *FEMS Microbiology Ecology*, vol. 63, no. 2, pp. 205–221, 2008.

[35] V. J. Paul, R. W. Thacker, K. Banks, and S. Golubic, "Benthic cyanobacterial bloom impacts the reefs of South Florida (Broward County, USA)," *Coral Reefs*, vol. 24, no. 4, pp. 693–697, 2005.

[36] L. Charpy, K. A. Palinska, B. Casareto et al., "Dinitrogen-fixing cyanobacteria in microbial mats of two shallow coral reef ecosystems," *Microbial Ecology*, vol. 59, no. 1, pp. 174–186, 2010.

[37] I. Berman-Frank, P. Lundgren, and P. Falkowski, "Nitrogen fixation and photosynthetic oxygen evolution in cyanobacteria," *Research in Microbiology*, vol. 154, no. 3, pp. 157–164, 2003.

[38] C. Charpy-Roubaud, L. Charpy, and A. Larkum, "Atmospheric dinitrogen fixation by benthic communities of Tikehau lagoon (Tuamotu Archipelago, French Polynesia) and its contribution to benthic primary production," *Marine Biology*, vol. 139, no. 5, pp. 991–997, 2001.

[39] P. Hallock, "Global change and modern coral reefs: new opportunities to understand shallow-water carbonate depositional processes," *Sedimentary Geology*, vol. 175, no. 1–4, pp. 19–33, 2005.

[40] S. Albert, J. M. O'Neil, J. W. Udy, K. S. Ahern, C. M. O'Sullivan, and W. C. Dennison, "Blooms of the cyanobacterium Lyngbya majuscula in coastal Queensland, Australia: disparate sites, common factors," *Marine Pollution Bulletin*, vol. 51, no. 1–4, pp. 428–437, 2005.

[41] D. G. Nagle and V. J. Paul, "Chemical defense of a marine cyanobacterial bloom," *Journal of Experimental Marine Biology and Ecology*, vol. 225, no. 1, pp. 29–38, 1998.

[42] I. B. Kuffner, L. J. Walters, M. A. Becerro, V. J. Paul, R. Ritson-Williams, and K. S. Beach, "Inhibition of coral recruitment by macroalgae and cyanobacteria," *Marine Ecology Progress Series*, vol. 323, pp. 107–117, 2006.

[43] E. A. Titlyanov, I. M. Yakovleva, and T. V. Titlyanova, "Interaction between benthic algae (Lyngbya bouillonii, Dictyota dichotoma) and scleractinian coral Porites lutea in direct contact," *Journal of Experimental Marine Biology and Ecology*, vol. 342, no. 2, pp. 282–291, 2007.

[44] P. Garrett and H. Ducklow, "Coral diseases in Bermuda," *Nature*, vol. 253, no. 5490, pp. 349–350, 1975.

[45] W. H. Wilson, A. L. Dale, J. E. Davy, and S. K. Davy, "An enemy within? Observations of virus-like particles in reef corals," *Coral Reefs*, vol. 24, no. 1, pp. 145–148, 2005.

[46] S. K. Davy, S. G. Burchett, A. L. Dale et al., "Viruses: agents of coral disease?" *Diseases of Aquatic Organisms*, vol. 69, no. 1, pp. 101–110, 2006.

[47] F. Rohwer and R. V. Thurber, "Viruses manipulate the marine environment," *Nature*, vol. 459, no. 7244, pp. 207–212, 2009.

[48] R. G. Carlton and L. L. Richardson, "Oxygen and sulfide dynamics in a horizontally migrating cyanobacterial mat: black band disease of corals," *FEMS Microbiology Ecology*, vol. 18, no. 2, pp. 155–162, 1995.

[49] M. Sussman, D. G. Bourne, and B. L. Willis, "A single cyanobacterial ribotype is associated with both red and black bands on diseased corals from Palau," *Diseases of Aquatic Organisms*, vol. 69, no. 1, pp. 111–118, 2006.

[50] D. Rasoulouniriana, N. Siboni, B. D. Eitan, K. W. Esti, Y. Loya, and A. Kushmaro, "Pseudoscillatoria coralii gen. nov., sp. nov., a cyanobacterium associated with coral black band disease (BBD)," *Diseases of Aquatic Organisms*, vol. 87, no. 1-2, pp. 91–96, 2009.

[51] L. L. Richardson and K. G. Kuta, "Ecological physiology of the black band disease cyanobacterium Phormidium corallyticum," *FEMS Microbiology Ecology*, vol. 43, no. 3, pp. 287–298, 2003.

[52] L. L. Richardson, A. W. Miller, E. Broderick et al., "Sulfide, microcystin, and the etiology of black band disease," *Diseases of Aquatic Organisms*, vol. 87, no. 1-2, pp. 79–90, 2009.

[53] D. Stanić, S. Oehrle, M. Gantar, and L. L. Richardson, "Microcystin production and ecological physiology of Caribbean black band disease cyanobacteria," *Environmental Microbiology*, vol. 13, no. 4, pp. 900–910, 2011.

[54] I. B. Kuffner and V. J. Paul, "Effects of the benthic cyanobacterium Lyngbya majuscula on larval recruitment of the reef corals Acropora surculosa and Pocillopora damicornis," *Coral Reefs*, vol. 23, no. 3, pp. 455–458, 2004.

[55] J. M. O'Neil, "Grazer interactions with nitrogen-fixing marine Cyanobacteria: adaptation for N- acquisition?" in *Marine Cyanobacteria*, L. Charpy and A. W. D. Larkum, Eds., vol. 19, pp. 293–317, Monaco Musée Océanographique, Monaco, 1999.

[56] D. W. Klumpp and N. V. C. Polunin, "Partitioning among grazers of food resources within damselfish territories on a coral reef," *Journal of Experimental Marine Biology and Ecology*, vol. 125, no. 2, pp. 145–169, 1989.

[57] M. J. Marnane and D. R. Bellwood, "Marker technique' for investigating gut throughput rates in coral reef fishes," *Marine Biology*, vol. 129, no. 1, pp. 15–22, 1997.

[58] C. R. Wilkinson and P. W. Sammarco, "Nitrogen fixation on a coral reef: effects of fish grazing and damselfish territoriality. The Reef and Man," in *Proceedings of the 4th International Coral Reef Symposium*, p. 589, 1981.

[59] R. W. Thacker, D. G. Nagle, and V. J. Paul, "Effects of repeated exposures to marine cyanobacterial secondary metabolites on feeding by juvenile rabbitfish and parrotfish," *Marine Ecology Progress Series*, vol. 147, no. 1–3, pp. 21–29, 1997.

[60] R. W. Thacker, D. W. Ginsburg, and V. J. Paul, "Effects of herbivore exclusion and nutrient enrichment on coral reef macroalgae and cyanobacteria," *Coral Reefs*, vol. 19, no. 4, pp. 318–329, 2001.

[61] E. Cruz-Rivera and V. J. Paul, "Feeding by coral reef mesograzers: algae or cyanobacteria?" *Coral Reefs*, vol. 25, no. 4, pp. 617–627, 2006.

[62] E. Cruz-Rivera and V. J. Paul, "Chemical deterrence of a cyanobacterial metabolite against generalized and specialized grazers," *Journal of Chemical Ecology*, vol. 33, no. 1, pp. 213–217, 2007.

[63] V. J. Paul, K. E. Arthur, R. Ritson-Williams, C. Ross, and K. Sharp, "Chemical defenses: from compounds to communities," *Biological Bulletin*, vol. 213, no. 3, pp. 226–251, 2007.

[64] D. G. Nagle and V. J. Paul, "Production of secondary metabolites by filamentous tropical marine cyanobacteria: ecological functions of the compounds," *Journal of Phycology*, vol. 35, no. 6, pp. 1412–1421, 1999.

[65] K. Sharp, K. E. Arthur, L. Gu et al., "Phylogenetic and chemical diversity of three chemotypes of bloom-forming Lyngbya species (cyanobacteria: Oscillatoriales) from reefs of southeastern Florida," *Applied and Environmental Microbiology*, vol. 75, no. 9, pp. 2879–2888, 2009.

[66] A. Capper and V. J. Paul, "Grazer interactions with four species of Lyngbya in southeast Florida," *Harmful Algae*, vol. 7, no. 6, pp. 717–728, 2008.

[67] A. Méjean, C. Peyraud-Thomas, A. S. Kerbrat et al., "First identification of the neurotoxin homoanatoxin-a from mats of Hydrocoleum lyngbyaceum (marine cyanobacterium) possibly linked to giant clam poisoning in New Caledonia," *Toxicon*, vol. 56, no. 5, pp. 829–835, 2010.

[68] T. E. Bowman and L. J. Lancaster, "A bloom of the planktonic blue-green alga, Trichodesmium erythraeum, in the Tonga Islands," *Limnology and Oceanography*, vol. 10, pp. 291–292, 1965.

[69] Y. I. Sorokin, "Phytoplankton and planktonic microflora in the coral reef ecosystem," *Journal of General Biology*, vol. 40, pp. 677–688, 1979.

[70] N. Revelante and M. Gilmartin, "Dynamics of phytoplankton in the great barrier reef lagoon," *Journal of Plankton Research*, vol. 4, no. 1, pp. 47–76, 1982.

[71] G. B. Jones, C. Burdon-Jones, and F. G. Thomas, "Influence of Trichodesmium red tides on trace metal cycling at a coastal station in the Great Barrier Reef Lagoon," in *the International Symposium on Coastal Lagoons*, pp. 319–326, Bordeaux, France, 1982.

[72] P. R. F. Bell, "Status of eutrophication in the Great Barrier Reef Lagoon," *Marine Pollution Bulletin*, vol. 23, pp. 89–93, 1991.

[73] I. Muslim and G. Jones, "The seasonal variation of dissolved nutrients, chlorophyll a and suspended sediments at Nelly Bay, Magnetic Island," *Estuarine, Coastal and Shelf Science*, vol. 57, no. 3, pp. 445–455, 2003.

[74] V. Cheevaporn and P. Menasveta, "Water pollution and habitat degradation in the Gulf of Thailand," *Marine Pollution Bulletin*, vol. 47, no. 1–6, pp. 43–51, 2003.

[75] C. Dupouy, M. Petit, and Y. Dandonneau, "Satellite detected cyanobacteria bloom in the southwestern tropical Pacific: implication for oceanic nitrogen fixation," *International Journal of Remote Sensing*, vol. 9, no. 3, pp. 389–396, 1988.

[76] V. P. Devassy, P. M. A. Bhattathiri, and S. Z. Qasim, "Succession of organisms following trichodesmium phenomenon," *Indian Journal of Marine Sciences*, vol. 8, pp. 89–93, 1979.

[77] S. P. Hawser, G. A. Codd, D. G. Capone, and E. J. Carpenter, "A neurotoxic factor associated with the bloom-forming cyanobacterium Trichodesmium," *Toxicon*, vol. 29, no. 3, pp. 277–278, 1991.

[78] S. P. Hawser, J. M. O'Neil, M. R. Roman, and G. A. Codd, "Toxicity of blooms of the cyanobacterium *Trichodesmium* to zooplankton," *Journal of Applied Phycology*, vol. 4, no. 1, pp. 79–86, 1992.

[79] J. M. O'Neil and M. R. Roman, "Ingestion of the cyanobacterium Trichodesmium spp. by pelagic harpacticoid copepods Macrosetella, Miracia and Oculosetella," in *Ecology and Morphology of Copepods*, F. D. Ferrari and B. P. Bradley, Eds., vol. 292-293, pp. 235–240, Kluwer Academic Publishers, Dordrecht, The Netherlands, 1994.

[80] C. Guo and P. A. Tester, "Toxic effect of the bloom-forming *Trichodesmium sp.* (Cyanophyta) to the copepod *Acartia tonsa*," *Natural Toxins*, vol. 2, no. 4, pp. 222–227, 1994.

[81] A. P. Negri, O. Bunter, B. Jones, and L. Llewellyn, "Effects of the bloom-forming alga *Trichodesmium erythraeum* on the pearl oyster *Pinctada maxima*," *Aquaculture*, vol. 232, no. 1–4, pp. 91–102, 2004.

[82] A. S. Kerbrat, H. T. Darius, S. Pauillac, M. Chinain, and D. Laurent, "Detection of ciguatoxin-like and paralysing toxins in *Trichodesmium spp.* from New Caledonia lagoon," *Marine Pollution Bulletin*, vol. 61, no. 7–12, pp. 360–366, 2010.

[83] D. Karl, A. Michaels, B. Bergman et al., "Dinitrogen fixation in the world's oceans," *Biogeochemistry*, vol. 57-58, pp. 47–98, 2002.

[84] C. Lugomela, T. J. Lyimo, I. Bryceson, A. K. Semesi, and B. Bergman, "Trichodesmium in coastal waters of Tanzania: diversity, seasonality, nitrogen and carbon fixation," *Hydrobiologia*, vol. 477, pp. 1–13, 2002.

[85] J. M. O'Neil, P. M. Metzler, and P. M. Glibert, "Ingestion of super 15N2-labelled Trichodesmium spp. and ammonium regeneration by the harpacticoid copepod Macrosetella gracilis," *Marine Biololgy*, vol. 125, pp. 89–96, 1996.

[86] T. A. Villareal, "Abundance and photosynthetic characteristics of *Trichodesmium spp.* along the Atlantic Barrier Reef at Carrie Bow Cay, Belize," *Marine Ecology*, vol. 16, no. 3, pp. 259–271, 1995.

[87] L. Charpy, J. Blanchot, and L. Lo, "Cyanobacteria *Synechococcus spp.* contribution to primary production in a closed atoll lagoon (Takapoto, Tuamotu, French polynesia)," *Comptes Rendus de l'Academie des Sciences - Serie III*, vol. 314, no. 9, pp. 395–401, 1992.

[88] L. Charpy and J. Blanchot, "Prochlorococcus contribution to phytoplankton biomass and production of Takapoto atoll (Tuamotu archipelago)," *Comptes Rendus de l'Academie des Sciences - Serie III*, vol. 319, no. 2, pp. 131–137, 1996.

[89] L. Charpy and J. Blanchot, "Photosynthetic picoplankton in French Polynesian atoll lagoons: estimation of taxa contribution to biomass and production by flow cytometry," *Marine Ecology Progress Series*, vol. 162, pp. 57–70, 1998.

[90] L. Charpy and J. Blanchot, "Picophytoplankton biomass, community structure and productivity in the Great Astrolabe lagoon, Fiji," *Coral Reefs*, vol. 18, no. 3, pp. 255–262, 1999.

[91] B. E. Casareto, Y. Suzuki, K. Fukami, and Y. Yoshida, "Particulate organic carbon budget and flux in a fringing coral reef at Liyako Island, Okinawa, Japan in July 1996," in *Proceedings of the 9th International Coral Reef Symposium*, vol. 1, pp. 95–100, Bali, Indonesia, 2002.

[92] N. D. Crosbie and M. J. Furnas, "Abundance, distribution and flow-cytometric characterization of picophytoprokaryote

populations in central (17∘S) and southern (20∘S) shelf waters of the Great Barrier Reef," *Journal of Plankton Research*, vol. 23, no. 8, pp. 809–828, 2001.

[93] N. D. Crosbie and M. J. Furnas, "Net growth rates of picocyanobacteria and nano-/microphytoplankton inhabiting shelf waters of the central (17∘ S) and southern (20∘ S) Great Barrier Reef," *Aquatic Microbial Ecology*, vol. 24, no. 3, pp. 209–224, 2001.

[94] L. Charpy, "Importance of photosynthetic picoplankton in coral reef ecosystems," *Vie et Milieu*, vol. 55, no. 3-4, pp. 217–223, 2005.

[95] Y. Thomas, P. Garen, C. Courties, and L. Charpy, "Spatial and temporal variability of the pico- and nanophytoplankton and bacterioplankton in a deep Polynesian atoll lagoon," *Aquatic Microbial Ecology*, vol. 59, no. 1, pp. 89–101, 2010.

[96] C. Ferrier-Pagès and P. Furla, "Pico- and nanoplankton biomass and production in the two largest atoll lagoons of French Polynesia," *Marine Ecology Progress Series*, vol. 211, pp. 63–76, 2001.

[97] L. Charpy, "Phytoplankton biomass and production in two tuamotu atoll lagoons (French polynesia)," *Marine Ecology Progress Series*, vol. 145, no. 1–3, pp. 133–142, 1996.

[98] B. E. Casareto, L. Charpy, J. Blanchot, Y. Suzuki, K. Kurosawa, and Y. Ishikawa, "Phototrophic prokaryotes in Bora Bay, Miyako Island, Okinawa, Japan," in *Proceedings of the 10th International Coral Reef Symposium*, vol. 1, pp. 844–853, Okinawa, Japan, 2006.

[99] T. Ayukai, "Picoplankton dynamics in davies Reeflagoon, the great barrier reef, Australia," *Journal of Plankton Research*, vol. 14, no. 11, pp. 1593–1606, 1992.

[100] K. Tada, K. Sakai, Y. Nakano, A. Takemura, and S. Montani, "Size-fractionated phytoplankton biomass in coral reef waters off Sesoko Island, Okinawa, Japan," *Journal of Plankton Research*, vol. 25, no. 8, pp. 991–997, 2003.

[101] C. Ferrier-Pagès and J. P. Gattuso, "Biomass, production and grazing rates of pico- and nanoplankton in coral reef waters (Miyako Island, Japan)," *Microbial Ecology*, vol. 35, no. 1, pp. 46–57, 1998.

[102] F. Houlbrèque, B. Delesalle, J. Blanchot, Y. Montel, and C. Ferrier-Pagès, "Picoplankton removal by the coral reef community of La Prévoyante, Mayotte Island," *Aquatic Microbial Ecology*, vol. 44, no. 1, pp. 59–70, 2006.

[103] I. C. Biegala and P. Raimbault, "High abundance of diazotrophic picocyanobacteria (<3 μm) in a Southwest Pacific coral lagoon," *Aquatic Microbial Ecology*, vol. 51, no. 1, pp. 45–53, 2008.

[104] J. M. González, J.-P. Torréton, P. Dufour, and L. Charpy, "Temporal and spatial dynamics of the pelagic microbial food web in an atoll lagoon," *Aquatic Microbial Ecology*, vol. 16, no. 1, pp. 53–64, 1998.

[105] A. Sakka, L. Legendre, M. Gosselin, and B. Delesalle, "Structure of the oligotrophic planktonic food web under low grazing of heterotrophic bacteria: Takapoto Atoll, French Polynesia," *Marine Ecology Progress Series*, vol. 197, pp. 1–17, 2000.

[106] A. J. Pile, "Finding Reiswig's missing carbon: quantification of sponge feeding using dual-beam flow cytometry," in *Proceeding 8th International Coral Reef Symposium*, vol. 2, pp. 1403–1410, 1997.

[107] G. Yahel, A. F. Post, K. Fabricius, D. Marie, D. Vaulot, and A. Genin, "Phytoplankton distribution and grazing near coral reefs," *Limnology and Oceanography*, vol. 43, no. 4, pp. 551–563, 1998.

[108] G. Yahel, J. H. Sharp, D. Marie, C. Häse, and A. Genin, "In situ feeding and element removal in the symbiont-bearing sponge *Theonella swinhoei*: Bulk DOC is the major source for carbon," *Limnology and Oceanography*, vol. 48, no. 1, pp. 141–149, 2003.

[109] M. Ribes, R. Coma, and J. M. Gili, "Heterotrophic feeding by gorgonian corals with symbiotic zooxanthella," *Limnology and Oceanography*, vol. 43, no. 6, pp. 1170–1179, 1998.

8

Exploring Relationships between Demersal Resources and Environmental Factors in the Ionian Sea (Central Mediterranean)

G. D'Onghia, A. Giove, P. Maiorano, R. Carlucci, M. Minerva, F. Capezzuto, L. Sion, and A. Tursi

Department of Biology, University of Bari "Aldo Moro", Via E. Orabona 4, 70125 Bari, Italy

Correspondence should be addressed to G. D'Onghia, g.donghia@biologia.uniba.it

Academic Editor: Jakov Dulčić

The relationships between the abundance of demersal resources, environmental variables, and fishing pressure in the northwestern Ionian Sea in the last two decades were evaluated. Data on the density collected during seventeen trawl surveys carried out from 1985 to 2005 were used. The following species were considered: *Aristaeomorpha foliacea, Nephrops norvegicus,* and *Parapenaeus longirostris* for crustaceans; *Merluccius merluccius, Phycis blennoides,* and *Mullus barbatus* for teleost fish. The recruitment index was also considered for *N. norvegicus, P. longirostris, M. merluccius* and *Mullus barbatus*. Six candidate models were evaluated for each density and recruitment data set either combining fishing effort with global (NAO) and regional (SST and precipitation) climatic indices, or models separately involving fishing effort, NAO, or regional climatic indices as the only predictive variable. Model selection was carried out using an information-theoretical approach that applies Akaike's Information Criterion (AIC). High changes over time were observed for the density data and recruitment indices in each species. Apart from hake abundance and recruitment data, for which a clear positive relationship with the NAO index alone was detected, the changes observed in the other species seem to be the consequence of the interaction between bottom-up effects linked to changes in physical environment and top-down ones due to the fishing pressure.

1. Introduction

It is well known that the abundance of the aquatic population fluctuates over both spatial and temporal scales in relation to the variability of abiotic and biotic factors in the ecosystem as well as to human activities, such as fishing [1–3]. Larkin [4] reported that the management of marine fisheries according to ecosystem properties is essentially a question of distinguishing the impacts of fishing (top-down effects) from those of fluctuations in the environment (bottom-up effects), understanding the dynamics of species interactions and appreciating the way in which fishing fleets will respond to changes in the abundance of various stocks. However, Larkin [4] himself reported that, although changes in many marine ecosystem have no doubt occurred, efforts to model their holistic dynamics have not yet been successful due to the number of variables involved and the way in which only small errors in their estimation may lead to large effects in the simulated results.

Since 1985 demersal resources in Italian waters have mostly been monitored by scientists in the context of the GRUND project [5] and successively as part of MEDITS project [6]. In this respect, governments and stakeholders ask scientists to provide their best advice, in terms of indicators and reference points that should trigger management action. Most Mediterranean demersal resources show high fluctuations over time and are considered to be particularly susceptible to overexploitation [7]. Their management is based on "effort control" measures, including licence limitation, time-area closure, and gear restrictions [8].

At the end of the 1980s the Mediterranean Sea underwent a major change that encompassed atmospheric, hydrological, and ecological systems and that led to new regime conditions [9–11]. The most important and evident phenomenon that involved the change in circulation and thermo-haline properties of the water masses is the so-called Eastern Mediterranean Transient (EMT), corresponding to the shift in deep water formation in the eastern basin from its usual source in the southern Adriatic to a new source in the Aegean Sea [12–15]. Even though such a change included the whole Mediterranean, some basins such as the Ionian, Adriatic, and Levantine basin were primarily involved. During the EMT highly saline (>39.0 psu), warmer (around 15°C) and well-oxygenated intermediate waters flowed from the Aegean Sea through the Western Cretan Arc Straits. This flux interrupted the traditional path of Levantine Intermediate Waters (LIW) spreading northwards into the Ionian Sea and eastwards into the Levantine basin and affecting the dynamics of the upper, intermediate, and deep layers. During the EMT an uprising of the nutrients close to the euphotic zone occurred in the northern and eastern Ionian [12, 15]. This large modification in the eastern Mediterranean circulation lasted approximately 10 years, until 1997 when the general Ionian cyclonic circulation was restored [15, 16]. This change in circulation of the north Ionian gyre from cyclonic to anticyclonic and vice versa has been recently suggested as an internal process occurring between Adriatic and Ionian water masses on a decadal scale called Bimodal Adriatic-Ionian Oscillations (BiOS) [17].

The EMT-BiOS-related regime shift recorded in the early 1990s involved sea surface temperature, sea level pressure, surface and deep circulation of water masses, salinity and nutrient changes, affecting the primary production, and pelagic community as recorded by long-term studies on mucilage, red tides, plankton, jellies, and anchovies in the western and eastern Mediterranean basins [18–20]. In particular, changes in the standing stock of the autotrophic biomass and zooplankton abundance with consequent changes in the pelagic community structure were detected in both the Ionian Sea [21–24] and Adriatic and Ligurian seas [18, 19, 25]. The inversion of the surface currents in the Ionian Sea was also thought to influence the occurrence of species in the Adriatic Sea, including the first record of the scleractinian coral *Astroides calycularis* and the teleost fish *Sparisoma cretense* in this basin [26]. Civitarese et al. [18] reported a list of biological records in the Adriatic indicating that the presence of Atlantic and western Mediterranean species was concomitant with the anticyclonic north Ionian gyre and thus with advection of Modified Atlantic Waters (MAW) into the Adriatic, while the occurrence of Lessepsian species coincided with the cyclonic north Ionian gyre that advects eastern Mediterranean waters into the Adriatic.

The first finding of the tropical fish *Sphoeroides pachygaster* [27], *Elates ransonnetii* [28], and *Dysomma brevirostre* [29] in the Ionian Sea was suggested by Maiorano et al. [30] as a consequence of the water masses variations occurring in the Ionian basin during EMT-BiOS. Moreover, these authors reviewed the changes of demersal species abundance and faunal assemblage structure over the last twenty years. The causes of these changes were suggested to be linked to the interaction between the bioecology of the species and environmental factors as well as fishing pressure in the study areas. However, no direct relationships between environmental variables and species abundance changes were examined. With this regard, this work represents an attempt to evaluate the relationships between the abundance of demersal resources and environmental variables and fishing pressure in the north-western Ionian Sea in the last two decades.

2. Material and Methods

2.1. Data Collection. The geographic area considered in this work is the north-western Ionian Sea, along the Italian coasts (Figure 1). The time series data utilized cover the period 1985–2005. On the basis of Conversi et al. [19] who report the significant change of several climatic indicators (SST, SLP, NAO index, and NHT) in the years 1986–1988, the sea surface temperature (SST) of the Ionian Sea was selected as an indicator of environmental changes in this basin. Temperature is one of the primary factors in determining the distribution pattern and abundance changes in aquatic populations [31–36]. SST data were obtained using the AVHRR (advanced very high-resolution radiometer) (http://podaac.jpl.nasa.gov/). Since the variations of the North Atlantic Oscillation (NAO) influence the weather on a large scale over the North Atlantic and Europe and have a strong impact on the oceanic conditions, including the Mediterranean region, the winter NAO index was taken from http://www.cgd.ucar.edu/. The precipitation values (in mm), often correlated to the NAO apart from local weather conditions, were recorded from http://www.ilmeteo.it/. The values of SST, NAO index, and autumn precipitation recorded between 1985 and 2005 are presented in Figure 2.

Fishing effort is an important variable influencing the abundance of living resources [1–3]. However, this is a complex variable that is difficult to quantify because it is influenced by different factors. The main fisheries of the north-western Ionian Sea are Gallipoli, Taranto, and Crotone (Table 1) (http://www.irepa.org/). Different fishing gears are used to exploit the demersal resources; however, trawling is the main technique in terms of catch and gross tonnage (GT). The trawlers are equipped with otter trawl nets. Fishing generally occurs from Monday to Friday during daylight hours only. Trawlers generally work on daily trips. Commercial hauls are carried out at different depths, from coastal waters to 700–800 m. The management of the fisheries is based on effort control and mesh size regulation. Since 1988 a 30–45-day "closed season" for trawling has been implemented in late summer-early autumn, as a management measure imposed by the Italian government [30, 37]. As in most Mediterranean fisheries data on effort are almost absent and there is a low level of enforcement of the legislation [8]. In the present study the number of days-at-sea per year has been used as a measure of fishing effort. This may be a fair measure of fishing effort for demersal gears [38]. Nonworking days were made up of days closed by management, days with wind speeds >15 knots, public holidays, Saturdays, and Sundays. The wind intensity values were obtained from http://www.ilmeteo.it/. The trend of days-at-sea over time is presented in Figure 3.

Data on the demersal resources used in this work were collected during seventeen trawl surveys carried out in the north-western Ionian Sea, from 1985 to 2005, as part of

TABLE 1: Number of vessels and mean gross tonnage (GT) for different gears recorded in the main fisheries (Gallipoli, Taranto, and Crotone) of the north-western Ionian Sea.

Fisheries	Trawling		Longline		Gillnet		Purse seine	
	N. vessels	Mean GT	N. vessels	Mean GT	N. vessels	Mean GT	N. vessels	Mean GT
Gallipoli	75	11.61	16	8.22	313	3.58	—	—
Taranto	53	9.27	2	9.43	118	2.65	6	8.92
Crotone	95	18.55	16	9.31	262	2.71	—	—

a national study project on the assessment of demersal resources [5]. All the trawl surveys were carried out in autumn. A trawl net with a stretched mesh of 40 mm in the codend was used and the experimental hauls were carried out according to a random-stratified sampling design at depths between 10 and 800 m [30, 37]. The Scanmar Sonar System was applied on the trawl net in order to estimate the horizontal net opening during each experimental haul [39]. Abundance data were standardised to the swept surface unit [40]. Thus, density (N/km^2) indices were computed for each species and survey. The following species were considered: *Aristaeomorpha foliacea*, *Parapenaeus longirostris*, and *Nephrops norvegicus* for crustaceans; *Merluccius merluccius*, *Phycis blennoides*, and *Mullus barbatus* for teleost fish. These species were chosen because they are of great commercial interest and comprise the bulk of the commercial trawl catch in the north-western Ionian fisheries. Apart from *M. barbatus* and, to a lesser extent, *M. merluccius* and *P. blennoides*, they are exclusively captured by trawling (Figure 4). They are also among the most studied commercial species in this Mediterranean basin. Since the recruitment of *P. longirostris*, *N. norvegicus*, *M. merluccius*, and *M. barbatus* occurs in the study area during autumn, the recruitment index (RI) for these species was calculated as the abundance (N/km^2) of specimens younger than 1 year (0 + group) [41, 42] collected during each survey. Although the shrimp *Aristeus antennatus* is one of the most abundant deep-water species in the north-western Ionian Sea, it has not been considered in this study due to its very wide depth distribution that often makes both adult stock and recruitment less available to trawling [43, 44].

2.2. Data Processing. With the purpose of trying to gain insight into which combination of the environmental variables considered and fishing effort could determine the observed changes in the abundance data, multiple linear regression analyses were carried out. For each density and recruitment data set (response variables), six candidate models were evaluated. Besides the global model, we selected models that combined fishing effort with global (NAO) or regional (SST and precipitation) climatic indices, and models involving NAO or regional climatic indices, separately, as the only predictive variable fishing effort. In order to reduce the effect of extreme abundance data, these were log-transformed.

Model selection was carried out in the framework of information-theoretic approach [45], which applies Akaike's Information Criterion (AIC) to select the most parsimonious model, that is, the one that achieves the most proper trade-off between bias and variance. Considering the small size of the sample with respect to the number of estimated parameters in each model, the second-order information criterion (AIC_c) was applied and AIC differences (Δ_i) and Akaike weights (w_i) were also based on AIC_c estimates. The best approximating model for each data set was selected by ranking of Akaike weights. For further comparison between the candidate models in terms of variation explained, adjusted R^2 values are also reported. Regression analyses were performed using SPSS version 18 [46].

3. Results

Fluctuating changes over time were observed for the density in crustacean (Figure 5(a)) and fish species (Figure 5(b)) and recruitment indices (Figure 6).

A summary of the six candidate models applied to explain the relationships between the density index and RI of each species and the environmental variables and fishing effort is reported in Tables 2(a) and 2(b), respectively. Table 3 reports the estimated regression coefficients and the relative standard errors in the models selected on the basis of ranking of Δ_i AIC_c.

Ranking of Δ_i for *A. foliacea* density data led to the selection of the model involving fishing effort and NAO as predictive variables, inversely and directly related to density indices, respectively. Comparable values of AIC_c and w_i were also estimated for the model including the fishing effort as the only predictive variable.

For *N. norvegicus* data, the selected model highlights an inverse relationship with fishing effort and a dependence on the regional climatic factors (inversely to SST and directly to precipitation).

The best approximating model for *P. longirostris* is that including the fishing effort together with global (NAO) and regional (SST and precipitation) climatic indices. This model indicated inverse relationships between density and fishing effort and between density and SST and positive relationships between density and NAO and between density and precipitation.

The best approximating model for *M. merluccius* was chosen to be that consisting of a direct regression with the NAO index.

For *P. blennoides* density data, comparable values of AIC_c and w_i were estimated for the model including as the only predictive variable the fishing effort (inverse) or the NAO index (direct). However, in both cases very low R^2 values were obtained.

TABLE 2: (a) Summary of the six candidate models for density data of the species investigated in the Ionian Sea, including AIC_c, Δ_i values for AIC_c, Akaike weights (w_i), based on AIC_c, and adjusted R^2 values. Models are ordered in terms of Δ_i for AIC_c. Predictive variables are: 1: fishing effort, 2: NAO index, 3: SST and 4: precipitation. (b) Summary of the six candidate models for recruitment index of the species investigated in the Ionian Sea, including AIC_c, Δ_i values for AIC_c, Akaike weights (w_i), based on AIC_c, and adjusted R^2 values. Models are ordered in terms of Δ_i for AIC_c. Predictive variables are: 1: fishing effort, 2: NAO index, 3: SST and 4: precipitation.

(a)

Dependent variable	Model	AIC_c	Δ_i AIC_c	w_i	Adjusted R^2
Density index (ln) *Aristaeomropha foliacea*	{1 2}	9.168	0.000	0.408	0.434
	{1}	9.381	0.213	0.367	0.397
	{1 2 3 4}	11.728	2.560	0.113	0.389
	{1 3 4}	12.059	2.891	0.096	0.352
	{3 4}	16.315	7.147	0.011	0.116
	{2}	18.228	9.060	0.004	−0.048
Density index *Nephrops norvegicus*	{1 3 4}	167.605	0.000	0.442	0.498
	{1}	169.025	1.420	0.217	0.402
	{1 2 3 4}	169.329	1.724	0.187	0.465
	{1 2}	170.446	2.841	0.107	0.380
	{3 4}	172.140	4.535	0.046	0.316
	{2}	178.854	11.249	0.002	−0.067
Density index (ln) *Parapenaeus longirostris*	{1 2 3 4}	5.772	0.000	0.555	0.235
	{1 2}	7.776	2.004	0.204	0.263
	{1 3 4}	8.510	2.738	0.141	0.171
	{3 4}	10.470	4.698	0.053	0.136
	{1}	11.131	5.359	0.038	0.162
	{2}	13.832	8.060	0.010	0.017
Density index *Merluccius merluccius*	{2}	203.555	0.000	0.454	0.306
	{1 2 3 4}	204.465	0.910	0.288	0.357
	{1 2}	205.453	1.898	0.176	0.261
	{3 4}	208.339	4.784	0.042	0.125
	{1 3 4}	209.139	5.584	0.028	0.121
	{1}	210.826	7.271	0.012	−0.064
Density index *Phycis blennoides*	{1}	181.760	0.000	0.329	−0.060
	{2}	181.827	0.067	0.318	−0.064
	{3 4}	183.444	1.684	0.142	−0.115
	{1 2}	183.738	1.979	0.122	−0.134
	{1 3 4}	184.990	3.230	0.065	−0.169
	{1 2 3 4}	186.990	5.230	0.024	−0.266
Density index (ln) *Mullus barbatus*	{1}	−1.962	0.000	0.455	0.098
	{1 2}	−0.847	1.115	0.260	0.083
	{2}	0.610	2.571	0.126	−0.049
	{1 3 4}	1.947	3.909	0.064	−0.035
	{3 4}	2.121	4.082	0.059	−0.092
	{1 2 3 4}	3.147	5.109	0.035	−0.070

(b)

Dependent variable	Model	AIC_c	Δ_i AIC_c	w_i	Adjusted R^2
Recruitment index (ln) *Parapenaeus longirostris*	{1}	15.095	0.000	0.447	0.476
	{1 3 4}	16.546	1.451	0.216	0.480
	{1 2}	16.653	1.558	0.205	0.453
	{1 2 3 4}	17.601	2.506	0.128	0.467
	{3 4}	25.136	10.041	0.003	0.099
	{2}	27.166	12.070	0.001	−0.066

(b) Continued.

Dependent variable	Model	AIC_c	$\Delta_i AIC_c$	w_i	Adjusted R^2
Recruitment index *Nephrops norvegicus*	{1 3 4}	151.903	0.000	0.465	0.481
	{1 2 3 4}	153.319	1.416	0.229	0.457
	{1}	154.145	2.242	0.152	0.350
	{3 4}	155.191	3.288	0.090	0.342
	{1 2}	155.976	4.073	0.061	0.311
	{2}	161.916	10.013	0.003	−0.026
Recruitment index *Merluccius merluccius*	{2}	184.842	0.000	0.539	0.293
	{1 2}	186.792	1.950	0.203	0.245
	{1 2 3 4}	187.136	2.294	0.171	0.290
	{3 4}	189.820	4.978	0.045	0.098
	{1 3 4}	191.050	6.208	0.024	0.072
	{1}	191.765	6.923	0.017	−0.062
Recruitment index (ln) *Mullus barbatus*	{2}	8.891	0.000	0.324	−0.070
	{1}	9.017	0.126	0.305	−0.079
	{3 4}	10.371	1.480	0.155	−0.124
	{1 2}	10.674	1.783	0.133	−0.149
	{1 3 4}	12.355	3.464	0.057	−0.235
	{1 2 3 4}	13.961	5.070	0.026	−0.335

For *M. barbatus*, the best approximating model from ranking of Δ_i and w_i was that involving the fishing effort as the only inverse predictive variable, but the value of R^2 estimated was also very low in this case.

With respect to recruitment data, the best approximating model for *P. longirostris* contemplates an inverse relationship with fishing effort.

For *N. norvegicus* recruits, the same dependence on fishing effort and regional climatic indices as the whole population was found.

The best approximating model for *M. merluccius* recruitment data testified a direct relationship with NAO.

For *M. barbatus* data, similar results were found for two models including only one predictive variable each (NAO and fishing effort), but both showed very low R^2 values, as in the case of the whole population.

4. Discussion and Conclusions

At the end of the 1980s the Ionian Sea experienced a strong modification of its water mass properties and thermohaline circulation [9, 12, 14, 15, 18]. The present results highlight that in the period 1985–2005 the SST in the Ionian did not show the increasing trend that characterized the Mediterranean as a whole [47]. On the contrary, at the end of the 1980s, the reversal of the Ionian gyre caused a greater input of colder and less saline Atlantic waters into this basin [18]. Only at the end of EMT did the SST values begin to match the surface waters of the whole Mediterranean [48].

All the demersal species considered in this study exhibited oscillating changes in their population and recruitment abundance.

The regime shift recorded at the end of the 1980s has had direct effects on several marine ecosystem components

encompassing more trophic levels and different Mediterranean basins (see [11, 18], and references therein [19, 24, 25, 49, 50]). The environmental change observed in the Ionian Sea at the end of 1980s could have affected recruitment directly, through its effects on the survival rate of individuals from eggs to juveniles, or indirectly, through bottom-up effects of the changes in zooplanktonic species [51]. The fluctuations observed in the abundance indices of the whole sampled populations most probably reflect the changes in recruitment, which represent the bulk of the stock in many demersal species [8].

In this respect, even though at the end of the 1980s the abrupt changes of SST, salinity, and density as well as nutrients of the Ionian water masses together with large scale climatic drivers (NAO) could have had both direct and indirect effects on the recruitment and abundance of the demersal resources, the role played by the fishing pressure in the area seems to be important. In fact, the analysis carried out through the model selection criteria detected significant relationships between the density of some species and both climatic factors and fishing effort. In all these species an increase in their abundances was significantly related to the decrease in the working days recorded over the study period.

Although population mechanisms act in very-high-dimension natural systems of great complexity which make it difficult to disentangle environmental from fishing effects as well as the nature of their interactions [52–54], there is no doubt that a decrease in the fishing pressure can favour an increase in the abundance of fish and shellfish stocks [2, 3]. However, the effect due to the reduction of fishing varies greatly between the different species. In this respect, the effect observed in *A. foliacea*, *P. longirostris*, and *N. norvegicus* seems to be realistically causal. In fact, these crustacean stocks show overfishing conditions in both the

TABLE 3: Estimated regression coefficients and relative standard errors in the models selected on the basis of AIC_c, Δ_i for AIC_c and Akaike weights (w_i) values. Predictive variables are: 1: fishing effort, 2: NAO index, 3: SST and 4: precipitation.

Dependent variable	Predictive variable	β	Standard error
Density index (ln) A. foliacea	(constant)	13.552	2.293
	1	−0.035	0.010
	2	0.214	0.154
Density index N. norvegicus	(constant)	2127.840	660.323
	1	−2.693	1.091
	3	−58.976	30.678
	4	0.355	0.495
Density index (ln) P. longirostris	(constant)	16.037	5.567
	1	−0.015	0.009
	2	0.183	0.126
	3	−0.253	0.258
	4	0.002	0.004
Density index M. merluccius	(constant)	548.557	93.731
	2	122.402	43.096
Density index P. blennoides	(constant)	369.053	369.330
	1	−0.489	1.572
Density index (ln) M. barbatus	(constant)	10.631	1.662
	1	−0.012	0.007
Recruitment index (ln) P. longirostris	(constant)	16.073	2.745
	1	−0.046	0.012
Recruitment index N. norvegicus	(constant)	1434.513	416.091
	1	−1.497	0.687
	3	−45.165	19.331
	4	0.036	0.312
Recruitment index M. merluccius	(constant)	232.864	54.057
	2	68.711	24.855
Recruitment index (ln) M. barbatus	(constant)	6.916	0.352
	2	−0.058	0.148

Ionian Sea [30, 37, 55–58] and Italian waters [7]. Due to the high vulnerability to trawling of both adults and juveniles in all three species, a reduction in the working days of this type of fishing could reliably favour stock recovery and, thus, an abundance increase. However, Capezzuto et al. [59] revealed that the increase in the abundance of A. foliacea verified mostly in the period 2000–2004 correlated significantly with the increase in temperature and salinity detected from 1995 to 2005 between 200 and 800 m in the Ionian Sea. This fact can be explained according to Ghidalia and Bourgois [60] and Bombace [61] by a hydrological hypothesis: A. foliacea is preferentially distributed in warmer and high-salinity waters. Temperature and salinity between 200 and 800 m were not considered in the present work.

In N. norvegicus an inverse relationship between density and both fishing effort and SST was detected while a direct relationship with precipitation came out. With reference to the first driver, since the catch of Norway lobster in the Ionian Sea mostly consists of juveniles [58], this relationship also seems to be realistic for recruitment. In respect to the SST, although the influence of the temperature on the recruitment

has also been proved for other Mediterranean demersal species [62, 63], probably temperature alone cannot explain the change in the N. norvegicus abundance observed in the Ionian Sea. Maynou and Sardà [64] found significant relationships between the catch rate of this crustacean and some environmental factors, such as atmospheric pressure, cloud cover, and sea-weather conditions. The precipitation effect detected in this work could be related to cloud cover observed by Maynou and Sardà [64]. These authors hypothesized that the light intensity reaching the seabed has a primary influence on the activity rates of N. norvegicus and thus on its vulnerability to trawling. Ungaro and Gramolini [65] found a possible influence of the water temperature at sea bottoms on the spatial distribution of P. longirostris in the southern Adriatic. However, these authors, in agreement with Abelló et al. [66], reported that spatial differences in abundance could also be due to the fishery exploitation. Recently, Ligas et al. [67] reported an increasing trend of P. longirostris abundance correlated to a rise in SST, a corresponding decrease of wind circulation and to the reduction of fishing effort.

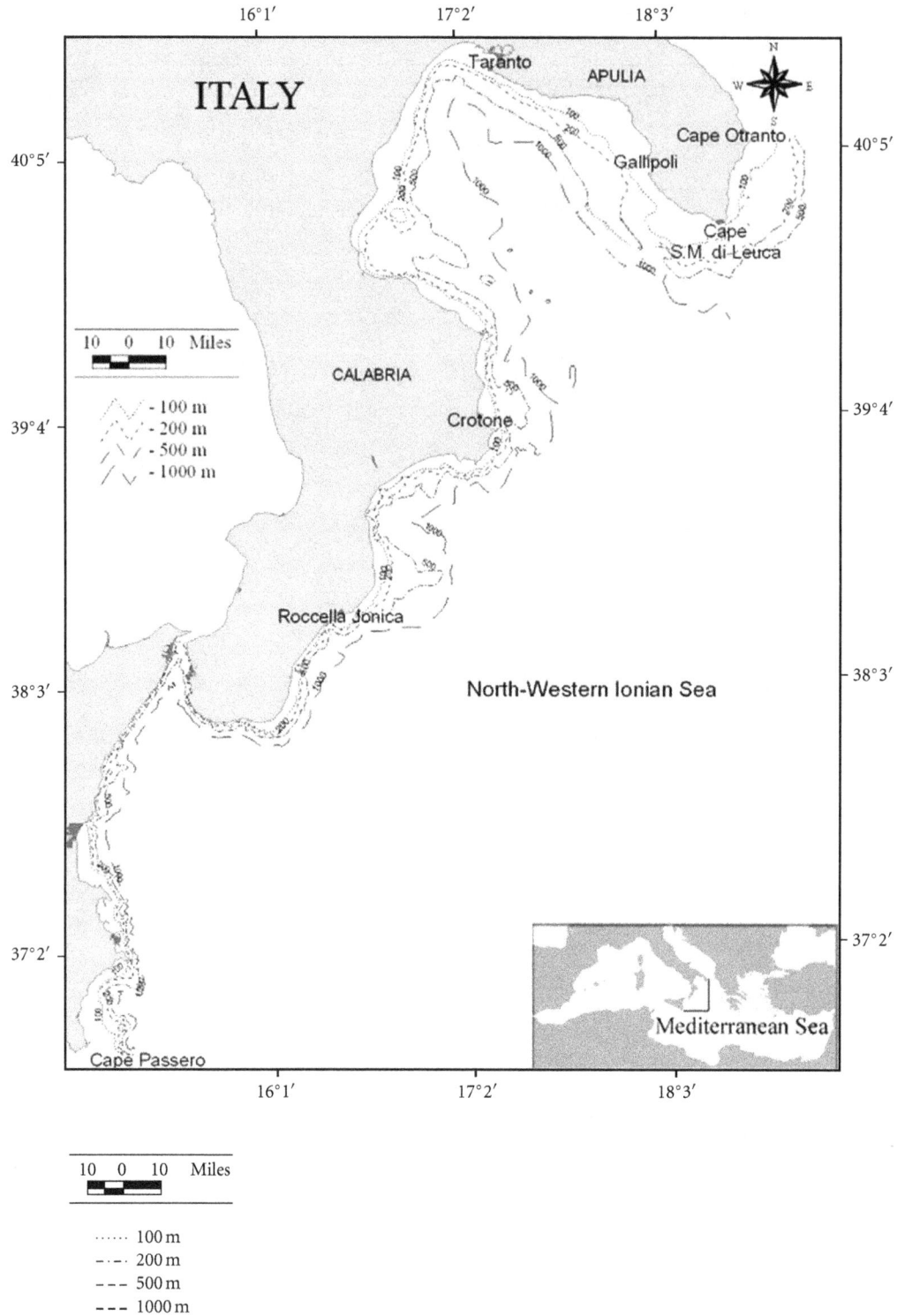

FIGURE 1: Map of the study area.

In *M. merluccius*, both for the whole sampled population and recruits, a positive relationship with the NAO index was detected. The existence of positive relationships between ecological processes in the sea and NAO was observed in many studies, even though the underlying mechanisms often remain unexplained ([18, and references therein], [19, and references therein] [51, 68–70]). The changes in environmental factors (atmospheric pressure, temperature, wind, precipitation, etc.) linked to the NAO can cause deep modifications to the ecosystem influencing, both directly and indirectly, individual organisms and populations as well as interspecies relationships and trophic webs [51, 70, 71].

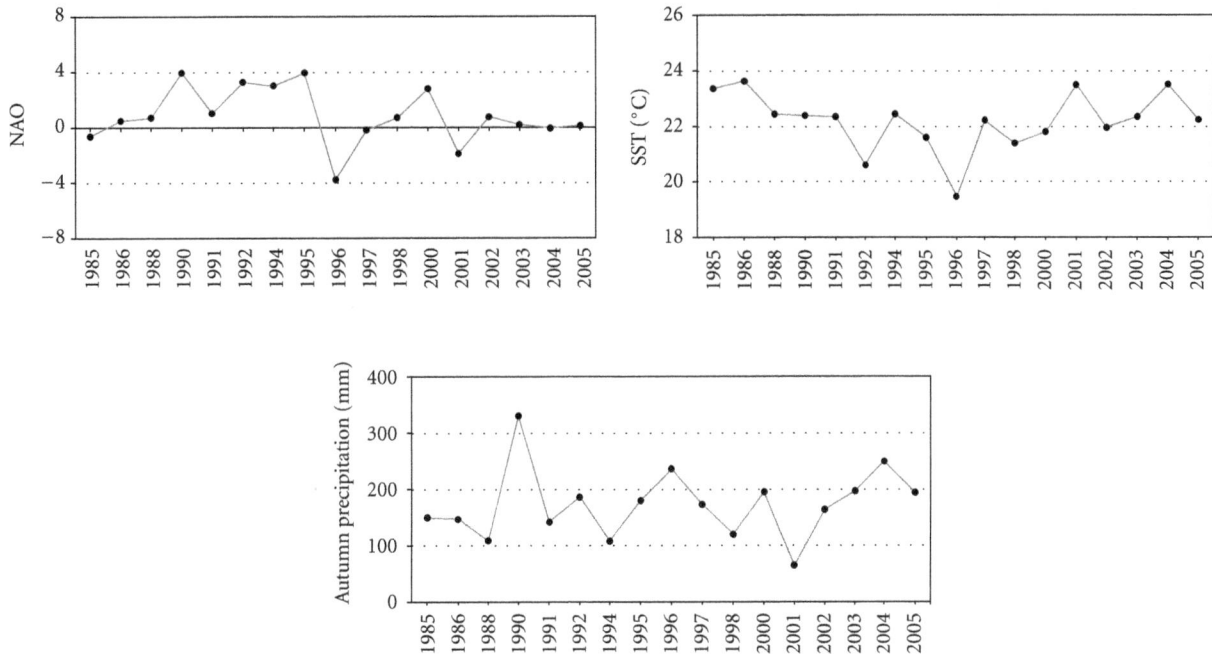

FIGURE 2: NAO index and Sea Surface Temperature (SST) and autumn precipitation recorded in the period 1985–2005 for the north-western Ionian Sea.

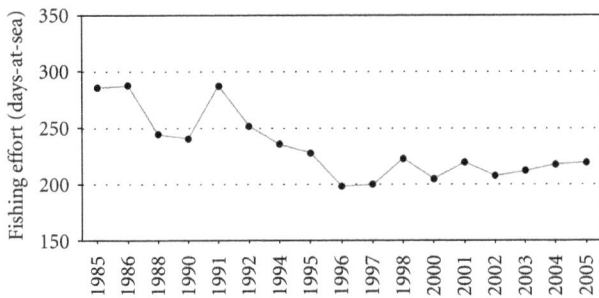

FIGURE 3: Fishing effort expressed in terms of the number of workdays in the north-western Ionian Sea from 1985 to 2005.

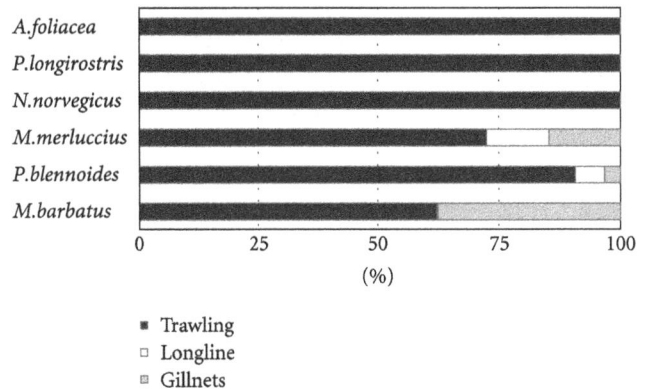

FIGURE 4: Percentage of commercial catches in some demersal species obtained with different gears in the north-western Ionian Sea.

The effects at the first levels of the trophic chain, that is on very short-lived species, are probably easier to identify ([18, and references therein] [25, 68]) while those on fish and other high-trophic level organisms are far from being well understood. In the Mediterranean, Lloret et al. [72] reported that the recruitment of continental shelf species, among which, *E. cirrhosa*, *M. merluccius,* and *M. poutassou*, was positively correlated with the river run-off and/or wind mixing index. Orsi Relini et al. [73] found a positive relationship between the summer biomass of *E. cirrhosa* and winter NAO index. A direct influence of NAO on the water mass circulation and consequent effects on distribution of *M. merluccius* was established by Abella et al. [74] from the Tyrrhenian to Ligurian Sea and on distribution of *M. merluccius* and *Aristeus antennatus* by Massutí et al. [75] off the Balearic Islands.

Maynou [76] found a positive correlation between the average annual NAO index and the annual catches of the shrimp *Aristeus antennatus* in six ports along the Catalonia coast. This author hypothesized that the NAO-induced environmental variability could favour secondary production enhancing the food supply for *A. antennatus*. The greater food availability for this shrimp would strengthen the reproductive potential of a particular year class resulting in increased catches 1 to 3 years later.

A similar hypothesis could also be suggested for hake in the Ionian Sea in agreement with the fact that the changes during 1987-1988 affected the features of the plankton communities. Indeed, an enrichment of the epipelagic system, in terms of chlorophyll *a* concentrations, primary production rates, ciliate abundance and zooplankton biomass, was

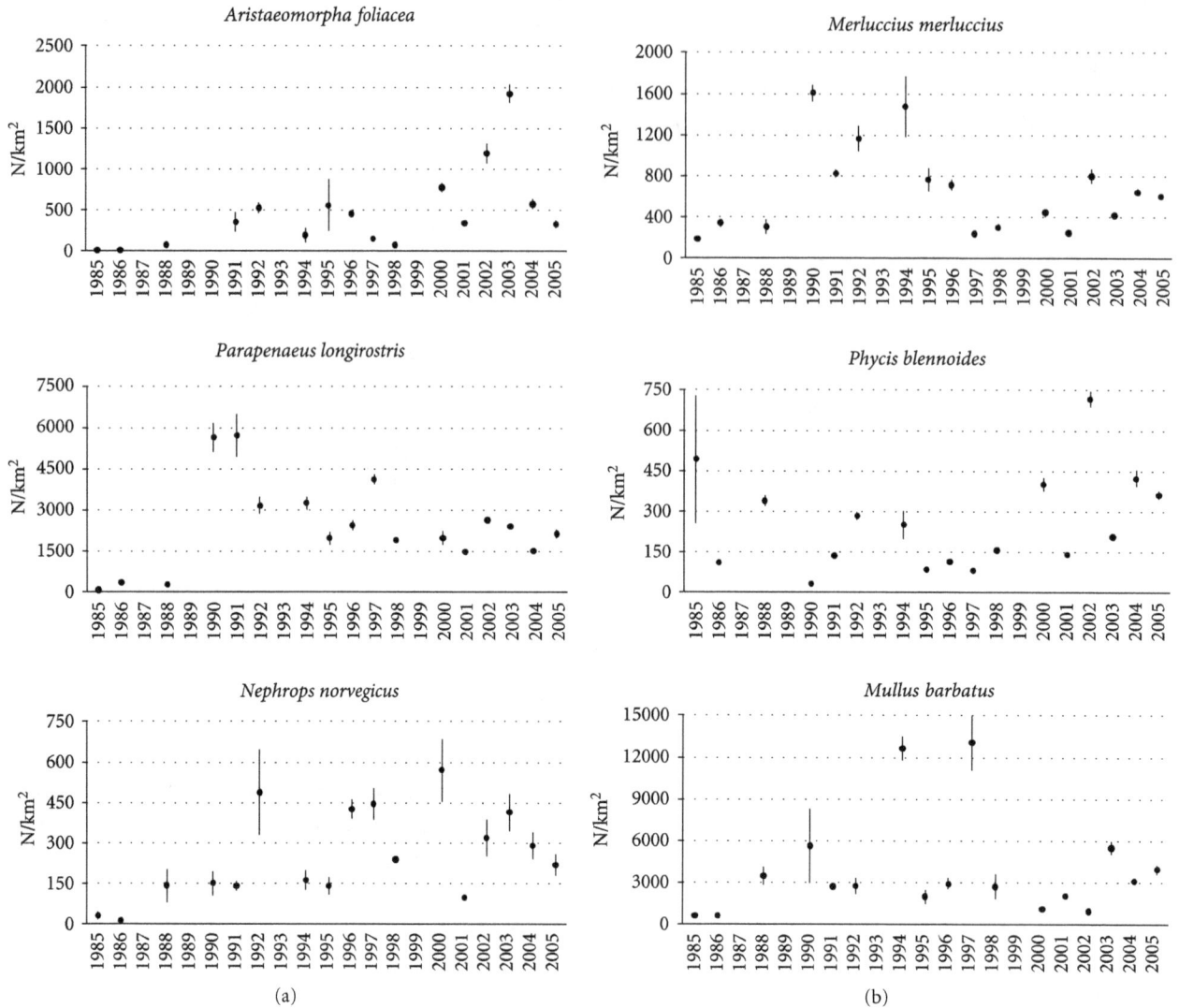

FIGURE 5: (a) Density indices with standard deviation in crustacean species caught from 1985 to 2005 in the north-western Ionian Sea. (b) Density indices with standard deviation in fish species caught from 1985 to 2005 in the north-western Ionian Sea.

detected during the years of cyclonic circulation of the water masses along the north-western Ionian coasts [18, 21–24]. Hakes feed actively in neritic water during their yearly life stages and the higher abundance of their potential prey can favour the recruitment success according to the "match-mismatch hypothesis" [1]. The marked increase in density observed in this fish during 1990 can be considered a consequence of successful recruitment. Nursery areas of this species have been identified along the north-western Ionian Sea [41, 42]. Thus, the period of higher food availability, indirectly related to positive NAO index, would lead to enhancement of the recruitment in $M.$ $merluccius$. The remarkable presence of juveniles in the catches of hake plays more in favour of environmental factors than fishing pressure as main cause of their abundance increase.

In the stock of $M.$ $barbatus$ of the Strait of Sicily it was shown that, for a given spawning stock level, higher recruitment levels correspond to positive SST anomalies [62]. In the stock of $M.$ $barbatus$ in the Ionian Sea the fishing

effort and NAO index seem to play a comparable role on its abundance changes. The same was detected for $P.$ $blennoides$. However, the very low R^2 values estimated in the models for these two fishes would indicate that the set of explanatory variables selected does not completely explain the variance in the abundance data of these species. In fact, mostly for $M.$ $barbatus$, the abundance in the Ionian Sea could also be affected by small vessels mostly working with gillnets the effort of which has not been considered in the present work.

In conclusion, apart from hake abundance and recruitment data, for which a clear positive relationship was detected with the NAO index alone, the changes observed in the other species appear to be an interaction between bottom-up effects linked to changes in physical environment and top-down ones due to fishing. In agreement with Gislason et al. [77], the effects of fishing seem to manifest themselves within a background of natural environmental variations that are the major agent of change in the system. Analyzing the changes in different species, in contrasting

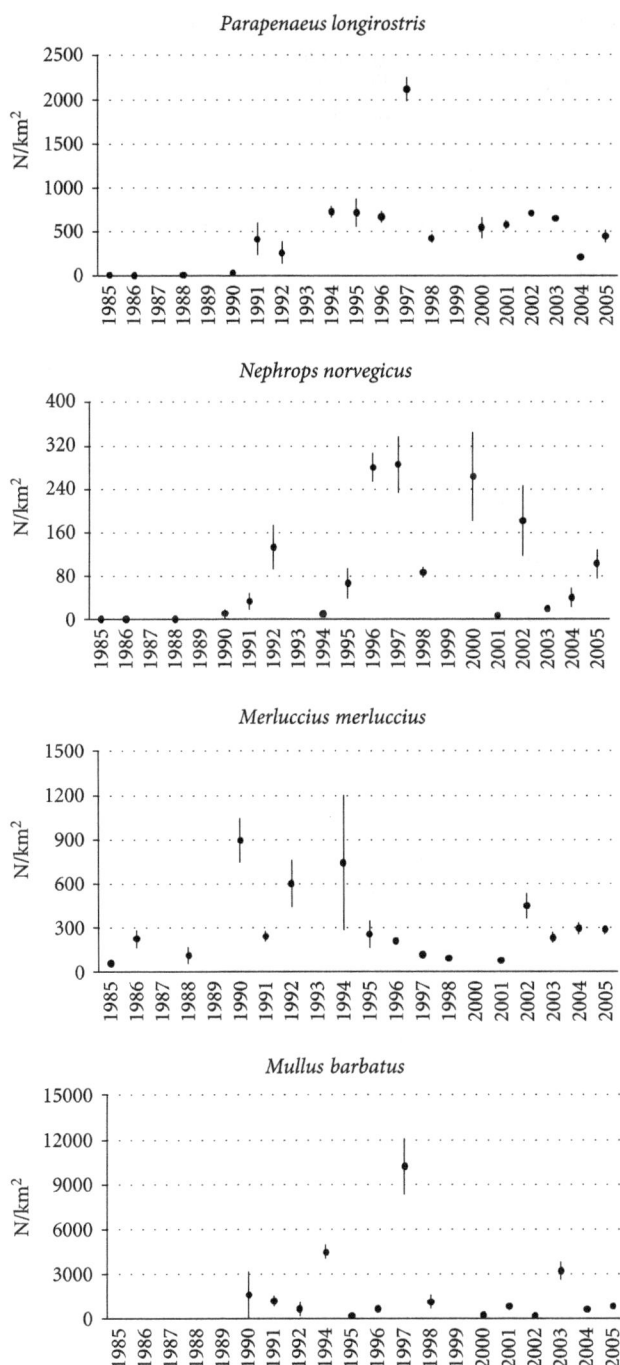

FIGURE 6: Recruitment indices (N/km^2) with standard deviation of demersal species caught from 1985 to 2005 in the north-western Ionian Sea.

environments and subject to different fishing pressures, is expected to shed light on the relative effects of these factors and the way they interact [78].

Acknowledgments

This study benefited from data collected during the trawl surveys financed by the Italian Government (MiPAAF) as part of the programme "Assessment of Demersal Resources." Statistical analyses were supported by a grant to the University of Bari (Local Research Unit CoNISMa) as part of the VECTOR project (vector.conismamibi.it).

References

[1] D. H. Cushing, *Climate and Fisheries*, Academic Press, London, UK, 1982.

[2] R. L. Haedrich and S. M. Barnes, "Changes over time of the size structure in an exploited shelf fish community," *Fisheries Research*, vol. 31, no. 3, pp. 229–239, 1997.

[3] S. J. Hall, *The effects of Fishing on Marine Ecosystems and Communities*, Fish Biology and Aquatic Resources, Series 1, Blackwell Sciences, 1999.

[4] P. A. Larkin, "Concepts and issues in marine ecosystem management," *Reviews in Fish Biology and Fisheries*, vol. 6, no. 2, pp. 139–164, 1996.

[5] G. Relini, "Valutazione delle risorse demersali," *Biologia Marina Mediterranea*, vol. 5, supplement 3, pp. 3–19, 1998.

[6] J. A. Bertrand, L. Gil de Sola, C. Papaconstantinou, G. Relini, and A. Souplet, "An international bottom trawl survey in the Mediterranean: The MEDITS programme," *IFREMER Actes de Colloques*, vol. 26, pp. 76–93, 2002.

[7] G. Relini, J. Bertrand, and A. Zamboni, "Synthesis of the knowledge on bottom fishery resources in Central Mediterranean, Italy and Corsica," *Biologia Marina Mediterranea*, vol. 6, supplement 1, p. 868, 1999.

[8] J. Lleonart and F. Maynou, "Fish stock assessments in the Mediterranean: state of the art," *Scientia Marina*, vol. 67, no. 1, pp. 37–49, 2003.

[9] CIESM, "The Eastern Mediterranean climatic transient, its origin, evolution and impact on the ecosystem," *CIESM Workshop Monographs*, vol. 10, p. 88, 2000.

[10] CIESM, "Tracking long-term hydrological change in the Mediterranean Sea," *CIESM Workshop Monographs*, vol. 16, p. 134, 2002.

[11] CIESM, "Climate warming and related changes in Mediterranean marine biota," *CIESM Workshop Monographs*, vol. 35, p. 152, 2008.

[12] B. Klein, W. Roether, B. B. Manca et al., "The large deep water transient in the Eastern Mediterranean," *Deep-Sea Research I*, vol. 46, no. 3, pp. 371–414, 1999.

[13] A. Lascaratos, W. Roether, K. Nittis, and B. Klein, "Recent changes in deep water formation and spreading in the Eastern Mediterranean Sea: a review," *Progress in Oceanography*, vol. 44, no. 1–3, pp. 5–36, 1999.

[14] P. Malanotte-Rizzoli, B. B. Manca, M. R. D'Alcala et al., "The Eastern Mediterranean in the 80s and in the 90s: the big transition in the intermediate and deep circulations," *Dynamics of Atmospheres and Oceans*, vol. 29, no. 2–4, pp. 365–395, 1999.

[15] B. B. Manca, L. Ursella, and P. Scarazzato, "New development of eastern mediterranean circulation based on hydrological observations and current measurements," *Marine Ecology*, vol. 23, supplement 1, pp. 237–257, 2002.

[16] G. L. E. Borzelli, M. Gačič, V. Cardin, and G. Civitarese, "Eastern mediterranean transient and reversal of the Ionian Sea circulation," *Geophysical Research Letters*, vol. 36, no. 15, Article ID L15108, 2009.

[17] M. Gačič, G. L. Eusebi Borzelli, G. Civitarese, V. Cardin, and S. Yairi, "Can internal processes sustain reversals of the

ocean upper circulation? The Ionian Sea example," *Geophysical Research Letters*, vol. 37, no. 9, pp. 1–5, 2010.

[18] G. Civitarese, M. Gačić, M. Lipizer, and G. L. Eusebi Borzelli, "On the impact of the Bimodal Oscillating System (BiOS) on the biogeochemistry and biology of the Adriatic and Ionian Seas (Eastern Mediterranean)," *Biogeosciences*, vol. 7, no. 12, pp. 3987–3997, 2010.

[19] A. Conversi, S. F. Umani, T. Peluso, J. C. Molinero, A. Santojanni, and M. Edwards, "The mediterranean sea regime shift at the end of the 1980s, and intriguing parallelisms with other european basins," *PLoS One*, vol. 5, no. 5, Article ID e10633, 2010.

[20] I. Vilibić, S. Matijević, and J. Šepić, "Changes in the Adriatic oceanographic properties induced by the Eastern Mediterranean Transient," *Biogeosciences Discussions*, vol. 9, pp. 927–956, 2012.

[21] R. Casotti, A. Landolfi, C. Brunet et al., "Composition and dynamics of the phytoplankton of the Ionian Sea (eastern Mediterranean)," *Journal of Geophysical Research C*, vol. 108, no. 9, 19 pages, 2003.

[22] F. D'Ortenzio, M. Ragni, S. Marullo, and M. Ribera d'Alcalà, "Did biological activity in the Ionian Sea change after the Eastern Mediterranean Transient? Results from the analysis of remote sensing observations," *Journal of Geophysical Research C*, vol. 108, no. 9, 20 pages, 2003.

[23] R. Koppelmann and H. Weikrt, "Deep-sea zooplankton ecology of the Eastern Mediterranean State-of-the-art and perspective," *CIESM Monographs*, vol. 23, pp. 47–53, 2003.

[24] M. G. Mazzocchi, D. Nervegna, G. D'Elia, I. Di Capua, L. Aguzzi, and A. Boldrin, "Spring mesozooplankton communities in the epipelagic Ionian Sea in relation to the Eastern Mediterranean Transient," *Journal of Geophysical Research C*, vol. 108, no. 9, pp. 1–12, 2003.

[25] A. Conversi, T. Peluso, and S. Fonda-Umani, "Gulf of Trieste: a changing ecosystem," *Journal of Geophysical Research C*, vol. 114, no. 7, Article ID C03S90, 2009.

[26] C. N. Bianchi, "Biodiversity issues for the forthcoming tropical Mediterranean Sea," *Hydrobiologia*, vol. 580, no. 1, pp. 7–21, 2007.

[27] A. Tursi, G. D'Onghia, and A. Matarrese, "Firs finding of *Sphoeroides pachygaster* (Müller & Troshel, 1848) in the Ionian Sea (Middle-Eastern Mediterranean)," *Cybium*, vol. 16, pp. 171–172, 1992.

[28] F. Mastrototaro, R. Carlucci, F. Capezzuto, and L. Sion, "First record of dwarf flathead *Elates ransonnetii* (Platycephalidae) in the Mediterranean Sea (North-Western Ionian Sea)," *Cybium*, vol. 31, no. 3, pp. 393–394, 2007.

[29] L. Sion, D. Battista, F. Mastrototaro, and R. Carlucci, "New findings of pignosed arrowtooth eel *Dysomma brevirostre* (Synaphobranchidae) in the Western Ionian Sea (Mediterranean Sea)," *Cybium*, vol. 32, no. 2, pp. 189–190, 2008.

[30] P. Maiorano, L. Sion, R. Carlucci et al., "The demersal faunal assemblage of the north-western Ionian Sea (central Mediterranean): current knowledge and perspectives," *Chemistry and Ecology*, vol. 26, (suppl), pp. 219–240, 2010.

[31] D. H. Gushing and R. R. Dickson, "The biological response in the sea to climate changes," *Advances in Marine Biology*, vol. 14, pp. 1–122, 1977.

[32] C. C. Coutant, "Compilation of temperature preference data," *Journal of the Fisheries Research, Board of Canada*, vol. 34, pp. 739–745, 1977.

[33] B. J. Rothschild, *Dynamics of Marine Fish Populations*, Harvard University Press, Cambridge, Mass, USA, 1986.

[34] A. J. Southward, G. T. Boalch, and L. Maddock, "Fluctuations in the herring and pilchard fisheries of Devon and Cornwall linked to change in climate since the 16th century," *Journal of Marine Biological Association of the the United Kingdom*, vol. 68, no. 3, pp. 423–445, 1988.

[35] C. Ravier and J. M. Fromentin, "Are the long-term fluctuations in Atlantic bluefin tuna (*Thunnus thynnus*) population related to environmental changes?" *Fisheries Oceanography*, vol. 13, no. 3, pp. 145–160, 2004.

[36] T. Brunel and J. Boucher, "Long-term trends in fish recruitment in the north-east Atlantic related to climate change," *Fisheries Oceanography*, vol. 16, no. 4, pp. 336–349, 2007.

[37] A. Tursi, A. Matarrese, G. D'Onghia, P. Maiorano, and M. Panza, "Sintesi delle ricerche sulle risorse demersali del Mar Ionio (da Capo d'Otranto a Capo Passero) realizzate nel periodo 1985–1997," *Biologia Marina Mediterranea*, vol. 5, supplement 3, pp. 120–129, 1998.

[38] S. P. R. Greenstreet, F. B. Spence, A. M. Shanks, and J. A. McMillan, "Fishing effects in northeast Atlantic shelf seas: patterns in fishing effort, diversity and community structure. II. Trends in fishing effort in the North Sea by UK registered vessels landing in Scotland," *Fisheries Research*, vol. 40, no. 2, pp. 107–124, 1999.

[39] L. Fiorentini, G. Cosimi, A. Sala, and V. Palumbo, "Caratteristiche e prestazioni delle attrezzature a strascico impiegate per la valutazione delle risorse demersali in Italia," *Biologia Marina Mediterranea*, vol. 1, supplement 2, pp. 115–134, 1994.

[40] D. Pauly, "Some simple methods for the assessment of tropical fish stocks," *FAO Fisheries Technical Paper, 1983*.

[41] R. Carlucci, G. Lembo, P. Maiorano et al., "Nursery areas of red mullet (*Mullus barbatus*), hake (*Merluccius merluccius*) and deep-water rose shrimp (*Parapenaeus longirostris*) in the Eastern-Central Mediterranean Sea," *Estuarine, Coastal and Shelf Science*, vol. 83, no. 4, pp. 529–538, 2009.

[42] M. Murenu, A. Cau, F. Colloca et al., "Mapping the potential locations of the European hake (*Merluccius merluccius*) nurseries in the Italian waters," in *GIS/Spatial Analyses in Fishery and Aquatic Sciences*, T. Nishida, P. J. Kailola, and C. E. Hollingworth, Eds., vol. 4, FAO, 2010.

[43] F. Sardà, G. D'Onghia, C. Y. Politou, J. B. Company, P. Maiorano, and K. Kapiris, "Deep-sea distribution, biological and ecological aspects of *Aristeus antennatus* (Risso, 1816) in the western and central Mediterranean Sea," *Scientia Marina*, vol. 68, no. 3, pp. 117–127, 2004.

[44] G. D'Onghia, P. Maiorano, F. Capezzuto et al., "Further evidences of deep-sea recruitment of *Aristeus antennatus* (Crustacea: Decapoda) and its role in the population renewal on the exploited bottoms of the Mediterranean," *Fisheries Research*, vol. 95, no. 2-3, pp. 236–245, 2009.

[45] K. P. Burnham and D. R. Anderson, *Model Selection and Multimodel Inference: A Practical Information-Theoretic Approach*, Springer, New York, NY, USA, 2nd edition, 2002.

[46] SPSS Inc., *PASW Statistics Base 18*, SPSS Inc., Chicago, Ill, USA, 2009.

[47] B. BuongiornoNardelli, R. Santoleri, S. Marullo, and M. Guarracino, "La temperatura superficiale del Mar Mediterraneo negli ultimi 21 anni: analisi delle misure satellitari. Clima e cambiamenti climatici," *Le attività del CNR*, pp. 345–348, 2007.

[48] B. B. Manca, V. Ibello, M. Pacciaroni, P. Scarazzato, and A. Giorgetti, "Ventilation of deep waters in the Adriatic and Ionian Seas following changes in thermohaline circulation of the Eastern Mediterranean," *Climate Research*, vol. 31, no. 2-3, pp. 239–256, 2006.

[49] J. C. Molinero, F. Ibanez, P. Nival, E. Buecher, and S. Souissi, "North Atlantic climate and northwestern Mediterranean plankton variability," *Limnology and Oceanography*, vol. 50, no. 4, pp. 1213–1220, 2005.

[50] A. Santojanni, E. Arneri, V. Bernardini, N. Cingolani, M. Di Marco, and A. Russo, "Effects of environmental variables on recruitment of anchovy in the Adriatic Sea," *Climate Research*, vol. 31, no. 2-3, pp. 181–193, 2006.

[51] N. C. Stenseth, G. Ottersen, J. W. Hurrell, and A. Belgrano, *Marine Ecosystem and Climate Variation*, Oxford University Press, 2004.

[52] B. E. Skud, "Dominance in fishes: the relation between environment and abundance," *Science*, vol. 216, no. 4542, pp. 144–149, 1982.

[53] C. M. Duarte and J. N. Cebrian Marba, "Uncertainty of detecting sea change," *Nature*, vol. 356, no. 6366, p. 190, 1992.

[54] S. I. Rogers, D. Maxwell, A. D. Rijnsdorp, U. Damm, and W. Vanhee, "Fishing effects in northeast Atlantic shelf seas: patterns in fishing effort, diversity and community structure. IV. Can comparisons of species diversity be used to assess human impacts on demersal fish faunas?" *Fisheries Research*, vol. 40, no. 2, pp. 135–152, 1999.

[55] G. D'Onghia, P. Maiorano, A. Matarrese, and A. Tursi, "Distribution, biology, and population dynamics of *Aristaeomorpha foliacea* (Risso, 1827) (Decapoda, Natantia, Aristeidae) in the north-western Ionian Sea (Mediterranean Sea)," *Crustaceana*, vol. 71, no. 5, pp. 518–544, 1998.

[56] G. D'Onghia, A. Matarrese, P. Maiorano, and F. Perri, "Valutazione di *Parapenaeus longirostris* (Lucas, 1846) (Crustacea, Decapoda) nel Mar Ionio," *Biologia Marina Mediterranea*, vol. 5, supplement 2, pp. 273–283, 1998.

[57] R. Carlucci, G. D'Onghia, L. Sion, P. Maiorano, and A. Tursi, "Selectivity parameters and size at first maturity in deep-water shrimps, *Aristaeomorpha foliacea* (Risso, 1827) and *Aristeus antennatus* (Risso, 1816), from the North-Western Ionian Sea (Mediterranean Sea)," *Hydrobiologia*, vol. 557, no. 1, pp. 145–154, 2006.

[58] F. Capezzuto, R. Carlucci, P. Maiorano et al., "Distribuzione spazio-temporale del reclutamento di *Nephrops norvegicus* (LINNAEUS, 1758) nel Mar Ionio," *Biologia Marina Mediterranea*, vol. 16, supplement 1, pp. 190–193, 2009.

[59] F. Capezzuto, R. Carlucci, P. Maiorano et al., "The bathyal benthopelagic fauna in the NW Ionian Sea: structure, patterns and interactions," *Chemistry and Ecology*, vol. 26, (suppl), pp. 199–217, 2010.

[60] W. Ghidalia and F. Bourgois, "Influence de la temperature et de l'éclairement sur la distribution des crevettes des moyennes et grandes profondeurs," *Stud. Rev. General Fisheries Council for the Mediterranean, FAO*, vol. 16, pp. 1–53, 1961.

[61] G. Bombace, "Considerazioni sulla distribuzione delle popolazioni di livello batiale con particolare riferimento a quelle bentonectoniche," *Pubblicazioni della Stazione Zoologica di Napoli*, vol. 39, pp. 7–21, 1975.

[62] D. Levi, M. G. Andreoli, A. Bonanno et al., "Embedding sea surface temperature anomalies into the stock recruitment relationship of red mullet (*Mullus barbatus* L. 1758) in the Strait of Sicily," *Scientia Marina*, vol. 67, supplement 1, pp. 259–268, 2003.

[63] V. Bartolino, F. Colloca, P. Sartor, and G. Ardizzone, "Modelling recruitment dynamics of hake, *Merluccius merluccius*, in the central Mediterranean in relation to key environmental variables," *Fisheries Research*, vol. 92, no. 2-3, pp. 277–288, 2008.

[64] F. Maynou and F. Sardà, "Influence of environmental factors on commercial trawl catches of *Nephrops norvegicus* (L.)," *ICES Journal of Marine Science*, vol. 58, no. 6, pp. 1318–1325, 2001.

[65] N. Ungaro and R. Gramolini, "Possible effect of Bottom Temperature on Distribution of *Parapenaeus longirostris* (Lucas, 1846) in the Southern Adriatic (Mediterranean Sea)," *Turkish Journal of Fisheries and Aquatic Sciences*, vol. 6, pp. 109–116, 2006.

[66] P. Abelló, A. Abella, A. Adamidou, S. Jukic-Peladic, P. Maiorano, and M. T. Spedicato, "Geographical patterns in abundance and population structure of *Nephrops norvegicus* and *Parapenaeus longirostris* (Crustacea: Decapoda) along the European Mediterranean coasts," *Scientia Marina*, vol. 66, no. 2, pp. 125–141, 2002.

[67] A. Ligas, S. De Ranieri, D. Micheli et al., "Analysis of the landings and trawl survey time series from the Tyrrhenian Sea (NW Mediterranean)," *Fisheries Research*, vol. 105, no. 1, pp. 46–56, 2010.

[68] J. M. Fromentin and B. Planque, "*Calanus* and environment in the eastern North Atlantic. II. Influence of the North Atlantic Oscillation on *C. finmarchicus* and *C. helgolandicus*," *Marine Ecology Progress Series*, vol. 134, no. 1–3, pp. 111–118, 1996.

[69] G. Ottersen and N. C. Stenseth, "Atlantic climate governs oceanographic and ecological variability in the Barents Sea," *Limnology and Oceanography*, vol. 46, no. 7, pp. 1774–1780, 2001.

[70] N. C. Stenseth, A. Mysterud, G. Ottersen, J. W. Hurrell, K. S. Chan, and M. Lima, "Ecological effects of climate fluctuations," *Science*, vol. 297, no. 5585, pp. 1292–1296, 2002.

[71] K. F. Drinkwater, "The response of Atlantic cod (*Gadus morhua*) to future climate change," *ICES Journal of Marine Science*, vol. 62, no. 7, pp. 1327–1337, 2005.

[72] J. Lloret, J. Lleonart, I. Solé, and J. M. Fromentin, "Fluctuations of landings and environmental conditions in the north-western Mediterranean sea," *Fisheries Oceanography*, vol. 10, no. 1, pp. 33–50, 2001.

[73] L. Orsi Relini, A. Mannini, F. Fiorentino, G. Palandri, and G. Relini, "Biology and fishery of *Eledone cirrhosa* in the Ligurian Sea," *Fisheries Research*, vol. 78, no. 1, pp. 72–88, 2006.

[74] A. Abella, F. Fiorentino, A. Mannini, and L. Orsi Relini, "Exploring relationships between recruitment of European hake (*Merluccius merluccius* L. 1758) and environmental factors in the Ligurian Sea and the Strait of Sicily (Central Mediterranean)," *Journal of Marine Systems*, vol. 71, no. 3-4, pp. 279–293, 2008.

[75] E. Massutí, S. Monserrat, P. Oliver et al., "The influence of oceanographic scenarios on the population dynamics of demersal resources in the western Mediterranean: hypothesis for hake and red shrimp off Balearic Islands," *Journal of Marine Systems*, vol. 71, no. 3-4, pp. 421–438, 2008.

[76] F. Maynou, "Environmental causes of the fluctuations of red shrimp (*Aristeus antennatus*) landings in the Catalan Sea," *Journal of Marine Systems*, vol. 71, no. 3-4, pp. 294–302, 2008.

[77] H. Gislason, M. Sinclair, K. Sainsbury, and R. O'boyle, "Symposium overview: incorporating ecosystem objectives within fisheries management," *ICES Journal of Marine Science*, vol. 57, no. 3, pp. 468–475, 2000.

[78] T. Rouyer, J. M. Fromentin, F. Ménard et al., "Complex interplays among population dynamics, environmental forcing, and exploitation in fisheries," *Proceedings of the National Academy of Sciences of the United States of America*, vol. 105, no. 14, pp. 5420–5425, 2008.

9

Shell Shape Analysis and Spatial Allometry Patterns of Manila Clam (*Ruditapes philippinarum*) in a Mesotidal Coastal Lagoon

Nathalie Caill-Milly,[1] Noëlle Bru,[2] Kélig Mahé,[3] Catherine Borie,[1] and Frank D'Amico[4]

[1] Laboratoire Ressources Halieutiques Aquitaine, IFREMER, FED 4155 MIRA, 1 allée du Parc Montaury, 64600 Anglet, France
[2] Laboratoire de Mathématiques et de leurs Applications de Pau, UMR CNRS 5142, FED 4155 MIRA,
 UNIV PAU & PAYS ADOUR, 64000 Pau, France
[3] Pôle de Sclérochronologie, IFREMER Centre Manche-Mer du Nord, 150 Quai Gambetta, 62200 Boulogne-sur-Mer, France
[4] UMR ECOBIOP, FED 4155 MIRA, UNIV PAU & PAYS ADOUR, Campus Montaury, 64600 Anglet, France

Correspondence should be addressed to Nathalie Caill-Milly, nathalie.caill.milly@ifremer.fr

Academic Editor: Robert A. Patzner

While gradual allometric changes of shells are intrinsically driven by genotype, morphometrical shifts can also be modulated by local environmental conditions. Consequently the common use of a unique dimension (usually length) to assess bivalves' growth may mask phenotypic differences in valve shape among populations. A morphometric exhaustive study was conducted on Manila clam, *Ruditapes philippinarum*, by acquiring data in the French Arcachon Bay (intrasite phenotypic variability) and by comparing with other sites in the literature (intersite phenotypic variability). 2070 shells were subsampled, weighted, and automatically measured using TNPC software. Some ratios' values indicate a relatively round and globular shape shell in comparison with other sites confirming poor conditions for some individuals. Among adult clams, three main morphological groups were identified and discussed according to spatial considerations. Allometric relations for pairs of shell descriptors were determined by testing classical linear and piecewise regression models on log-transformed relation of Huxley. A significant shape change correlated to size was observed; it corresponds to the second year of life of the clam. Relationships between density, disease, and shell shape are demonstrated and discussed related to other potential factors affecting shell shape. Finally, consequences on population regulation are addressed.

1. Introduction

Growth of individuals is commonly assessed by correlating the evolution of the largest dimension of the individuals along time. For bivalves, this dimension can be the valve's length as for cockle, clam, mussel, razor shell, the valve's height as for oyster, scallop [1]. The shape changes of the shell are induced by the differential growth vectors operating at distinct locations around the mantle edge [2], organ that plays a key role in the shell secretion. This highlights the need of taking into account several allometric ratios.

Those gradual allometric changes are clearly driven by genotype and occur during ontogeny; they are usually associated with the conservation concept of physiological favorable surface area to volume ratios [1]. At the same time, those morphometrical shifts can also be modulated by local environmental conditions [3–12].

So, the convenient approach which consists in considering growth through a unique dimension can mask phenotypic plasticity responses in valve shape. While bivalves allocate a significant portion of their total energy budget to shell growth [13], apparent disparities concerning length could occur among individuals even if they dedicate the same amount of energy to shell growth. Shell increments could indeed shift to other dimensions such as height, width, and thickness of the valve. This concern is important because it gives new arguments to suggest that deficient growth in length is not always due to problems of energy input deficit (e.g., phytoplankton) but could also be related to intrinsic considerations leading to specific morphological patterns.

In order to investigate the question of growth in the fullest possible way for ecological and commercial purposes, an exhaustive morphometric study is proposed by acquiring data in Arcachon Bay (intrasite phenotypic variability) and

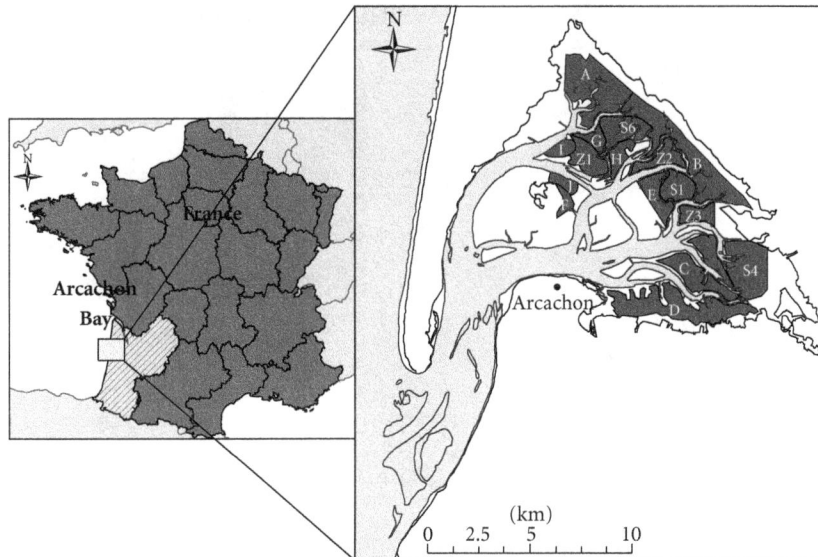

FIGURE 1: Maps showing the studied site Arcachon Bay (France) and localization of the sampled strata (Sources: ESRI, BD Carthage, Ifremer—M. Lissardy).

by comparing with other sites in the literature (intersite phenotypic variability). The chosen biological model is Manila clam, *Ruditapes philippinarum* (Adams and Reeve, 1850) which supports an important commercial fishing activity with around 550 to 1.000 tons per year for the Arcachon Bay located in the southwest coast of France, hoisting it at the first rank of French production sites.

Various ecological factors are identified for their effect on bivalve shell shape: wave impact, trophic conditions, water depth, density... [2, 14]. In this study, we decided to focus on two factors integrating a populational dimension and be susceptible to influence the growth and the shell shape: density and brown muscle disease—emerging pathology. Several studies have already been conducted on density dependence of the Manila clam's shell shape but without consensus [15–18]. One of the supposed effects is that the dorso-ventral shell axis (height) should be the most affected by space competition since buried clams present their anteroposterior axis (length) almost perpendicular to the substrat [18]. Concerning pathology, we considered it challenging to study relationships between the prevalence of BMD affection which is known for having negative impact on the functioning of the adductor muscles involved in the shell's closure [19] and shell shape. Those features were preferred to other potential factors such as perkinsosis or brown ring disease since their low impacts on the Arcachon Bay population were demonstrated [20]. For other environmental modulators, (such as type of sediment, shore level, etc) data at sufficient fine scale are not currently defined.

2. Materials and Methods

2.1. Study Site. Arcachon Bay is a 156 km² semi sheltered lagoon in the southwest coast of France (Figure 1). Mostly composed of tidal flats (110 km² within the inner lagoon);

this mesotidal system is characterized by a sediment composition ranging from mud to muddy sands and colonized by vast *Zostera noltii* seagrass meadows. Both influenced by external neritic waters and by continental inputs [20], this bay presents a semidiurnal macrotidal rhythm. Temperature and salinity gradients within the bay are controlled by the characteristics of these water masses as well as the slow renewal of water by tides [21]. With an average salinity of 30 (source Archyd network) which is higher than freshwater salinity and lower than seawater one's, the investigated area is considered as brackish water [22].

2.2. Origin of the Data. *R. philippinarum* shells were collected in Arcachon Bay during the last biomass survey in 2010. Henceforth biennial, this field survey is carried out with a standardised protocol (stratified random sampling) with 14 strata located at intertidal level (excluding channels) investigated for a total of 490 sampling stations. Sediments, core of 0.25 m² (0.5 m × 0.5 m) on a 0.2 m depth, were sampled with a Hamon grab at the high tide and filtered onboard with running water over a set of three sieves with 2, 1, and 0.5 cm mesh sizes. On the whole collected shells, 2070 shells were randomly subsampled.

2.3. Shells Preparation and Morphology Descriptors of Individuals. In laboratory, all the shells were first cleaned and dried at 38°C for 48 h. Then, the valves were separated and analyses on two high-resolution pictures (lateral and ventral views) were performed with the TNPC software (Digital Processing for Calcified Structures, http://www.TNPC.fr) on left valve. To describe the morphology of the individuals, seven classical parameters were retained:

(i) length (L), defined as the longest distance from front edge to back edge (mm). It is the reference length obtained from lateral view;

TABLE 1: Morphometric variables for arcachon bay.

Descriptor	Formulae	Minimum	Maximum	Mean	Ecart-type
Elongation index	$= H/L$	0.66	0.86	0.75	0.02
Compacity index	$= W/L$	0.19	0.38	0.28	0.02
Convexity index	$= W/H$	0.35	0.51	0.37	0.03
Circle index lateral view	$= AL/0.25 \times L^2 \times \pi$	0.65	0.85	0.73	0.02
Circle index ventral view	$= 2 \times AV/0.25 \times LVent^2 \times \pi$	0.36	0.70	0.48	0.04
Reference Ellipse index lateral view	$= AL/0.25 \times L^2 \times \pi$	2.63	3.43	2.94	0.10
Reference Ellipse index ventral view	$= 2 \times AV/0.25 \times LVent^2 \times \pi$	0.73	1.40	0.95	0.08
Weight ratio 1	$= SM/L$	0.06	1.76	0.57	0.27
Weight ratio 2	$= SM/H$	0.08	2.17	0.76	0.35
Weight ratio 3	$= SM/W$	0.23	5.80	2.02	0.88

(ii) ventral length (*LVent*), as the longest distance from front edge to back edge (mm). It is the reference length obtained from ventral view;

(iii) height (*H*), as distance from the umbo to edge (mm). It is obtained from lateral view;

(iv) width (*W*), as the longest distance of the valve in a lateral plane across the valve (mm). It is obtained from ventral view;

(v) weight or shell mass (*SM*), as dry mass of the left valve (mg);

(vi) area lateral (AL), as area of the left valve projection (mm^2) from lateral view;

(vii) area ventral (AV), as area of the left valve projection (mm^2) from ventral view.

These seven parameters consist in linear or surface measures directly obtained from the shape analysis (lateral or ventral views) and weight measure of each left valve. Respective accuracies of almost 1.10^{-3} millimeter and 0.1 mg are associated to linear and weight measures. Taken separately, those parameters do not allow describing the shell morphology because of a size effect. So not only to characterize the general morphology tendencies at the bay and at the strata scales, but also to examine allometric patterns for this population, 10 shape descriptors were defined from these parameters (Table 1). Such indices are supposed to give synthetic information on growth [23].

(i) Elongation index, Compactness index, and Convexity index to consider the coupled ratios between the 3 dimensions of the valve: a compact and convex bivalve has its width greater than 50% of its height which means a Convexity index greater than 0.5 [24];

(ii) Circle index lateral view (CIL) and Circle index ventral view (CIV) to compare the valves' forms to circular ones: the first one is the area of the valve (lateral view) related to the surface of a circle with *L* as

diameter. The second one is twice the area of the valve (ventral view) related to the surface of a circle with *LVent* as diameter. Those two indices are complementary, respectively, to Elongation and Compactness index. For both of them, if the value is close to 1, it means that the shape tends to be circular.

(iii) Ellipse index lateral view (EIL) and Ellipse index ventral view (EIV) are also considered to compare the valves' forms with elliptical ones. The first one is the area of the valve (lateral view) related to the surface of an ellipse with *L* and 0.5 *L* as diagonals. The second one is twice the area of the valve (ventral view) related to the surface of an ellipse with *LVent* and 0.5 *LVent* as diagonals. Those indices are also complementary to Elongation and Compactness indices since they provide information on the more or less flattened form of the shell.

(iv) Three descriptors involving the weight and successively the 3 dimensions (*L*, *H*, and *W*) with weight ratio 1, weight ratio 2, and weight ratio 3.

For *Ruditapes philippinarum*, globularity is described as a result of a faster increase of width related to length [7]. In terms of indices retained above, this character can be defined for clams presenting high Compactness, CIV, and EIV indices.

According to Hamai [25], Ohba [3], and Eagar et al. [9], morphometric characteristics could reveal more or less favorable environmental conditions. For Manila clam, a low Elongation index is observed to localities presenting clams with high condition index [23] meaning favorable environmental conditions. Consequently, CIL index close to 1 and EIL index greater than 1 could indicate unfavorable environmental conditions.

Moreover, Watanabe and Katayama [23] linked those morphometrics characteristics to commercial considerations for *Ruditapes philippinarum*: the less palatable individuals are the ones with fatter and rounder shell shape that is, high value of Elongation index and CIL index close to 1.

In other words, the best shape for ecological and economical purposes needs to be considered as a combination of several characteristics: at least low H/L ratio combined with low values of CIL and EIL.

2.4. Multivariate Analysis

2.4.1. Shell Shape Analysis. In this subsection, all samples were considered. After brief descriptive statistics, linear regressions were performed on the logarithmically transformed data on parameters and shape descriptors versus size. It allows to account for significant shape changes correlated to size and is based on the classical equations for allometry ($Y = aX^k$) proposed by Huxley [26]. Focusing on the coefficient named k (allometric exponent) provides information about differences of growth rates between the two considered descriptors, in particular when it concerns width versus length it can reflect globularity as defined above. This parameter allows summarizing the growth rate and so it is compared to other areas.

When problems of "bad fitting" (presence of a shift between observed and predicted data) were detected using for example the corresponding residuals of this classical model, an alternative model built on piecewise regressions [27] was tested. It is also called segmented regressions model. It provides regression analysis on both sides of an automatically determined breakpoint. This methodology was used by Katsanevakis et al. [28] for bivalves' species. The choice of the "best" model was undertaken by using the AIC (Akaike Information Criterion; [29]) score; the lowest AIC score gives the one to select. The breakpoint is a parameter which can be viewed as an indicator of fast ecological changes or linked to marked events [28]. The regression equations before and after this breakpoint illustrate the shift of the shell growth to different morphological patterns during ontogeny.

2.4.2. Spatial Allometry Patterns. In this part, clams longer than 30 mm (i.e., 661 clams) were retained to consider spatial variability of allometric descriptors. This choice was decided in order to avoid size effect and to consider whether a link with two of the shape potential drivers (density or the brown muscle disease—coded BMD) could be detectable (since the smallest clam presenting this pathology was 26 mm long). To interpret the allometric variability in terms of density, two classes were defined: below 100 clams m^{-2} ("low density"); above 100 clams m^{-2} ("high density"). Concerning pathology, proportion of clams infected by BMD within the sample was coded into 3 classes: no clams infected ("No BMD"); below 3.7% of infected clams ("Low BMD"); above 3.7% of infected clams ("High BMD"). Normalized principal components analysis (NPCA) was conducted in order to consider the relations between the synthetic shape descriptors and spatial considerations (strata encoded by letters, see Figure 1), density and proportion of clams infected by BMD (added as supplementary factors). This methodology is widely used for such investigations [30, 31].

Calculations were carried out under R (http://cran.r-project.org/web/packages/Rcmdr/index.html) and the following packages: factomineR for NPCA, stats for linear regressions, and SiZer for piecewise regressions (http://cran.r-project.org/web/packages/SiZer/index.html; [32]).

3. Results

3.1. Shell Shape in Arcachon Bay. The range of studied population length was between 10.4 and 45.0 mm (mean: 27.4 mm; standard deviation: 5.21 mm). Other statistics on the morphometric variables are summarized in Table 1. The mean shell pattern (average values) was described by an Elongation index of 0.75, a Compactness index of 0.28, and a Convexity index of 0.37. By comparison with traditional forms (circular or elliptical), valve shape was better described as circular when it was seen from the side view and elliptical when seen from ventral view. For the three weight ratios, the values were ranging between 0.57 and 2.02 and presented higher variability within the considered population than the other morphometric variables (relative standard deviations around 45%).

3.2. Allometric Patterns. Classical linear regressions (Figures 2(a)–2(g); see also Supplementary Materials available online at doi: 10.1155/2012/281206) revealed isometry for pattern such H related to L (called H to L allometric relation), characterized by an allometric exponent (k) equal to 1.00 indicating that an increase of the length did not induce changes of growth rate of the height. For other (W versus L, H versus L, SM versus L, SM versus H, and SM versus W) allometric relations, k exceeded 1.00 which means a positive allometry for these pairs of variables. Higher k values were obtained for relation involving the weight. No significant relations were obtained for the circle or ellipse indices in relation to the length. For H to L allometric relation, classical and piecewise regressions had similar goodness of fit (AIC scores close to -2360). For SM to L allometric relation, AIC score with classical model amounted to 859, whereas it is established to 854 with piecewise regression, this second modeling was considered to be better adjusted to our data for this ratio. The results of these two piecewise regressions (H to L, SM to L) showed breakpoints corresponding, respectively, to $L = 15.7$ mm and $L = 19.6$ mm.

3.3. Spatial Allometry Patterns, Density, and Disease Effects. For clams longer than 30 mm, the shape was correctly summarized by the first three dimensions of the NPCA (93.8% of the variance) described in Table 2, hereafter called Character 1, Character 2, and Character 3.

Heavy shells (regarding classical linear dimensions) with high values of width related to length were found in strata A, C, and S1. Those globular shells were associated with low-density levels (average 48 clams m^{-2}) and high proportions of clams infected by BMD (average 12% of the sample) (see Figures 3 and 4). Besides, those two factors were significantly discriminated on the first axe which was characterized by the variables SM/L, SM/H, W/L ratios, circle and reference ellipse lateral indices.

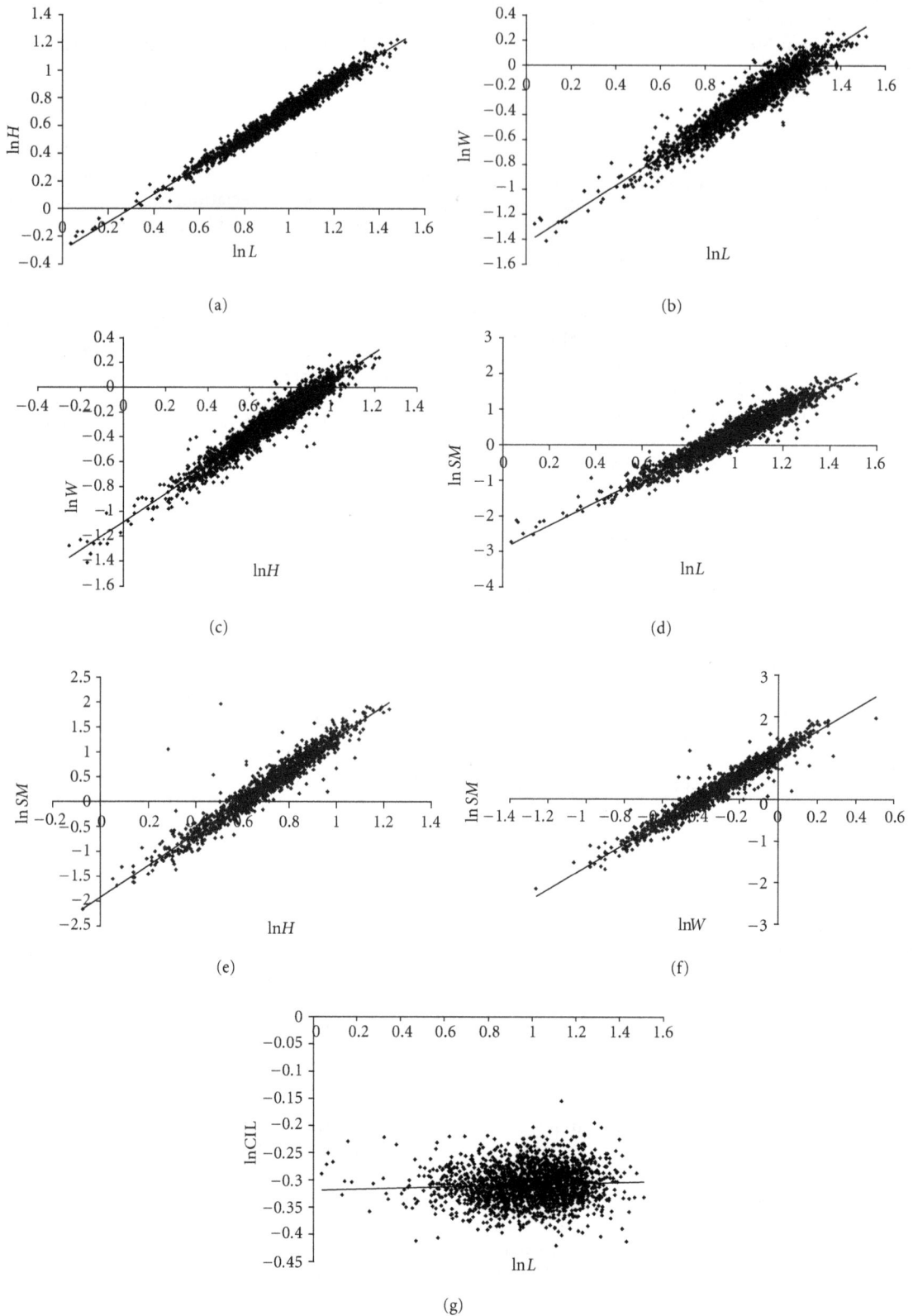

FIGURE 2: Allometric relations for pairs of parameters and shape descriptors (logarithmically transformed data).

TABLE 2: NPCA synthetic descriptors.

Selected principal components	Description	Shape tendency
Character1	Individuals presenting high values for *SM/L*, *SM/H*, *W/L* ratios, high indices related to ventral circle and reference ellipse	Heavy shells regarding linear measures and high values of width related to length-globular individuals
Character2	*L*, *H/L*, EIL, CIL and two of the weight ratios (*SM/H* and *SM/W*): the higher the indices of related mass and length, the less the clam presents a round form (lateral view)	Heavy shells regarding linear dimensions but associated to high values of lengths
Character3	Shells with high *H/L* ratio and lateral indices, by opposition to low *W/L* and *W/H* ratios	Round shells (lateral view) and little width related to length and height

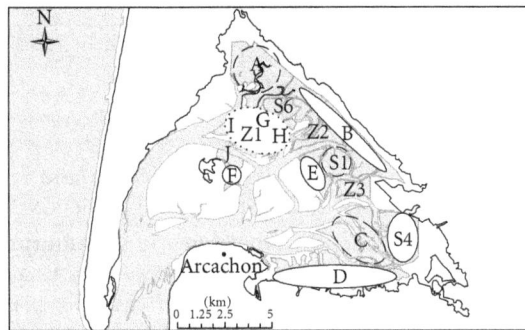

⌐⌐⌐ Relative heavy and globular clams associated with high BMD and low denisty

◯ Relative light clams with D case (particularly round (front) clams)

⋯⋯ Relative heavy and elongated clams

FIGURE 3: Spatial vizualisation of the allometry patterns related to density and disease.

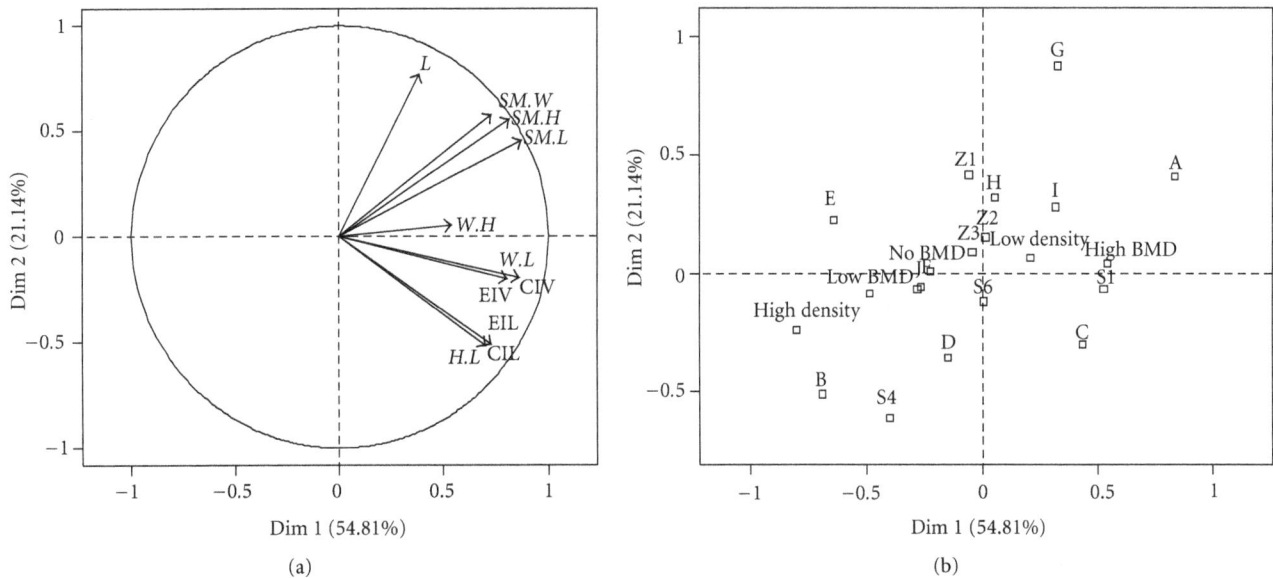

(a)

(b)

FIGURE 4: NPCA on morphometric variables (individuals and variables graphs) with strata, density, and BMD as supplementary factors.

TABLE 3: Comparison of morphometric results with other sites including Europe.

Site/Area	Mean morphometrics results	Length (S. D.) indications of the studied population (mm)	Sources
Japan	$\overline{H/L}$ from 0.70 to 0.75 $\overline{W/L}$ from 0.40 to 0.45	Range 10.8 (0.60) to 11.5 (0.63)	Deduced from Watanabe and Katayama [23]—Figure 3
China	$\overline{H/L}$ = 0.69 $\overline{W/L}$ = 0.44	Mean Length 20.35 (1.35)	Deduced from Fan et al. [33]—Table 1
Spain (bottom culture)	$\overline{H/L}$ from 0.58 to 0.73 $\overline{W/L}$ from 0.41 to 0.48	Range 13.7 (0.26) to 26.6 (0.43)	Deduced from Cigarría and Fernández [18]—Table 1
Tunisia	$\overline{H/L}$ = 0.69 $\overline{W/L}$ = 0.44	Mean Length 35.0 (3.82)	Deduced from Ben Ouada et al. [30]—Table 1
Italy	$\overline{H/L}$ = 0.71 $\overline{W/L}$ = 0.50	Mean Length 39.2 (5.7)	Deduced from Geri et al. [34]—Table 1
France (Barfleur)	$\overline{H/L}$ from 0.68 to 0.71 $\overline{W/L}$ from 0.40 to 0.48	Range 19 to 33	Deduced from Gérard [7]—Figures 32 and 36

Note: after having checked for valves symmetry on a subsample, the W/L ratio (obtained by TNPC system) for Arcachon was multiplied by two in order to allow for realistic comparison with published ratios.

Heavy shells (regarding classical linear dimensions) but associated to high values of lengths were found in strata G, H, Z1, and Z3.

Those two types of strata differ from strata B, E, and S4 for which individuals presented low weight related to length, height, or width.

Stratum D was illustrated by circular shells (lateral view) and little width related to length and height by opposition to strata F, S1, and Z3.

Spatial visualization of those results revealed a typology defined by 3 main areas which differ in particular by their distance to the ocean connection (Figure 3).

4. Discussion

4.1. Shell Shape: Profile and Main Characteristics according to Some Populational Descriptors. This current study investigates the morphometric traits of *Ruditapes philippinarum* within Arcachon Bay using ratios classically employed for calcified structures. Observed values for those ratios (Table 1) establish that this population is characterized by relatively round ($H/L = 0.75$, CIL = 0.73) and globular shells ($W/L = 0.28$, EIV = 0.95). Those results confirm the professional observations on the existence of a peculiar morphological pattern (so-called globular or "*boudeuses*" clams by French fishermen) which is described by a more compact form as usually observed.

Those conventional morphometric variables allow for comparison between the Arcachon Bay population and other ones from remote sites (deduced from [7, 18, 23, 30, 33, 34]). It emphasizes that shells are characterized by a much less elongated and more globular form than in other sites including Europe (Table 3) highlighting so a less favorable environment for the development of the clam and also individuals less attractive for economical point of view. These results are still consistent if we consider only the smallest individuals as Cigarría and Fernández [18], Watanabe and Katayama [23] did (data not shown).

Dependence of shell morphology on local environmental conditions has been indicated by Costa et al. [12] reminding inter alia the phenotypic plasticity in valve shape with a reference to the work of Kwon et al. [11] on transplantation results. They correspond more globally to the concept proposed by Lucas [8] in which ecological conditions seem to strongly influence both bivalves' morphology and physiology. The spatial patterns observed at the bay scale advocate for the existence of specific local environmental conditions that could be related for example to the hypsometric level or to continental input. Unfortunately in this present study, we couldn't take into account those kinds of factors.

Our study focuses on two factors that may affect the observed shape descriptors within the bay: i/density and ii/presence of an emergent disease, the BMD. Relationships among low density (below 100 clams m^{-2}), absence or low presence of BMD (proportion of clams infected by BMD within the sample ranging from 0 to 3.7%) and clams presenting high SM/L, SM/H, and W/L ratios are highlighted using the observed values in Arcachon Bay.

The question of density dependence of the Manila clam's shell shape appears to have been considered in the literature but with no consensus. In our study, concomitance between high densities and propensity to be elongated is depicted. This is comparable to the results obtained by Cigarría and Fernández [18] in Eo Estuary (Spain) for this same species but under lowest densities conditions for Arcachon Bay (maximal density in the present work: 358 clams m^{-2}; 1000 clams m^{-2} for Eo estuary). For the Spanish site, this height's density dependence was interpreted as a consequence of space competition since the height corresponds to the horizontal occupation in the sediment (because of the position of the buried clam). Space competition issue was also reported for *Cerastoderma edule* [35, 36]. Alumno-Bruscia et al. [37] describe as well an elongation of the shell for high population density for *Mytilus edulis*, while it is interesting to note that they raised the question of a consequence of a real physical interference, food depletion, or a combination of both. With respect to Manila clam, different observations have been made. Clam density was identified among other factors to impact the suitability of lagoon's areas (at local scale) for clam cultivation through a model-based approach [38]. For high densities conditions, Ohba [15] observed an increase of length proportionately less important than the ones in height and width. Bourne and Adkins [16] reported a common happening of stunted clams for wild populations and Mitchell [17] stated for Manila clam that density in the Canadian beach determines the maximum size they will grow before stunting occurs. Otherwise, a competition for space was also suggested between *Ruditapes philippinarum* and three species (*Mactra veneriformis, Nihonotrypaea japonica, and Upogebia major*) by Tamaki et al. [39]. For the Arcachon population, low density levels are associated with heavy and globular character. A common unfavorable environmental factor (see below) could be considered to explain these observations. Because the densities remain much lower than other studied sites, we can address in the same time the question of an environmental reason going beyond density strictly seen as spatial interference as suggested by some authors. A possible limitation of the maximum carrying capacity due to all the filter feeding species and not only within the Manila clam's population could be a hypothesis. On intraspecific competition issues to explain growth deficit, it would also have been interesting to consider the biomass. Insofar as we selected here the individuals above 30 mm, we consider that the biomass and density are well correlated.

To the best of our knowledge, correlation between disease and morphology had not been studied. For the Arcachon population, relationships between high proportion of clams affected by BMD and globular form (associated to heavy clams regarding length and height) have been described. Recently highlighted by Dang et al. [19], this pathology affects the posterior adductor muscle and leads to a progressive calcification of this organ. Valve activity and clam mobility appear to be affected, including valves hermetic ability. Because causal relationship between the globular form and BMD has not been demonstrated, we can wonder if the disease could have impacted the globular form or if unfavorable environmental condition leading to specific shape patterns could have favored the affection development in specific site. The first hypothesis is supported by a significant discrimination of the BMD modalities on the first axe which is characterized by the variables *SM/L*, *SM/H*, *W/L* ratios, circle and reference ellipse lateral indices. It implies that the disease would develop a sufficiently long time to modify the shell shape. So far, the only available information is that the smaller infected recorded clam is 26 mm long (de Montaudouin, personal communication). For the second hypothesis, no argument is available up to now. Both of them are in accordance with the description of a decrease of the condition index (for Arcachon Bay) associated to the BMD pathology [40]. It is also consistent with the highlighted correlation between *H/L* and to a less extent *W/L* with nutritional condition indices described by

Watanabe and Katayama [23]. More generally, the observed globularity is in accordance with phenotypic changes under "unfavorable" conditions described by several authors [3, 9, 25].

Apart from density and disease effects, other factors have been proposed or demonstrated to impact the shell morphology. From an evolutionary point of view, defense against predator is considered as the most important function of the shell as reminded by Tokeshi et al. [41]. Considering different species of bivalves including a related species *Ruditapes variegatus*, these authors pointed out that the larger the shell, the more resistant the shell is regarding breakage by predators. For *M. balthica* in the North Sea, the hypothesis of a selective predation of the more globular shells has been proposed by Luttikhuizen et al. [42]. For avian predators, this form may mean a harder prey to swallow but also promotes a higher salt content which is according to Visser et al. [43] research, energetically costly to excrete. Those biological interpretations are applicable for fast predators but for slower ones, the capacity of moving away and burrowing deeper is considered as prevailing and is easier for flatter shells [42].

The main influence of predation on molluscan shell morphology has also been indicated by Watanabe and Katayama [23], but they attributed preferentially significant local differences in elongation and compactness indices to differences of nutritional conditions as explained above.

Other factors such as current velocity, water depth, or nature of the sediment have also been proposed for their influence on Venerids' shape [7, 44–46]. For Gérard [7], the nature of the sediment is of a great influence on the sharpness of the shell. For *Tapes rhomboides*, a related species, globular character was depicted in the Plymouth region (Great Brittany). Originally attributed by Holme [44] to an effect of pressure (related to water depth), this conclusion was challenged by Eagar [45] who focused on other physical conditions (muddy substrates and sheltered localizations). He made the physiological assumption that "obesity" could prevent the shell from sinking within the sediment and could provide stability. On other bivalves species (*Tellina tenuis, Donax vittatus, Macoma balthica,* and *Cerastoderma edule*), Trueman et al. [47] reported an effect of substrate of particle size and shell shape on the penetration of the bivalves' shells.

Up to now, those environmental factors are incompletely known for Arcachon Bay by comparison with other sites and should require further investigations.

4.2. Allometric Patterns: Synthetic Descriptors of Main Interest of Growth. For the first time, our study provides allometric data for the main exploited stock in France (see Table 1 and Figure 2). Allometric coefficient (k) is ranging between 1 (isometry for H to L) and 3 (positive allometry with higher coefficient for ratios taking into account the weight). Those values are consistent with other allometric patterns described for *Ruditapes philippinarum* and related species such as *Ruditapes decussatus, Ruditapes variegatus,* and *Tapes rhomboides* [7, 41, 48, 49]. In particular, considering that age is in turn reflected by the length like Eagar et al. [9] did, our results confirm that the globular character is more pronounced in ageing individuals.

Both linear and piecewise regressions models fit correctly the data for H to L and SM to L relationships, despite the fact that the piecewise model appears to be better in the case of SM to L. Both of them highlight discontinuities in the relative growth curves with marked breakpoints at length 15.7 mm and 19.6 mm. After the growth becomes faster for directions other than length and clams tend to be globular. Katsanevakis et al. [28] estimated that those changes in growth trajectories during ontogeny were worth being taking into account since they can be linked to marked events in the life history or fast ecological changes. For Manila clam, the identified breakpoints correspond to the second year of life of the clam; they match with the supposed size of maturity for this species. According to Holland and Chew [50], sexual maturation begins at 5 mm and spawning at 20 mm for Manila clam.

For *Venerupis senegalensis*, it is interesting to note that Eagar [45] observed a second breakpoint interpreted as a shell-limiting process for H to L and W to L for length class plotted against length of shell. The physiological explanation proposed by this author was a weaker efficacy of muscles when respiratory and food-collecting capacity per unit tissue decrease while the length increases. Those results were obtained with mean ratios per length classes, ignoring the individual variability information.

Spatial variability is shown and three main morphological groups of adult clams are identified. For those clams, similar shell shape appears to be grouped in the proximate strata G, H, and Z1; they could be seen as clams living in optimal conditions compared to the rest of the bay. The shape tendencies also appear to differ by their distance to the ocean connection (Figure 3). A high intertidal localization (involved in particular in the access time to food supply) could intervene but the necessary data were not considered in this work to address this point. This would be consistent with the observations of stunted clams especially in high intertidal areas and at higher clam densities done by Bourne and Adkins [16].

Morphometric investigations led by Ben Ouada et al. [30] on *Ruditapes decussatus* for sites along the Tunisian coast identified the existence of three phenotypes (globular, slender, truncated) and established a high polymorphism not only at between-population but also at within-population scales. This high variability within population was also genetically demonstrated [51, 52]. Nevertheless, Luttikhuizen et al. [42] established that shell shape variation was not randomly distributed over sites for *Macoma balthica*, and Costa et al. [12] indicated that "contribution of local adaptation to the morphological differentiation of population of clam is still poorly studied phenomenon."

4.3. Regulation Considerations. Currently four conservation measures are applied to regulate the fishing activity (number of licenses, minimum legal harvest size, fishing period, and no-take zones). Among those, the minimum legal harvest size is the only measure defined by European legislation; in practice applicable to the whole catches in Europe. Manila clam in Arcachon Bay presents a growth deficiency above 32 mm [20] and the present study reveals a different

morphology for the shell shape compared to other sites with a propensity to be globular. These characteristics are both driven by environmental factors and exploitation; do they reflect a situation of stress for this species or adaptation? Can they challenge the minimum legal harvest size for the benefit of a local one more adapted to this situation?

Conflict of Interests

All authors declare no conflict of interests.

Acknowledgments

The authors are grateful to Xavier de Montaudouin and Cindy Binias for supplying data on BMD, to Muriel Lissardy for providing localization map. They also wish to thank the anonymous reviewer and Dr. Robert A. Patzner for their constructive comments on the paper. University of Pau & Pays de l'Adour (UPPA-ED 211) provided financial support to N. Caill-Milly to attend the 5th EUROLAG/25–30 July 2011 in Aveiro (Portugal).

References

[1] E. Gosling, *Bivalve Molluscs: Biology, Ecology and Culture*, Fishing News Books, Oxford, UK, 2003.

[2] R. Seed, "Shell growth and form in the Bivalvia," in *Skeletal Growth of Aquatic Organisms*, D. C. Rhoads and R. A. Lutz, Eds., pp. 23–67, Plenum Press, New York, NY, USA, 1980.

[3] S. Ohba, "Ecological studies in the natural population of a clam, *Tapes japonica*, with special reference to seasonal variations in the size and structure of the population and to individual growth," *Biological Journal of Okayama University*, vol. 5, pp. 13–42, 1959.

[4] R. Seed, "Factors influencing shell shape in the mussel, *Mytilus edulis* L," *Journal of the Marine Biological Association of the United Kingdom*, vol. 48, pp. 561–584, 1968.

[5] G. D. Rosenberg, *Patterned growth of the bivalve Chione undatella (Sowerby) relative to the environment [Dissertation thesis]*, Departement of Geology, University of California, 1972.

[6] R. A. Brown, R. Seed, and R. J. O. O'Connor, "A comparison of relative growth in *Cerastoderma* (= Cardium) *edule, Modiolus modiolus* and *Mytilus edulis* (Mollusca: Bivalvia)," *Journal of the Zoological Society of London*, vol. 179, pp. 297–315, 1976.

[7] A. Gérard, *Recherches sur la variabilité de diverses populations de Ruditapes decussatus et Ruditapes philippinarum (Veneridae, Bivalvia) [Dissertation thesis]*, Université de Bretagne Occidentale, 1978.

[8] A. Lucas, "Adaptations écophysiologiques des bivalves aux conditions de culture," *Bulletin de la Société d'Ecophysiologie*, vol. 6, pp. 27–35, 1981.

[9] R. M. C. Eagar, N. M. Stone, and P. A. Dickson, "Correlations between shape, weight and thickness of shell in four populations of *Venerupis rhomboides* (Pennant)," *Journal of Molluscan Studies*, vol. 50, pp. 19–38, 1984.

[10] H. P. Stirling and I. Okumus, "Growth, mortality and shell morphology of cultivated mussel (*Mytilus edulis*) stocks cross planted between two Scottish sea lochs," *Marine Biology*, vol. 119, no. 1, pp. 115–124, 1994.

[11] J. Y. Kwon, J. W. Park, Y. H. Lee, Y. K. Hong, and Y. J. Chang, "Morphological variation and genetic relationship among

populations of shortnecked clam *Ruditapes philippinarum* collected from different habitats," *Journal of Fisheries Science and Technology*, vol. 2, pp. 98–104, 1999.

[12] C. Costa, J. Aguzzi, P. Menesatti, F. Antonucci, V. Rimatori, and M. Mattoccia, "Shape analysis of different populations of clams in relation to their geographical structure," *Journal of Zoology*, vol. 276, no. 1, pp. 71–80, 2008.

[13] A. J. S. Hawkins and B. L. Bayne, "Physiological interrelations and the regulation of production," in *The Mussel Mytilus: Ecology, Physiology, Genetics and Culture*, E. M. Gosling, Ed., pp. 171–222, Elsevier Science, Amsterdam, The Netherlands, 1992.

[14] C. M. Yonge and J. I. Campbell, "On the heteromyarian conditions in the Bivalvia with special reference to *Dreissena polymorpha* and certain Mytilacea," *Transactions of the Royal Society of Edinburgh*, vol. 68, pp. 21–43, 1968.

[15] S. Ohba, "Effect of population density on mortality and growth in an experimental culture of bivalve, *Venerupis semidecussata*," *Biological Journal of Okayama University*, vol. 112, pp. 169–173, 1956.

[16] N. Bourne N and B. Adkins, "Savary Island clam study," *Canadian Manuscript Report of Fisheries and Aquatic Sciences*, pp. 69–95, 1848.

[17] D. Mitchell, "Effect of seed density on Manila clam growth and production at a British Columbia clam farm," *Bulletin of Aquaculture Association of Canada*, vol. 92, no. 4, pp. 29–32, 1992.

[18] J. Cigarría and J. Fernández, "Manila clam (*Ruditapes philippinarum*) culture in oyster bags: influence of density on survival, growth and biometric relationships," *Journal of the Marine Biological Association of the United Kingdom*, vol. 78, no. 2, pp. 551–560, 1998.

[19] C. Dang, X. De Montaudouin, P. Gonzalez, N. Mesmer-Dudons, and N. Caill-Milly, "Brown muscle disease (BMD), an emergent pathology affecting Manila clam *Ruditapes philippinarum* in Arcachon Bay (SW France)," *Diseases of Aquatic Organisms*, vol. 80, no. 3, pp. 219–228, 2008.

[20] C. Dang, *Dynamique des populations de palourdes japonaises (Ruditapes philippinarum) dans le bassin d'Arcachon, conséquences sur la gestion des populations exploitées [dissertation thesis]*, University of Bordeaux I, Bordeaux, France, 2009.

[21] M. Plus, D. Maurer, J. Y. Stanisière, and F. Dumas, "Caractérisation des composants hydrodynamiques d'une lagune mésotidale, le bassin d'Arcachon," *Rapport Ifremer*, RST/LER/AR/06. 007, 2006.

[22] D. S. Mc Lusky and M. Elliott, *The Estuarine Ecosystem—Ecology, Threats and Management*, Oxford University Press, New York, NY, USA, 3rd edition, 2004.

[23] S. Watanabe and S. Katayama, "Relationships among shell shape, shell growth rate, and nutritional condition in the manila clam (*Ruditapes philippinarum*) in Japan," *Journal of Shellfish Research*, vol. 29, no. 2, pp. 353–359, 2010.

[24] N. I. Selin, "Shell form, growth and life span of *Astarte arctica* and *A. borealis* (Mollusca: Bivalvia) from the subtidal zone of northeastern Sakhalin," *Russian Journal of Marine Biology*, vol. 33, no. 4, pp. 232–237, 2007.

[25] I. Hamai, "A study of one case in which environmental conditions produce different types in *Meretrix meretrix* (L.)," *Science Reports of Tôhoku University*, vol. 10, pp. 485–498, 1935.

[26] J. S. Huxley, *Problems of Relative Growth*, Methuen, London, UK, 1932.

[27] J. D. Toms and M. L. Lesperance, "Piecewise regression: a tool for identifying ecological thresholds," *Ecology*, vol. 84, no. 8, pp. 2034–2041, 2003.

[28] S. Katsanevakis, M. Thessalou-Legaki, C. Karlou-Riga, E. Lefkaditou, E. Dimitriou, and G. Verriopoulos, "Information-theory approach to allometric growth of marine organisms," *Marine Biology*, vol. 151, no. 3, pp. 949–959, 2007.

[29] Y. Sakamo, M. Ishiguro, and G. Kitagawa, *Akaike Information Criterion Statistics*, D. Reidel, 1986.

[30] H. Ben Ouada, M. N. Medhioub, A. Medhioub, H. Jammoussi, and M. Beji, "Variabilité morphométrique de la palourde *Ruditapes decussatus* (Linné, 1758) le long des côtes tunisiennes," *Haliotis*, vol. 27, pp. 43–55, 1998.

[31] D. Wildish, H. Akagi, B. Hatt et al., "Population analysis of horse mussels of the inner Bay of Fundy based on estimated age, valve allometry and biomass," *Canadian Technical Report of Fisheries and Aquatic Sciences*, Vol. 2257, 1998.

[32] D. L. Sonderegger, H. Wang, W. H. Clements, and B. R. Noon, "Research communications research communications Using SiZer to detect thresholds in ecological data," *Frontiers in Ecology and the Environment*, vol. 7, no. 4, pp. 190–195, 2009.

[33] D. Fan, A. Zhang, Z. Yang, and X. Sun, "Observations on shell growth and morphology of the bivalve *Ruditapes philippinarum*," *Chinese Journal of Oceanology and Limnology*, vol. 25, no. 3, pp. 322–329, 2007.

[34] G. Geri, G. Parisi, P. Lupi et al., "Caratteristiche corporee in due specie di vongole [*Tapes decussatus* L. e. *Tapes semidecussatus* (Reeve) di taglia commercial," *Zootecnica e Nutrizione Animale*, vol. 22, pp. 103–118, 1996 (Italian).

[35] K. T. Jensen, "The presence of the bivalve *Cerastoderma edule* affects migration, survival and reproduction of the amphipod *Corophium volutator*," *Marine Ecology Progress Series*, vol. 25, pp. 269–277, 1985.

[36] X. De Montaudouin and G. Bachelet, "Experimental evidence of complex interactions between biotic and abiotic factors in the dynamics of an intertidal population of the bivalve *Cerastoderma edule*," *Oceanologica Acta*, vol. 19, no. 3-4, pp. 449–463, 1996.

[37] M. Alunno-Bruscia, E. Bourget, and M. Fréchette, "Shell allometry and length-mass-density relationship for *Mytilus edulis* in an experimental food-regulated situation," *Marine Ecology Progress Series*, vol. 219, pp. 177–188, 2001.

[38] C. M. Spillman, D. P. Hamilton, and J. Imberger, "Management strategies to optimise sustainable clam (*Tapes philippinarum*) harvests in Barbamarco Lagoon, Italy," *Estuarine, Coastal and Shelf Science*, vol. 81, no. 2, pp. 267–278, 2009.

[39] A. Tamaki, A. Nakaoka, H. Maekawa, and F. Yamada, "Spatial partitioning between species of the phytoplankton-feeding guild on an estuarine intertidal sand flat and its implication on habitat carrying capacity," *Estuarine, Coastal and Shelf Science*, vol. 78, no. 4, pp. 727–738, 2008.

[40] C. Dang and X. DeMontaudouin, "Brown muscle disease and Manila clam *Ruditapes philippinarum* dynamics in Arcachon Bay, France," *Journal of Shellfish Research*, vol. 28, no. 2, pp. 355–362, 2009.

[41] M. Tokeshi, N. Ota, and T. Kawai, "A comparative study of morphometry in shell-bearing molluscs," *Journal of Zoology*, vol. 251, no. 1, pp. 31–38, 2000.

[42] P. C. Luttikhuizen, J. Drent, W. Van Delden, and T. Piersma, "Spatially structured genetic variation in a broadcast spawning bivalve: quantitative vs. molecular traits," *Journal of Evolutionary Biology*, vol. 16, no. 2, pp. 260–272, 2003.

[43] G. H. Visser, A. Dekinga, B. Achterkamp, and T. Piersma, "Ingested water equilibrates isotopically with the body water pool of a shorebird with unrivaled water fluxes," *American Journal of Physiology*, vol. 279, no. 5, pp. R1795–R1804, 2000.

[44] N. A. Holme A, "Shell form in *Venerupis rhomboides* (Pennant)," *Journal of the Marine Biological Association of the United Kingdom*, vol. 41, pp. 705–722, 1961.

[45] R. M. C. Eagar, "Shape and function of the shell: a comparison of some living and fossil bivalve mollusks," *Biological Reviews of the Cambridge Philosophical Society, Part II*, vol. 53, pp. 169–210, 1978.

[46] J. Kakino, "Relationship between growth of Japanese little neck clam *Ruditapes philippinarum* and current velocity on Banzu tidal flat, Tokyo Bay," *Bulletin of Chiba Prefectural Fisheries Experimental Station*, 1996 (Japanese).

[47] E. R. Trueman, A. R. Brand, and P. Davis, "The effect of substrate and shell shape on the burrowing of some common bivalves," *Proceedings of the Malacologica Society of London*, vol. 37, pp. 97–109, 1966.

[48] F. Laruelle, J. Guillou, and Y. M. Paulet, "Reproductive pattern of the clams, *Ruditapes decussatus* and *R. philippinarum* on intertidal flats in Brittany," *Journal of the Marine Biological Association of the United Kingdom*, vol. 74, no. 2, pp. 351–366, 1994.

[49] M. B. Gaspar, M. N. Santos, P. Vasconcelos, and C. C. Monteiro, "Shell morphometric relationships of the most common bivalve species (Mollusca: Bivalvia) of the Algarve coast (southern Portugal)," *Hydrobiologia*, vol. 477, pp. 73–80, 2002.

[50] D. A. Holland and K. K. Chew, "Reproductive cycle of the Manila clam (*Venerupis japonica*) from Hood Canal, Washington," *Proceedings of the National Shell-Fisheries Association*, vol. 64, pp. 53–58, 1974.

[51] D. Moraga, *Polymorphisme enzymatique de populations naturelles et expérimentales de la palourde européenne Ruditapes decussatus (L.) (Veneridae, Bivalvia) [dissertation thesis]*, Université de Bretagne Occidentale, 1984.

[52] P. Jarne, P. Berrebi, and O. Guélorget, "Variabilité génétique et morphométrique de cinq populations de la palourde *Ruditapes decussatus*," *Oceanologia Acta*, vol. 11, pp. 401–407, 1988.

10

Detection of Bioactive Compounds in the Mucus Nets of *Dendropoma maxima*, Sowerby 1825 (Prosobranch Gastropod Vermetidae, Mollusca)

Anne Klöppel,[1] Franz Brümmer,[1] Denise Schwabe,[2] and Gertrud Morlock[3]

[1] *Department of Zoology, Biological Institute, University of Stuttgart, Pfaffenwaldring 57, 70569 Stuttgart, Germany*
[2] *Animal Evolutionary Ecology, Zoological Institute, University of Tübingen, Auf der Morgenstelle 28, 72076 Tübingen, Germany*
[3] *Institute of Food Chemistry, University of Hohenheim, Garbenstraße 28, 70599 Stuttgart, Germany*

Correspondence should be addressed to Anne Klöppel; anne.kloeppel@bio.uni-stuttgart.de

Academic Editor: Horst Felbeck

The sessile suspension-feeding wormsnail *Dendropoma maxima*, Sowerby 1825 (Vermetidae) secretes a mucus net to capture planktonic prey. The nets are spread out over the corals and often have remarkable deleterious effects on them like changes in growth form and pigmentation shifts not uncommonly resulting in tissue necrosis. Until now, there is no explanation for this phenomenon although the indication as well as theories about its genesis is mentioned in several publications. Vermetids are well studied concerning the intraspecific competition with neighboring individuals but not in their interaction with other taxa like corals or fish. We did extensive *in situ* video recording and observed that fish avoided the plankton-load nets although several specialized taxa are known to be molluscivores, mucivores, and/or feed on plankton. As many molluscs use chemical weapons to combat feeding pressure and to defend themselves against predators, we screened empty and plankton-load mucus nets for potential bioactive metabolites. Bioactivity testing was performed with a recently developed system based on a chromatographic separation (high-performance thin-layer chromatography (HPTLC)) and a bioassay with luminescent bacteria *Vibrio fischeri*. Thus, we found at least two active compounds exclusively accumulated by the wormsnails themselves. This is the first record of bioactive properties in the whole family of Vermetidae.

1. Introduction

The vermetid gastropod *Dendropoma maxima*, Sowerby 1825 (Vermetidae, Littorinimorpha, Gastropoda) is a dominant, very abundant encrusting species of outer tropical reefs and widespread throughout the IndoPacific [1–3]. In the Red Sea, *D. maxima* populations can attain substantial densities, a phenomenon largely restricted to that area [4, 5]. Wormsnails mainly live in the infralittoral and upper-circalittoral between the breaker zone and the outer reef edges facing the current [1, 2, 6]. To combat the competition for space and nutrients, wormsnails developed particularly strong substrate preferences. In the Red Sea, substrates are living corals like hydrozoan *Millepora* spp. (Figure 1(a)) and *Porites* spp., dead coral rock next to the reef edge, or they can be found on the nonliving reef flat substratum [2, 7].

Vermetids are a rather peculiar and poorly studied family. Although they are common inhabitants of coral reefs and rocky shores, little is known about the basic biology of most species [5]. Wormsnails have an irregularly uncoiled calcareous shell, cemented to or embedded in the substrate [2, 3, 6, 8]. Growth rate of the shell is rather fast and enables the vermetid to escape overgrowth by corals and to maintain access to food [9]. Often, only the proteinaceous operculum is seen forming a plug to the accessible shell aperture whose diameter is a good index of body size [5, 10].

Aspects of the feeding ecology of vermetids have been examined in a number of studies (e.g., [3, 6, 8, 9, 11]). Wormsnails are active and passive suspension feeders [9]. Next to ctenidial cilia filter feeding, what is marginally in *D. maxima*, vermetids excrete a sticky mucus net (containing mucopolysaccharides and mucins (glycosylated proteins))

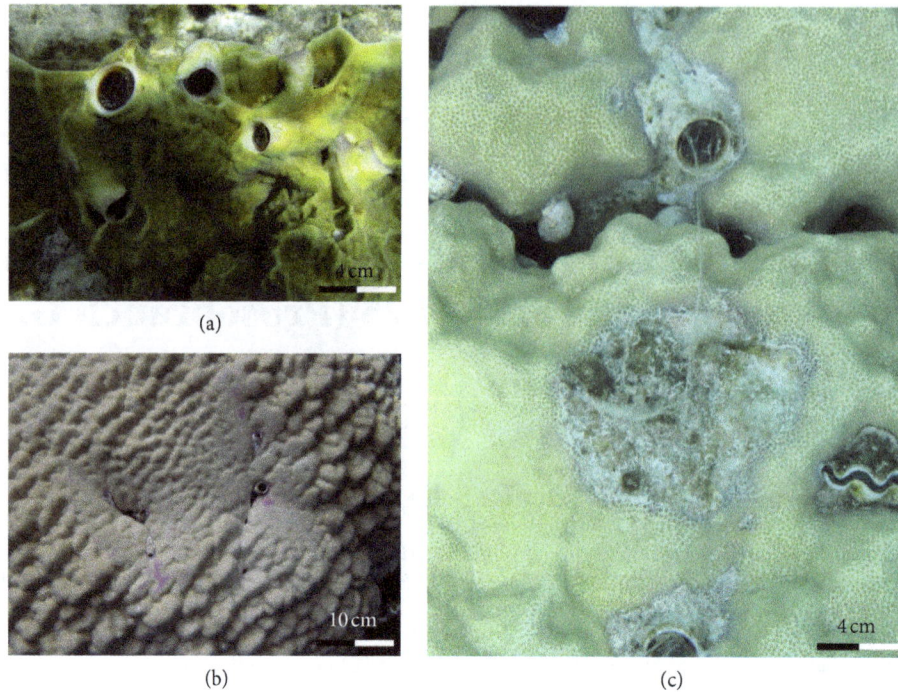

FIGURE 1: Deleterious effects of the vermetid gastropod *Dendropoma maxima* on associated corals. Several individuals growing inside the hydrozoan fire coral *Millepora platyphylla* that minimizes the contact area between coral tissue and nets because of its erected and branching growth form (a). Morphology and pigmentation shift of the massive coral *Porites* sp. where nets get in touch with the coral tissue. The naturally uneven surface exhibits a more even, flatted one (b). Dead coral tissue of a mucus net affected area (here: dimension approximately 20.1 cm^2) of *Porites* sp. overgrown by algae (c). The highly particle-enriched veils are negatively buoyant and descend on the coral tissue underneath. Almost the whole contact surface between net and coral tissue is degenerated.

well adapted to capture near-bed small planktonic organisms consisting mainly of copepods, rhizopods, meiobenthos, and detritus [3, 6, 8, 11, 12].

Analyzing the interspecific interaction between wormsnails and their living coral substratum disclosed that *D. maxima* depresses coral growth in the region of infestation (Colgan in [9]). Zvuloni et al. [1] reported that the presence of *D. maxima* in the Gulf of Aqaba (Red Sea) was associated with morphological anomalies and reduced size of corals they live inside. This suggests either that wormsnails harm corals or that they recruit preferentially to degraded coral habitats [13].

The extruded webs seem to be more detrimental on surrounding organisms than the vermetids' body. They often contact neighboring substrates including corals [2, 14]. Field experiments clearly demonstrated deleterious effects of vermetid nets on corals [13]. Since the underlying mechanisms are still unknown [13], theories speak of abrasive effects on the coral polyps, the competition for substrate and planktonic food, or the presence of bioactive secondary metabolites [1, 15]. Presumably, there is an additional effect of surrounding invertebrate colonies and associated fish. For example, the guard crab *Trapezia serenei* ameliorates the strong negative effect of *D. maxima* on pocilloporid coral growth rate by dislodging and consuming the vermetid mucus (as they do with sediments) [16].

Fishes are notable among vertebrates for having comparatively broad and flexible diets [17, 18]. In the Red Sea, many mucoid (e.g., *Labroides* spp., *Chaetodon* spp.) and molluscivore (e.g., *Anampses* spp., *Bodianus* spp., *Cheilinus* spp.) reef fish are known. Until now there is no publication about any predator—besides the guard crab *Trapezia serenei* as suggested in [16]—that feeds on the wormsnail's mucus even nets are additionally loaded with plankton, mainly copepods, what should be of great interest for plankton feeding reef inhabitants. Already in 1973, it was assumed that vermetid mucus must be distasteful to other animals to function as an effective feeding inhibitor [19].

Molluscs are a dominant group within aquatic invertebrates producing bioactive substances. Among them, opisthobranchs, especially nudibranchs, are mostly preyed on [20]. As molluscs often lack sufficient morphological protective mechanisms, they have adopted an impressive array of defensive strategies [21, 22]. Passive and active chemical defensive (mucus) secretion is known in numerous marine molluscs [23, 24]. Besides mechanical protection, gastropod mucus may contain specific products to render the animal poisonous, distasteful or irritating, or some combination of these known from many mucus producing invertebrates [19, 23, 25]. Experiments with carnivorous fishes as potential predators have indicated the possession of chemical defense mechanisms in opisthobranch gastropods, including

Detection of Bioactive Compounds in the Mucus Nets of Dendropoma maxima, Sowerby 1825
(Prosobranch Gastropod Vermetidae, Mollusca)

105

the secretion of strongly acidic or noxious substances with ichthyotoxic and antibacterial properties [26–28]. Many of these deterrents are diet-derived or built up *in vivo* from dietary precursors (e.g., sponges for dorid nudibranchs) and not synthesized *de novo* [22, 29–31].

Until now, nothing is known either about predators feeding on the plankton-load nets or any interaction between ambient marine organisms (besides corals) and the mucus veils of Vermetidae. No studies have been done analyzing the toxic potential of mucus net ingredients on corals associated with vermetid gastropods, on surrounding reef inhabitants or their potential application in human medicine. Marine invertebrates are one of the most efficient sources for bioactive secondary metabolites with pharmaceutical properties [24, 32–35]. Some have entered preclinical and clinical trials or even made it to the commercial sector [34, 36–38].

Therefore, we made video observations to document the interaction reef fish vermetids concerning the capability as food source and screened the nets for bioactive substances. A recently optimized system (high-performance thin-layer chromatography (HPTLC)/bioassay) for bioactivity screening in invertebrates was used [39].

2. Material and Methods

2.1. Study Area. Field work was carried out in the Gulf of Aqaba (Red Sea) along the Egyptian eastern and western coastline near Dahab (Suleiman Reef), 100 km north of Sharm El Sheikh and at Mangrove Bay (Sharm Fugani), 30 km south of El Quseir (Egypt) during the period from April 15 to December 15, 2010. *Dendropoma maxima* individuals are very common on the reef crest using corals (*Millepora dichotoma*, *Millepora platyphylla* (Figure 1(a)), *Porites lutea*, and *Porites nodifera*) as their main substrate, and on the rear reef flat, where organisms are often attached to blocks of dead or live coral rubble. The latter feature makes *D. maxima* very suited for aquaria experiments. Their habitat is always characterized by intense wave action, strong current, and high abundance of reef fish.

2.2. Documentation and Video Recording. Filming experiments were done in front of several wormsnail populations on the reef flat, edge, and slope, covering more than 12 hours at different day times. Filming was done with a digital camera in an underwater case (PowerShot G9, Canon, Germany). The camera was attached to a small flexible tripod and positioned near a wormsnail community. After focusing, the system was left alone for half an hour in order to not interfere with the organisms. The video material was evaluated by identifying the fish species [40, 41] and counting their net attacks or bites.

2.3. In Situ and Ex Situ Sampling of Mucus Nets. In situ sampling was focused on the shallow subtidal zone and done via snorkeling. All individuals exhibited an aperture width of about 15 mm. Secreted mucus nets were collected with a metal wire ring positioned on the outside perimeter of the shell's aperture. Prior to collection, the plankton-load net was wrested from the wormsnail by twisting the metal device. Removing the nets stimulated the wormsnails

to start producing a new one within less than a minute. The combined veils were transferred to the laboratory and extracted immediately. Additionally, plankton was caught with a 15 μm pore-sized plankton net and sediments were taken next to the wormsnails' *in situ* habitat.

Individuals, all of about 15 mm aperture width, used for *ex situ* net extraction were collected from horizontal surfaces at depths between 0.5 and 1.5 m. Only animals attached to lose coral blocks were transferred to the aquaria without getting exposed to air and brought back after one day and final plankton feeding. Before inserting, the coral blocks were cleaned with a wire brush to get rid of attached microfauna and flora that could diffuse in the ambient water. The plankton-load net produced in situ was removed. The ex situ tank was flushed with filtered seawater at 25°C (<15 μm pore size) to get rid of most of the suspended particles. Unfiltered seawater contained a high density of visible detrital particles; the filtered seawater appeared clear. The aquarium was equipped with a perforated acrylic glass rack on which coral blocks were positioned. This facilitates a controlled current flow from below and sidewise induced by pumps. The vermetids were triggered to produce the mucus nets floating upwards by this external stimulus. A ventilation-air-stone positioned under the rack provided small air bubbles that attach to the translucent mucus strands rendering the nets visible [8]. The back side of the aquarium was covered with a black plate to enhance the contrast. Collection was done according to *in situ* sampling. Clear nets from different individuals were combined and covered with solvent.

2.4. Extraction of Mucus Nets. Mucus nets (collected *in situ* and *ex situ*) from different individuals were combined in reaction tubes and extracted exhaustively over night in methanol, ethanol, diethyl ether, or ethyl acetate (~ 5 veils mL^{-1} solvent). After centrifugation, the supernatant was transferred in HPTLC prevalent vials. Sediment and plankton samples were weighed and extracted according to the mucus webs (~500 mg mL^{-1} solvent).

2.5. High-Performance Thin-Layer Chromatography (HPTLC). Chromatography was performed on silica gel 60 F$_{254}$ plates (Merck, Darmstadt, Germany). All plates were prewashed by development with methanol, then dried at 100°C for 15 min, and stored protected in a desiccator. Samples were applied on the plate as 4 mm bands, 9 mm apart, 10 mm from the lower edge, and 15 mm from the left edge by means of the Automatic TLC Sampler 4 (ATS4, CAMAG, Switzerland). The application volume of the mucus extracts ranged from 20 to 80 μL per band. Plates were developed with a twelve-step gradient by use of the Automated Multiple Development System (AMD2, CAMAG). The gradient was based on methanol, isopropyl acetate, and *n*-hexane (all technical grade and distilled before, BASF, Ludwigshafen, Germany) using 4 mm increments for the successively ascending migration distances. The drying time of the last AMD2 step was increased to be 15 min to avoid residual solvent interferences with the subsequent bioassay. Plates were documented by use of the

DigiStore 2 Documentation System (CAMAG) at 254 nm, 366 nm, and with white light illumination (reflectance and transmission mode). Data processing for all instrumentation was performed with the software platform winCATS (CAMAG). Digital evaluation (transfer from image into analogue curve) was processed with separate software (VideoScan, CAMAG). The VideoScan settings used were: "track borders and slopes" was selected as integration mode, the *Savitsky-Golay* filter width was 11, and the offset was set to "yes." The track borders were set outside the fraction to enable background subtraction of the mean area of both borderline pixels.

2.6. Bioluminescence Assay with Vibrio fischeri.

The developed plate was automatically immersed at a speed of 3.5 cm s^{-1} and with an immersion time of 1 s, by means of the TLC Chromatogram Immersion Device III (CAMAG), into the luminescent bacteria (*Vibrio fischeri*) suspension, which was prepared according to the Bioluminex assay protocol (culture medium and buffer from ChromaDex, Boulder, CO, USA). For imaging, the HPTLC plate was placed in the compartment of the BioLuminizer (CAMAG), a dark chamber with a cooled 16-bit high-resolution CCD camera on top, specially designed for HPTLC-bioluminescence detection. In the compartment, the HPTLC plate was covered by a glass plate keeping the bacteria moist for a prolonged time. Images were captured with an exposure time of 30 s over a period of 30 min. This allowed the study any time-dependent changes.

3. Results

3.1. Observations on Coral-Wormsnail Interaction.

Observations made at Sharm Fugani (Mangrove Bay, South Sinai, Western Coast, Red Sea) and Suleiman Reef (Dahab, South Sinai, Eastern Coast, Red Sea) showed clear structural deformations of surrounding coral tissue inhabiting *Dendropoma maxima*. Less affected were fire corals of the genus *Millepora* (Figure 1(a)), due to the erected and branching growth form and the limited contact area between coral tissue and mucus webs. Most affected were massive corals of the genus *Porites*. This vermetid-coral association was always characterized by the wormsnail's mucus net covering parts of the coral block. Veils of different individuals often merge together forming huge sheets spread on the reef's surface. Some veils were so thick and dense that they trap gas bubbles, which get in touch with coral surface. Different states of host-coral degeneration were observable, beginning with pigment modification and surface structure flattening (Figure 1(b)) until the exposure of dead coral rock was overgrown by algae (Figure 1(c)).

3.2. Video Observation on Fish-Wormsnail Interaction.

The evaluation of the video material does not show any fish biting or interested in the nets, despite the full load of plankton and bacteria accumulated within the web; even though amongst the identified reef fish were well known mucoid and plankton feeding organisms known to consume, for example, nutritious coral mucus aggregates. Observed reef fish sharing wormsnail habitats are listed in Table 1.

3.3. Sampling of Mucus Nets and Screening for Bioactive Metabolites (In Situ and Ex Situ).

Preliminary tests preferentially showed unpolar and less polar components accumulated inside the nets. Therefore, ethyl acetate was chosen as extracting agent. Dipping in a *Vibrio fischeri* suspension leads to an inverse correlation between the concentration of bioactivity-reducing substances and the bioluminescence of bacteria.

Mucus nets with plankton sampled *in situ* and extracted with ethyl acetate showed at least four *V. fischeri* toxic substances (M1, M3–M5) detected by bioluminescence inhibition (black spots on luminescent background). Additionally, there was one metabolite (M2) enhancing the bioluminescence indicated by a white spot on the HPTLC plate (Figure 2(a)). Because of different application volumes, the more concentrated extracts led to a stronger decrease of luminescence on the plate. When nets are particle enriched (*in situ*), it is not possible to say whether the wormsnail or the captured plankton is responsible for product synthesis.

Inside the aquaria (*ex situ*), wormsnails retained producing their mucus nets which were almost invisible due to the lack of adherent particles. Defecation was frequent (2–4 times h^{-1}) and ejected as pellets in batches leading to the cleanup of all facilities as well as the contamination of veils which had to be discarded. Wormsnails kept on generating slime nets for approximately 24 hours, until they stopped and could not be motivated again without additional feeding. During the final analysis, the wormsnails secreted a small bolus of mucus that accumulated on the uppermost lip of the operculum.

Bioactive metabolites, detected *in situ*, could also be found in extracts from nets collected *ex situ* (Figure 2(b)). The digital detection of the inverted (grey-scale) bioluminescence signals showed the same intensity of bioluminescence inhibition concerning the most nonpolar substances (M4 and M5) between equal concentrated extracts of *in situ* and *ex situ* sampling (Figure 3). All other metabolites disclosed stronger effects in the *in situ* extracts compared to the *ex situ* pendants. The ratio of bioluminescence inhibition between metabolite five and one is M1 > M5 in the *in situ* extracts including plankton and M1 < M5 in *ex situ* extracts excluding plankton (Figure 3). Same results were obtained for M4. Thus, M4 and M5 seemed to (exclusively) synthesize or accumulate from the wormsnail itself whereas M1 is of planktonic origin. Despite using filtered seawater, the *ex situ* extracts were not completely free of plankton. This was due to inserting some microfauna and flora attached to the coral blocks the wormsnails live inside. The inserted particles could have been trapped in the veils in small amounts leading to grayish spots (M1) also in the *ex situ* extracts. Extracts of surrounding sediments and plankton caught by a 15 μm plankton net only exhibited the polar or further non polar bioactive substances eluting later on the HPLTC plate (data not shown). This is the first record of bioactivity in the whole family of Vermetidae.

4. Discussion

4.1. First Record of Bioactivity in the Mucus Nets of Vermetids.

This study is the first report of bioactive substances in mucus

Detection of Bioactive Compounds in the Mucus Nets of Dendropoma maxima, Sowerby 1825
(Prosobranch Gastropod Vermetidae, Mollusca)

107

TABLE 1: List of reef fish and their preferred feeding mode that share the same habitat like wormsnails.

Family	Genus/species	Feeding
Acanthuridae	*Acanthurus sohal*	Algae, zooplankton
Acanthuridae	*Naso elegans*	Brown algae
Acanthuridae	*Naso unicornis*	Brown algae
Acanthuridae	*Zebrasoma xanthurum*	Algae
Apogonidae	*Apogon* spp.	Fishes, crustaceans, zooplankton
Balistidae	*Rhinecanthus assasi*	Benthic invertebrates
Blenniidae	*Cirripectes castaneus*	Benthic algae
Blenniidae	*Ecsenius dentex*	Thread algae
Blenniidae	*Ecsenius gravieri*	Thread algae
Chaetodontidae	*Chaetodon auriga*	Polychaetes, anemones, coral polyps and mucus, algae
Chaetodontidae	*Chaetodon austriacus*	Coral polyps and mucus, cnidarian tentacles
Chaetodontidae	*Chaetodon fasciatus*	Cnidarian polyps and mucus, invertebrates, algae
Chaetodontidae	*Chaetodon paucifasciatus*	Coral polyps and mucus, invertebrates, algae
Chaetodontidae	*Heniochus intermedius*	Zooplankton, benthic invertebrates
Cirrhitidae	*Paracirrhites forsteri*	Fishes, crustaceans
Holocentridae	*Sargocentron diadema*	Snails, polychaetes, crustaceans
Labridae	*Cheilinus lunulatus*	Benthic invertebrates like molluscs and crustaceans
Labridae	*Gomphosus caeruleus*	Small fish, benthic invertebrates
Labridae	*Halichoeres hortulanus*	Benthic invertebrates
Labridae	*Labroides dimidiatus*	Fish parasites, skin and mucus
Labridae	*Thalassoma lunare*	Small fish, benthic invertebrates
Labridae	*Thalassoma rueppelii*	Small fish, benthic invertebrates
Mullidae	*Parupeneus* spp.	Fishes, benthic invertebrates
Ostraciidae	*Ostracion cyanurus*	Sessile invertebrates, sponges, algae
Ostraciidae	*Tetrasomus gibbosus*	Benthic, sessile invertebrates
Pomacanthidae	*Pygoplites diacanthus*	Sponges, tunicates
Pomacentridae	*Abudefduf* spp.	(zoo)plankton, algae
Pomacentridae	*Amblyglyphidodon* spp.	Zooplankton, drifting organic material
Pomacentridae	*Chromis* spp.	Plankton, algae, (omnivores)
Pomacentridae	*Chromis viridis*	Plankton
Pomacentridae	*Dascyllus aruanus*	Zooplankton, benthic invertebrates, algae
Pomacentridae	*Dascyllus trimaculatus*	Algae, planktonic crustaceans (copepods)
Pomacentridae	*Pomacentrus trilineatus*	Invertebrates, algae
Pomacentridae	*Stegastes nigricans*	Algae, gastropods, sponges, copepods
Pseudochromidae	*Pseudochromis* spp.	Crustaceans, worms, zooplankton
Scaridae	*Chlorurus gibbus*	Benthic grazer of algae
Scaridae	*Hipposcarus harid*	Thread algae
Scaridae	*Scarus* spp.	Thread algae, benthic algae
Serranidae	*Cephalopholis* sp.	Fishes, crustaceans
Serranidae	*Epinephelus fasciatus*	Fishes, crustaceans
Serranidae	*Epinephelus tauvina*	Fishes
Serranidae	*Pseudanthias* spp.	(zoo)plankton
Tetraodontidae	*Arothron diadematus*	Invertebrates, algae
Tetraodontidae	*Arothron hispidus*	Invertebrates (sponges, tunicates, worms, crustaceans, corals, molluscs, echinoderms), algae

And some unidentified Labridae, Serranidae, Pomacentridae, and Acanthuridae

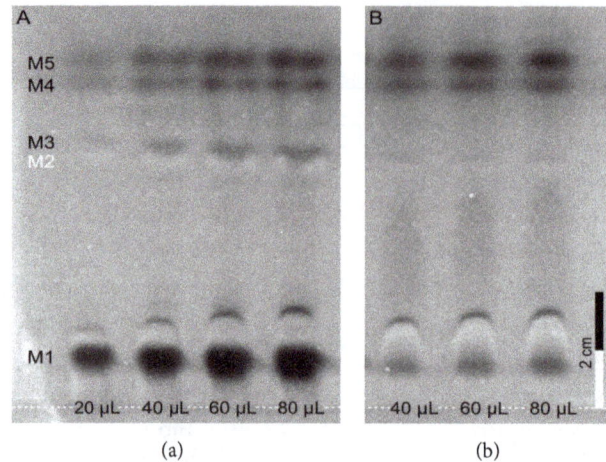

FIGURE 2: HPTLC plate image (detail) obtained from the screening for bioactive secondary metabolites in ethyl acetate extracts (~5 veils mL^{-1} solvent) of *in situ* (with plankton, (a)) and *ex situ* (without plankton, (b)) mucus nets from *Dendropoma maxima*. Detection was performed after the plate was immersed into a bioluminescent *Vibrio fischeri* suspension. M1–M5: bioactive metabolites with decreasing polarity; dotted line displays zone of application. Concentrations on plate: 20–80 μL/band. Dark spots denote luminescence inhibition (M1, M3–M5); white spots (M2) denote luminescence enhancement. $N = 5$.

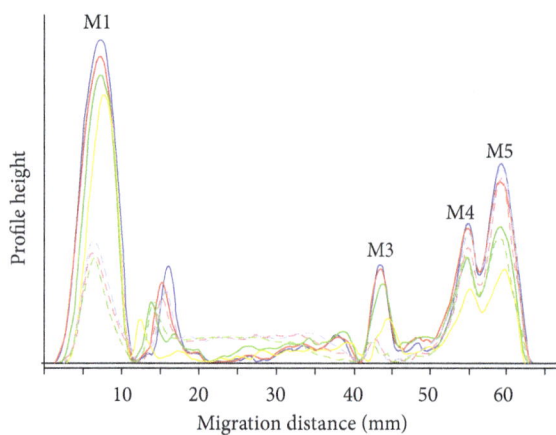

FIGURE 3: Digital detection of the inverted (grey-scale) bioluminescence signal of respective tracks. Each extract was made of ~5 nets educed in 1 mL ethyl acetate. M1–M5: bioactive metabolites with decreasing polarity. Lines bright coloured = nets with plankton sampled *in situ*, lines light coloured = nets without plankton sampled *ex situ*. Concentrations on plate: blue: 80 μL/band, red: 60 μL/band, green: 40 μL/band, and yellow: 20 μL/band. Due to equal intensities of M4 and M5 and the inverse ratio of M1/M4 and M1/M5 in both extracts (*in situ*: M4 and M5 < M1, *ex situ*: M4 and M5 > M1) of same concentration, M4 and M5 seem to exclusively be synthesised or accumulated by the wormsnail itself whereas M1 is of planktonic origin. $N = 3$.

nets of the family Vermetidae. Although it was already presumed in 1973 by Coles and Strathmann [19] and despite the already known occurrence of toxic compounds in many mucus releasing invertebrates, no study has been completed on the detection of bioactive secondary metabolites in vermetids. The compounds could be either of gastropod's origin synthesized by the wormsnail itself, accumulated via passive suspension feeding, or built up starting from diet-derived

precursors. Here we suggest that M4 and M5 are exclusively produced by the vermetid itself due to equal bioluminescence inhibition of these metabolites and the inverse ratio of M1 and M5 (or M1 and M4) between *in situ* and *ex situ* extracts (Figure 3). Traces of M1 to M3 along the HPTLC plate of clean extracts are presumably due to small amounts of plankton transferred to the *ex situ* tanks, for example, microfauna and flora attached to the coral block, that got caught in the mucus veils. In case the bioactive metabolites would not be accumulated by the wormsnail but by the plankton, the ratio of substance amounts—and therefore the intensity of bioluminescence inhibition—between M1 and M4 as well as M5 and all the other metabolites should be constantly equal. This requires the equality of the qualitative and quantitative composition of the remaining planktonic organisms in the *ex situ* tank and the *in situ* habitat. The secretion of mucus agglomerates from animals that kept in aquaria for more than 24 hours was also observed by Yonge [42]. These attributes led to the conclusion that *D. maxima* uses the bolus to consolidate filtered particles prior to ingestion. Since filter feeding is possible but negligible in *D. maxima* and it is fully adapted to mucus-trap feeding continuously nourishing on slime nets, the bolus seems to be an artifact of *ex situ* conditions.

4.2. Trophic Significance and Biotoxicity/Ichthyotoxicity of Mucoidal Material. The significance of mucoidal material of either fish or invertebrate origin as food source for tropical marine fishes is well known (e.g., [43–45]). External mucus slime provides the predator with a rich, continuously renewed resource of energy and amino acids that may comprise more than 95% of their diet [46, 47]. Lipids, particularly wax esters, and mucins containing glucose, galactose, glucosamine, galactosamine, and arabinose were found to be a major, consistent component of coral mucus [43, 47]. The molluscan mucus is phylogenetically and functionally

Detection of Bioactive Compounds in the Mucus Nets of Dendropoma maxima, Sowerby 1825 (Prosobranch Gastropod Vermetidae, Mollusca)

109

different from coelenterate mucins but seems quite similar to them chemically [47]. Predators feeding on mucus are *Labroides* spp. cleaning wrasses (Labridae) or juvenile angelfish (Pomacanthidae) but also noncleaning fish like cichlids [44, 48]. Further families including mucoid reef fish are the Chaetodontidae (predominant feeding mode), Tetraodontidae, Balistidae, Monacanthidae, Pomacentridae and Scaridae [45]. Corallivores are well known to feed compulsively or facultatively on the nutritious slime of live coral polyps. Several corals (e.g., *Acropora* spp., *Porites* spp.) release large aggregates of denatured mucus that becomes suspended and floats in the water column. These flocks, additionally containing algae, occasional protozoa, organic debris, filamentous algae, and crustacean molts, as well as inorganic particles, are highly energy enriched and may be an important food source [19, 43, 47, 49, 50].

However, our study gives no information whether bioactive substances are responsible for coral tissue degeneration or have ichthyotoxic properties concerning habitat sharing reef fish. The antibiotic effect may also be used to reduce the degradation of plankton by bacteria while it is trapped within the net. Bioactive metabolites could be the explanation for mucoid and plankton feeding reef fish not being interested in the veils as noted and presumed before [19]. Our findings that several species of *Chromis* sp. ignore the mucus nets contradict with the studies completed by Johannes [49], who showed that these fish have a preference for mucus sheets in general. It is also possible that reef fish ignore the webs because of their dimension, as seen with macroscopic mucus aggregates from corals [19]. Further analyses are needed to study the correlation between the separated metabolites and their effect.

4.3. Effect of Vermetids on Coral Morphology and Survival. The deleterious influence of mucus nets on coral morphology leading to, for example, the flattened shape, even surface and air bubble trapping was also previously reported (e.g., [1, 13]). Flattened corals cannot block the horizontal component of the local currents and may enable the gastropods' webs to spread over the coral head. Thus, the mucus net may be secreted over a larger area [1]. A comparable strong coral degradation marked by the exposure of dead tissue overgrown by algae (Figure 1(c)) is also published in [13]. They revealed a negative correlation between the local density of vermetids and the percent cover of live coral. Further time-series recordings are planned to analyze whether the reported different stages of hostcoral interaction reflect a real gradient across time. In contrast to our studies, Zvuloni and coworkers [1] also observed branching corals, for example, the hydrozoan *Millepora* sp., dramatically affected by the gastropods mucus nets. This may end in the loss of the coral's typical terminal polyps [1]. Until now, there is no explanation for the observed deformations and degenerations of corals. The physiological effects and the exact mechanisms underlying vermetid-coral interactions remain unknown. The following hypotheses were presumed: (1) occluding mucus nets reduce the water flow around the corals and prevent them from feeding on plankton inducing competition pressure (Colgan in [9]), (2) mucus nets have an abrasive effect on coral

polyps (own suggestion), (3) *D. maxima* consumes the coral's (secondary) metabolites and/or their mucus (suggested by Fenner [51]), and (4) mucus nets contain chemicals that influence coral growth (suggested by Zvuloni and coworkers [1], own suggestion in combination with (2)).

4.4. Conclusion/Outlook. This is the first record of bioactive properties in mucus producing organisms belonging to the Vermetidae. *Ex situ* feeding experiments with isolated substances are in progress to study further potential bioactive properties (e.g., ichthyotoxicity) besides the antimicrobial effect. Additionally, analyses are ongoing to elucidate the structural formula of the acting compounds by HPTLC/HRMS. We suggest that other mucus net producing taxa (e.g., polychaetes) also contain bioactive substances what open a new field in the search for marine natural products with possible application in human medicine.

Conflict of Interests

There is no conflict of interests in the submitted paper.

Acknowledgments

The authors thank the DiveIn, Dahab, all the staff, especially Andreas Tischer, and Hans Lange for their support and providing equipment; the Dahab Marine Research Center (DMRC) and all interns "milking" the wormsnails, Stefan Rauchut, and Tim Cross; furthermore, special thanks to Nils Anthes, University of Tübingen for sparking the interest in wormsnails and providing the literature. The authors are thankful to Merck, Darmstadt, Germany, and CAMAG, Berlin, Germany, for support regarding plate material and instrumentation, respectively.

References

[1] A. Zvuloni, R. Armoza-Zvuloni, and Y. Loya, "Structural deformation of branching corals associated with the vermetid gastropod *Dendropoma maxima*," *Marine Ecology Progress Series*, vol. 363, pp. 103–108, 2008.

[2] R. N. Hughes and A. H. Lewis, "On the spatial distribution, feeding, and reproduction of the vermetid gastropod *Dendropoma maximum*," *Journal of Zoology*, vol. 172, pp. 531–547, 1974.

[3] M. G. Hadfield, E. A. Kay, M. U. Gillette, and M. C. Lloyd, "The vermetidae (Mollusca: Gastropoda) of the Hawaiian Islands," *Marine Biology*, vol. 12, no. 1, pp. 81–98, 1972.

[4] M. Zuschin and W. E. Piller, "Molluscan hard-substrate associations in the northern Red Sea," *Marine Ecology*, vol. 18, no. 4, pp. 361–378, 1997.

[5] N. E. Phillips and J. S. Shima, "Reproduction of the vermetid gastropod *dendropoma maximum* (Sowerby, 1825) in moorea, french polynesia," *Journal of Molluscan Studies*, vol. 76, no. 2, pp. 133–137, 2010.

[6] G. Ribak, J. Heller, and A. Genin, "Mucus-net feeding on organic particles by the vermetid gastropod *Dendropoma maximum* in and below the surf zone," *Marine Ecology Progress Series*, vol. 293, pp. 77–87, 2005.

[7] M. Zuschin and M. Stachowitsch, "The distribution of molluscan assemblages and their postmortem fate on coral reefs in the gulf of aqaba (northern red sea)," *Marine Biology*, vol. 151, no. 6, pp. 2217–2230, 2007.

[8] I. Kappner, S. M. Al-Moghrabi, and C. Richter, "Mucus-net feeding by the vermetid gastropod *Dendropoma maxima* in coral reefs," *Marine Ecology Progress Series*, vol. 204, pp. 309–313, 2000.

[9] T. L. Smalley, "Possible effects of intraspecific competition on the population structure of a solitary vermetid mollusc," *Marine Ecology Progress Series*, vol. 14, pp. 139–144, 1984.

[10] R. Hughes, "Feeding behaviour of the sessile gastropod Tripsycha tulipa (Vermetidae)," *Journal of Molluscan Studies*, vol. 51, pp. 326–330, 1985.

[11] A. Gagern, T. Schürg, N. K. Michiels, G. Schulte, D. Sprenger, and N. Anthes, "Behavioural response to interference competition in a sessile suspension feeder," *Marine Ecology Progress Series*, vol. 353, pp. 131–135, 2008.

[12] C. R. Boettger, "Studien zur Physiologie der Nahrungsaufnahme festgewachsener Schnecken. Die Ernährung der Wurmschnecke *Vermetus*," *Biologisches Zentralblatt*, pp. 581–598, 1930.

[13] J. S. Shima, C. W. Osenberg, and A. C. Stier, "The vermetid gastropod *Dendropoma maximum* reduces coral growth and survival," *Biology Letters*, vol. 6, no. 6, pp. 815–818, 2010.

[14] J. E. Morton, "Form and function in the evolution of the Vermetidae," *Bulletin of the British Museum (Natural History). Zoology*, vol. 2, pp. 585–630, 1965.

[15] N. E. Chadwick and K. M. Morrow, "Competition among sessile organisms on coral reefs," in *Coral Reefs: An Ecosystem in Transition*, Z. Dubinsky and N. Stambler, Eds., pp. 347–372, US Government, 2011.

[16] A. C. Stier, C. S. McKeon, C. W. Osenberg, and J. S. Shima, "Guard crabs alleviate deleterious effects of vermetid snails on a branching coral," *Coral Reefs*, vol. 29, no. 4, pp. 1019–1022, 2010.

[17] R. H. Lowe-McConnell, *Fish Communities in Tropical Freshwaters*, Longman Press, London, UK, 1975.

[18] P. A. Larkin, "Interspecific competition and population control in freshwater fish," *Journal of the Fisheries Research Board of Canada*, vol. 13, pp. 327–342, 1956.

[19] S. L. Coles and R. Strathmann, "Observations on coral mucus "flocs" and their potential trophic significance," *Limnology and Oceanography*, vol. 18, pp. 673–678, 1973.

[20] M. L. Harmelin-Vivien and Y. Bouchon-Navaro, "Feeding diets and significance of coral feeding among Chaetodontid fishes in Moorea (French Polynesia)," *Coral Reefs*, vol. 2, no. 2, pp. 119–127, 1983.

[21] V. DiMarzo, A. Marin, R. R. Vardaro, L. DePetrocellis, G. Villani, and G. Cimino, "Histological and biochemical bases of defense mechanisms in four species of Polybranchioidea ascoglossan molluscs," *Marine Biology*, vol. 117, no. 3, pp. 367–380, 1993.

[22] G. Cimino, S. De Rosa, S. De Stefano, and G. Sodano, "The chemical defense of four Mediterranean nudibranchs," *Comparative Biochemistry and Physiology Part B*, vol. 73, no. 3, pp. 471–474, 1982.

[23] C. D. Derby, "Escape by inking and secreting: marine molluscs avoid predators through a rich array of chemicals and mechanisms," *Biological Bulletin*, vol. 213, no. 3, pp. 274–289, 2007.

[24] J. W. Blunt, B. R. Copp, M. H. G. Munro, P. T. Northcote, and M. R. Prinsep, "Marine natural products," *Natural Product Reports*, vol. 28, no. 2, pp. 196–268, 2011.

[25] M. S. Davies and S. J. Hawkins, "Mucus from marine molluscs," *Advances in Marine Biology*, no. 34, pp. 1–71, 1998.

[26] T. E. Thompson, "Defensive adaptations on opisthobranchs," *Journal of the Marine Biological Association of the United Kingdom*, vol. 39, pp. 123–134, 1960.

[27] J. R. Pawlik, "Marine invertebrate chemical defenses," *Chemical Reviews*, vol. 93, no. 5, pp. 1911–1922, 1993.

[28] L. Gunthorpe and A. M. Cameron, "Bioactive properties of extracts from Australian dorid nudibranchs," *Marine Biology*, vol. 94, no. 1, pp. 39–43, 1987.

[29] G. Cimino, A. Fontana, and M. Gavagnin, "Marine opisthobranch molluscs: chemistry and ecology in sacoglossans and dorids," *Current Organic Chemistry*, vol. 3, no. 4, pp. 327–372, 1999.

[30] C. Avila, "Natural products of opisthobranch molluscs: a biological review," *Oceanography and Marine Biology: An Annual Review*, vol. 33, pp. 487–559, 1995.

[31] G. Cimino, A. Crispino, V. Di Marzo, M. Gavagnin, and J. D. Ros, "Oxytoxins, bioactive molecules produced by the marine opisthobranch mollusc Oxynoe olivacea from a diet-derived precursor," *Experientia*, vol. 46, no. 7, pp. 767–770, 1990.

[32] G. J. Bakus, N. M. Targett, and B. Schulte, "Chemical ecology of marine organisms: an overview," *Journal of Chemical Ecology*, vol. 12, no. 5, pp. 951–987, 1986.

[33] D. Bhakuni and D. Rawat, *Bioactive Marine Natural Products*, Anamaya, New Delhi, India, 2005.

[34] N. Fusetani, *Drugs from the Sea*, Karger, Basel, Switzerland, 2000.

[35] P. Proksch, R. Ebel, R. A. Edrada, V. Wray, and K. Steube, "Bioactive natural products from marine invertebrates and associated fungi," in *Sponges (Porifera)*, W. E. G. Müller, Ed., pp. 117–142, Springer, Berlin, Germany, 2003.

[36] D. J. Newman and G. M. Cragg, "Marine natural products and related compounds in clinical and advanced preclinical trials," *Journal of Natural Products*, vol. 67, no. 8, pp. 1216–1238, 2004.

[37] P. Proksch and W. E. G. Muller, *Frontiers in Marine Biotechnology*, Horizon Bioscience, Wymondham, UK, 2006.

[38] A. M. S. Mayer, K. B. Glaser, C. Cuevas et al., "The odyssey of marine pharmaceuticals: a current pipeline perspective," *Trends in Pharmacological Sciences*, vol. 31, no. 6, pp. 255–265, 2010.

[39] A. Klöppel, W. Grasse, F. Brümmer, and G. E. Morlock, "HPTLC coupled with bioluminescence and mass spectrometry for bioactivity-based analysis of secondary metabolites in marine sponges," *Journal of Planar Chromatography*, vol. 21, no. 6, pp. 431–436, 2008.

[40] E. Lieske and R. Myers, *Coral Reef Guide Red Sea*, Franckh-Kosmos Verlags, GmbH & Co. KG, Stuttgart, Germany, 2004.

[41] R. Froese and D. Pauly, "FishBase. World Wide Web electronic publication. Version (04/2013)," 2013, http://www.fishbase.org/search.php.

[42] C. M. Yonge, *Notes on Feeding and Digestion in Pterocera and Vermetus, with a Discussion on the Occurence of the Crystalline Style in the Gastropoda*, vol. 1 of *Great Barrier Reef Expedition 1928-1929, Scientific Report*, 1932.

[43] A. A. Benson and L. Muscatine, "Wax in coral mucus: energy transfer from corals to reef fishes," *Limnology and Oceanography*, vol. 19, pp. 810–814, 1974.

[44] D. L. Gorlick, "Ingestion of host fish surface mucus by the Hawaiian cleaning wrasse, *Labroides phthirophagus* (Labridae), and its effect on host species preference," *Copeia*, vol. 4, pp. 863–868, 1980.

Detection of Bioactive Compounds in the Mucus Nets of Dendropoma maxima, Sowerby 1825
(Prosobranch Gastropod Vermetidae, Mollusca)

111

[45] A. J. Cole, M. S. Pratchett, and G. P. Jones, "Diversity and functional importance of coral-feeding fishes on tropical coral reefs," *Fish and Fisheries*, vol. 9, no. 3, pp. 286–307, 2008.

[46] K. O. Winemiller and H. Y. Yan, "Obligate mucus-feeding in a South American trichomycterid catfish (Pisces: Ostariophysi)," *Copeia*, vol. 2, pp. 511–514, 1989.

[47] H. W. Ducklow and R. Mitchell, "Composition of mucus released by coral reef coelenterates," *Limnology and Oceanography*, vol. 24, pp. 706–714, 1979.

[48] A. S. Grutter and R. Bshary, "Cleaner fish, *Labroides dimidiatus*, diet preferences for different types of mucus and parasitic gnathiid isopods," *Animal Behaviour*, vol. 68, no. 3, pp. 583–588, 2004.

[49] R. E. Johannes, "Ecology of organic aggregates in the vicinity of a coral reef," *Limnology and Oceanography*, vol. 12, pp. 189–195, 1967.

[50] S. Richman, Y. Loya, and L. B. Slobodkin, "The rate of mucus production by corals and its assimilation by the coral reef copepod *Acartia negligens*," *Limnology and Oceanography*, vol. 20, pp. 918–923, 1975.

[51] D. Fenner, "Is a mollusc that sculptures coral a parasite?" in *Proceedings of the 82nd Australian Coral Reef Society Conference*, Mission Beach, Australia, 2006.

11

Spirulina (Arthrospira): An Important Source of Nutritional and Medicinal Compounds

Abdulmumin A. Nuhu

Department of Chemistry, Ahmadu Bello University, P.M.B. 1069, Zaria, Kaduna, Nigeria

Correspondence should be addressed to Abdulmumin A. Nuhu; aanuhu@yahoo.com

Academic Editor: Horst Felbeck

Cyanobacteria are aquatic and photosynthetic organisms known for their rich pigments. They are extensively employed as food supplements due to their rich contents of proteins. While many species, such as *Anabaena* sp., produce hepatotoxins (e.g., microcystins and nodularins) and neurotoxins (such as anatoxin a), *Spirulina (Arthrospira)* displays anticancer and antimicrobial (antibacterial, antifungal, and antiviral) activities via the production of phycocyanin, phycocyanobilin, allophycocyanin, and other valuable products. This paper is an effort to collect these nutritional and medicinal applications of *Arthrospira* in an easily accessible essay from the vast literature on cyanobacteria.

1. Introduction

Cyanobacteria are ancient photosynthetic organisms that are found in various aquatic environments [1–3]. Their photosynthetic pigments confer different colors on them, but they are generally regarded as blue-green. Calling them algae is, however, a misnomer since they are truly prokaryotes that share most of the characteristics of eubacteria. Some of these organisms have nitrogen-fixing potential which makes them important in rice paddy waters [4].

Cyanobacteria form colonies [5] or live as individual cells [6]. They also form coccoid [7] or filamentous structures [8]. The filamentous colonies show the ability to differentiate into three different cell types [9]. Vegetative cells, the normal photosynthetic cells formed under favorable growth conditions; climate-resistant spores in harsh environmental conditions and a thick-walled heterocyst containing the enzyme nitrogenase for nitrogen fixation.

In the last 3.5 billion years, cyanobacterial morphology has been largely maintained as they are very resistant to contamination. Sigler et al. [10] have shown that cyanobacteria form monophyletic taxon. Culture-based morphological characteristics of endolithic cyanobacteria have been extensively described by Al-Thukair and Golubic [11]. Since characterization of microorganisms based on morphology is highly subjective and sometimes very speculative, the shift by genome-based characterization is now gaining momentum. Koksharova and Wolk [12] have presented a good review on the available genetic tools for cyanobacteria studies.

Cyanobacteria are very resistant as they produce protective compounds which shield them against harsh environmental conditions [13]. Some of these compounds also have strong insecticidal activities [14]. Toxic species, including *Anabaena* species, produce toxins such as microcystins and nodularins which are hepatotoxic, and neurotoxins such as anatoxin *a* [15, 16].

The Darling River cyanobacterial bloom of 1991 is a clear representation of the environmental hazard that such species pose [17]. However, some species of cyanobacteria possess the ability to produce substances with therapeutic activities such as anticancer and antimicrobial applications [18–22].

Among the myriads of cyanobacteria, *Arthrospira platensis* is a blue-green cyanobacterium that thrives in elevated alkaline pH [23]. *A. platensis* is recognized by its peculiar shape of cylindrical trichomes that are arranged in a left-handed helix throughout the filament [24]. The correct taxonomic definitions of *Arthrospira* have been revealed through the study of the ultrastructural details of its trichomes and 16S rRNA gene sequences [25]. An important ligation detection reaction, in combination with universal array, capable of identifying various cyanobacteria, including *Arthrospira*, in environmental samples, has been developed [26]. Good understanding of the ecology of this alkaliphilic organism is a catalyst to its mass production and commercial viability as food supplement. By the end of year 2009, its total annual production in Ordos Plateau of Mongolia was in excess of 700 t [27]. With retrospect, the Mexicans [28] and Kanenbu tribe of Chad [29] have been exploiting the protein potentials of *S. platensis* in their diets for long time now, and about 3000 metric tons of *S. platensis* is currently produced for commercial purposes [30]. A fed-batch process has been employed in the cultivation of *Arthrospira* [31], and different solid-liquid separation techniques give various degrees of recovery. Which technique is ultimately selected will depend on the cyanobacterial species, intended concentration of the finished product, and product quality [32]. Cultivation of *A. platensis* under different trophic modes was shown to affect the product yield [33]. High-value compounds from this organism have been put to assorted uses as cosmaceuticals, nutraceuticals, and as functional foods [34]. Phycocyanin and allophycocyanin, two of such important compounds, have been determined in *Spirulina* supplements and raw materials by a 2-wavelength spectrophotometric method [35]. Bioactivity and health functions of *Arthrospira* food supplements have been reviewed [36–38]. Specific functions that have been tested for compounds extracted from this organism are grouped under the following subheadings.

2. Nutritional Functions

Arthrospira (*Spirulina*) is among the richest sources of proteins. Its protein content is about 60–70% [39]. In a study that attempted using *Spirulina* as a protein supplement, it was observed that it can replace up to 40% of protein content in tilapia diets [40]. Rabelo et al. [41] have explained the development of cassava doughnuts enriched with *S. platensis* biomass.

Unlike many other cyanobacteria that have proven toxicity, no such property has been attributed to *Spirulina*. While testing for mutagenicity, acute, subchronic, and chronic toxicities and teratogenicity in animal experimentations, Chamorro et al. [42] have shown that *Spirulina* did not exhibit any potential for organ or system toxicity even though the doses given were elevated above those for expected human consumption. Rather, *Spirulina* was shown to protect fish from sublethal levels of some chemicals [43]. Likewise, dietary supplementation of *Spirulina* has helped in alleviating the incidence of anemia experienced during pregnancy and lactation. In the study conducted by Kapoor and Mehta [44], dietary supplementation of *S. platensis* was found to increase the iron storage of rats, better than achieved from the combination of casein and wheat gluten diets, during the first half of pregnancy and lactation. A review that treats the influence of different compounds from *Spirulina* on the immune system has been written [45].

3. Antioxidant Functions

Apart from its importance as a food additive for supplementary dietary proteins, there are also a lot of potentials for medical and therapeutic applications [46]. For example, *A. platensis* plays a hepatoprotective role [47]. This role, which has to do with the antioxidant activity of *Spirulina*, has been previously asserted by various researchers. The antioxidant activity of *Spirulina* is ascribed to the presence of two phycobiliproteins: phycocyanin and allophycocyanin, as determined by its action against OH radical generated from ascorbate/iron/H_2O_2 system. The activity was found to be proportional to the concentration of the phycobiliproteins and was mainly due to the phycocyanin content [48]. As an antioxidant effect, oxygen stress was inhibited by phycocyanin and phycocyanobilin from *Spirulina* leading to protection against diabetic nephropathy [49]. In an earlier experiment to determine the radical scavenging activity of C-phycocyanin isolate of *S. platensis*, an intraperitoneally administered C-phycocyanin was found to reduce the peroxide values of CCl_4-induced lipid peroxidation in rat liver microsomes [50]. Following a study conducted on 60 patients presenting with chronic diffuse disorders in the liver and on 70 experimental animals, Gorban' et al. [51] have found that *Spirulina* administration prevented the transformation of chronic hepatitis into hepatic cirrhosis. Recently, Paniagua-Castro et al. [52] have demonstrated the protective efficacy of *Arthrospira* against cadmium-induced teratogenicity in mice.

There are indications that these therapeutic potentials are not the exclusive rights of *S. platensis*. *Spirulina fusiformis* also has shown some free radical scavenging activities. In rats, Kuhad et al. [53] have found that radical scavenging activity of *S. fusiformis* did protect against nephrotoxicity resulting from oxidative and nitrosative stress of the aminoglycoside, gentamicin, an antibiotic commonly used for the treatment of Gram-negative bacterial infections. Pretreatment of mice with *Arthrospira maxima* effectively led to the reduction in liver total lipids, liver triacylglycerols, and serum triacylglycerols, thus protecting against Simvastatin-induced hyperlipidemia [54]. The hexane extract of *Spirulina* achieved an impressive 89.7% removal of arsenic from rat liver tissue, which is a better result than obtained with either alcohol or dichloromethane extract [55]. In a more recent finding, aqueous extract of *S. platensis* showed suppressive potency, through free radical scavenging activity, against cyclophosphamide-induced lipid peroxidation in goat liver homogenates [56].

As a nephroprotective activity, *S. platensis* extract counteracted the hyperoxaluria experimentally induced by the administration of sodium-oxalate to rats, through stabilization of antioxidant enzymes and glutathione metabolizing enzymes [57]. Protections against mercuric chloride-($HgCl_2$-) induced renal damage and oxidative stress were attributed to the administration of *A. maxima* to experimental mice [58]. Administration of *A. platensis* to rats also rendered protection against $HgCl_2$-induced testis injury and sperm quality deteriorations [59].

S. platensis biomass preparations have shown some corrective influences on atherosclerotic processes in 68 patients with ischemic heart disease (IHD) and atherogenic dyslipidemia. The patients' immunological states were altered, in addition to changes in lipid spectra [60]. Pretreatment of experimental animals with *Spirulina* has proved its cardio-protective function, this time against doxorubicin-induced toxicity, as evident from lower mortality, lower degree of lipid peroxidation, decreased ascites, and normalization of antioxidant enzymes, without compromising the antitumor activity of the drug, doxorubicin [61]. The contribution of reactive oxygen species (ROS) to brain injury in neurodegenerative conditions, such as Parkinson's disease, is hampered with proper administration of *A. maxima* supplement. Following a 40-day pretreatment with 700 mg/kg/day of this supplement, various indicators of toxicity in rat injected with a single dose of 6-hydroxydopamine, 6-OHDA (16 μg/2 μL), were decreased [62]. This is an indication of the neuroprotective effect of this supplement against the harmful effect of free radicals. *Arthrospira* supplement has also a radioprotective effect. This is demonstrated by its free radical scavenging function against gamma-irradiation-induced oxidative stress and tissue damage in rats [63]. Cell death through apoptosis is prevented or delayed by using a cold water extract of *S. platensis* [64]. Hence, it is suggested that the inclusion of cyanobacterial supplement in beverages and food products should be strongly considered.

4. Antitumor Functions

Strong evidences have shown that *S. platensis* is also imbued with antitumor and anticancer functions. In this regard, it was discovered that significant to full tumor regression was obtained with intravenous injection of Radachlorin, a new chlorine photosensitizer that was derived from *S. platensis* [65]. It was shown that hot-water extract of *S. platensis* facilitated enhanced antitumor activity of natural killer (NK) cells in rats [66]. Recently, complex polysaccharides from *Spirulina* have brought about suppression of glioma cell growth by downregulating angiogenesis via partial regulation of interleukin-17 production [67]. High production of tumor necrosis factor-α (TNF-α), in macrophages, was recorded in the presence of acidic polysaccharides from *A. platensis* [68]. Li et al. [69] have shown that with increased phycocyanin concentration, expression of CD59 proteins in HeLa cells was promoted while Fas protein that induces apoptosis was increased with an attendant decline in the multiplication of HeLa cells. These findings are an evidence for the multidimensional applications of phycocyanin content of *S. platensis*.

5. Antiviral Functions

Many compounds with antimicrobial activities have been isolated from different marine organisms, and a number of evidences are put forward for the antiviral activity of *Spirulina* [70, 71]. This antiviral activity, in a large part, is attributable to the richness of *S. platensis* in vital proteins, fatty acids, minerals, and other important constituents [72].

Previously, calcium spirulan (Ca-SP), a novel sulfated polysaccharide that was isolated from hot water extract of *S. platensis*, has shown antiviral activities against different enveloped viruses such as *Herpes simplex* virus type-I,

measles virus, HIV-I and influenza virus. This high sought for antiviral activity has been suggested to be due to the effect that chelation of calcium ions to sulfate groups has on molecular conformation [73]. Both extracellular and intracellular spirulan-like molecules from the polysaccharide fractions of *A. platensis* displayed significant antiviral activities against wide range of viruses, including human cytomegalovirus and HIV-I [74]. About 50% and 23% reductions in viral load were recorded for methanolic and aqueous extracts of *S. platensis*, respectively [75]. Reduction in viral load was attributed to inhibition of HIV-I replication in human T cells, langerhans cells, and peripheral blood mononuclear cells (PBMCs), with up to 50% reduction accorded to PBMCs [76]. Antiviral and immunostimulatory properties of *S. platensis* preparations were elicited through increased mobilization of macrophages, cytokine production, antibodies generation, accumulation of NK cells, and mobilization of B and T cells [77]. A recent study on the antiviral activity of *Spirulina* has resulted in the isolation of Cyanovirin-N (CV-N), a novel cyanobacterial carbohydrate-binding protein that inhibits HIV-I and other enveloped viral particles [78]. The Kanenbu tribe of Chad and most people in Korea and Japan, who consume *Spirulina* diet daily, have been shown to display lower cases of HIV/AIDS than their surrounding neighbors who do not take such diet. Therefore, it is expected that consistent intake of diets containing *Spirulina* can help in reducing the prevalence of HIV/AIDS [79]. Antiherpetic activities were noted for the crude extracts of *S. fusiformis* [80]. While Hernández-Corona et al. [81] have reported antiviral activity of *S. maxima* against HSV-2, Shalaby et al. reported similar activity for *S. platensis* against HSV-I [82].

6. Antibacterial Functions

Spirulina is not without antibacterial activity. In 3-week-old chicks injected with either *Escherichia coli* or *Staphylococcus aureus* suspensions, 0.1% *Spirulina* was found to enhance their bacterial clearance abilities, as shown by the improvement in the activities of different phagocytotic cells, including heterophils, thrombocytes, macrophages, and monocytes in the chickens [83]. Microalgal cultures of *A. platensis* have displayed significant antibacterial activity against six *Vibrio* strains: *Vibrio parahaemolyticus*, *Vibrio anguillarum*, *Vibrio splendidus*, *Vibrio scophthalmi*, *Vibrio alginolyticus*, and *Vibrio lentus* [84]. Antibacterial activity against *Streptococcus pyogenes* and/or *S. aureus* was proven for the phycobiliproteins isolated from *A. fusiformis* [85]. Purified C-phycocyanin from *S. platensis* markedly inhibited the growth of some drug resistant bacteria: *E. coli*, *Klebsiella pneumoniae*, *Pseudomonas aeruginosa*, and *S. aureus* [86]. This shows the potentials of compounds isolated from these cyanobacterial species in the fight against drug resistance.

7. Antifungal Functions

Recently, *Spirulina* has also exhibited antifungal activities [87]. Activity of 13 mm was recorded against *Candida glabrata* in the butanol extract of *Spirulina* sp. [88]. The immunostimulatory effect of *S. platensis* extract was tested in Balb/C mice infected with candidiasis [89]. In this experiment, pretreatment of the mice with 800 mg/kg of the extract for 4 days before intravenous inoculation with *C. albicans* resulted in increased production of cytokines TNF-α and interferon-gamma (IFN-γ), leading to increased survival time and better fungal clearance than in control groups. Glucosamine production was reduced by about 56% when the antifungal activity of the methanolic extract of *S. platensis* was tested against *Aspergillus flavus* [90]. Contrary to these findings, *S. platensis* grown in Zarrouk media, DB_1 media and papaya skin extract media did not show any antifungal activity [91]. In some instances, extracts from *S. platensis* may display a stimulatory effect toward cultured microorganisms. It was found by Gorobets et al. [92] that different doses of *S. platensis,* when added to culture fluid, displayed important stimulatory and inhibitory effects on the cultured microorganisms due to the presence of complex metabolites that were active in the prepared nutrient agar. Similarly, *S. platensis* biomass was used to maintain the counts of starter organisms in acidophilus-bifidus-thermophilus (ABT) milks at satisfactory levels during whole duration of storage. This is a novel opportunity for the production and maintenance of functional dairy foods [93].

8. Miscellaneous Functions

Many compounds produced from marine organisms, including cyanobacteria, have important protective functions against various allergic responses such as asthma, atopic dermatitis, and allergic rhinitis [94].

Powders of *S. platensis* have inhibited anaphylactic reaction resulting from antidinitrophenol IgE-induced histamine release or from TNF-α of rats [95]. *Spirulina* was also found effective against allergic rhinitis [96]. In an earlier human feeding study conducted in this regard, *Spirulina*-based dietary supplement was found effective in suppressing the level of interleukin- (IL-) 4 [23]. Zymosan-induced upsurge in the level of beta-glucuronidase of experimental mice was significantly reduced following the administration of phycocyanin [97]. This antiarthritic action may be due to the combination of various mechanisms such as free radical scavenging, inhibition of arachidonic acid metabolism as well as inhibition of TNF-α within the mice.

S. platensis has also neuroprotective ability. Its neuroprotective effect was demonstrated in adult Sprague-Dawley rats through a significant reduction in the volume of cerebral cortex infarction and increased poststroke locomotor activity. Hence, it is suggested that chronic treatment with *Spirulina* can reduce ischemic brain damage [98]. A report has shown that lead-induced increase in mast cells in rat ovary, during estrous cycle, is curtailed by using *Spirulina* at 300 mg/kg [99].

Another important compound synthesized by *Spirulina,* which equally has a lot of vital applications, is polyhydroxyalkanoates. These are polyesters produced by bacterial fermentation of sugars or lipids. According to Campbell et al. [100], *S. platensis* stores about 6% of its total dry weight and this value decreases during stationary phase of its growth profile. However, Jau et al. [101] have posited that this value can be increased to 10% when the organisms are grown under nitrogen-deficient, mixotrophic culture medium. The use of recombinant *E. coli*, due to its fast rate of growth and minimal nutrient requirements, to overproduce polyhydroxyalkanoates, has the potential of increasing the number of polyhydroxyalkanoates inclusions per cell [102]. Polyhydroxyalkanoates hold an assuring promise in therapeutic applications as drug carriers that display a release pattern similar to those of monolithic devices; an early rapid release followed by a prolonged, but slower release pattern. This type of drug-release behavior is normally required for depositing adequate concentration of drug of interest at the site of infection [103]. Among the polyhydroxyalkanoates, 4-hydroxybutyrate has long been advocated for the treatment of alcohol withdrawal syndrome in alcohol-dependent subjects [104].

9. Conclusions

In the foregoing essay, various nutritional and medicinal potencies have been attributed to metabolites from the cyanobacteria, *Spirulina (Arthrospira)* sp. In the present clamor for alternative medicine, these organisms serve as very viable potential sources of bioactive products with commercial imports. Therefore, more should be done in the study, culture, isolation, and purification of these organisms to enable beneficial harvest of their important inclusions.

References

[1] J. A. Downing, S. B. Watson, and E. McCauley, "Predicting cyanobacteria dominance in lakes," *Canadian Journal of Fisheries and Aquatic Sciences*, vol. 58, no. 10, pp. 1905–1908, 2001.

[2] T. Miyatake, B. J. MacGregor, and H. T. S. Boschker, "Depth-related differences in organic substrate utilization by major microbial groups in intertidal marine sediment," *Applied and Environmental Microbiology*, vol. 79, no. 1, pp. 389–392, 2013.

[3] L. . Charpy, B. E. Casareto, M. J. Langlade, and Y. Suzuki, "Cyanobacteria in coral reef ecosystems: a review," *Journal of Marine Biology*, vol. 2012, Article ID 259571, 9 pages, 2012.

[4] M. Singh, N. K. Sharma, S. B. Prasad, S. S. Yadav, G. Narayan, and A. K. Rai, "Freshwater cyanobacterium *Anabaena doliolum* transformed with ApGSMT-DMT exhibited enhanced salt tolerance and protection to nitrogenase activity, but changed its behavior to halophily," *Microbiology*, vol. 159, pp. 641–648, 2013.

[5] M. Zhang, X. Shi, Y. Yu, and F. Kong, "The acclimative changes in photochemistry after colony formation of the cyanobacteria *Microcystis aeruginosa*," *Journal of Phycology*, vol. 47, no. 3, pp. 524–532, 2011.

[6] M. Sabart, D. Pobel, E. Briand et al., "Spatiotemporal variations in microcystin concentrations and in the proportions of microcystin-producing cells in several *Microcystis aeruginosa* populations," *Applied and Environmental Microbiology*, vol. 76, no. 14, pp. 4750–4759, 2010.

[7] J. Kazmierczak, W. Altermann, B. Kremer, S. Kempe, and P. G. Eriksson, "Mass occurrence of benthic coccoid cyanobacteria and their role in the production of Neoarchean carbonates of

South Africa," *Precambrian Research*, vol. 173, no. 1–4, pp. 79–92, 2009.

[8] N. Morin, T. Vallaeys, L. Hendrickx, L. Natalie, and A. Wilmotte, "An efficient DNA isolation protocol for filamentous cyanobacteria of the genus Arthrospira," *Journal of Microbiological Methods*, vol. 80, no. 2, pp. 148–154, 2010.

[9] I. V. Kozhevnikov and N. A. Kozhevnikova, "Taxonomic studies of some cultured strains of Cyanobacteria (Nostocales) isolated from the Yenisei River basin," *Inland Water Biology*, vol. 4, no. 2, pp. 143–152, 2011.

[10] W. V. Sigler, R. Bachofen, and J. Zeyer, "Molecular characterization of endolithic cyanobacteria inhabiting exposed dolomite in central Switzerland," *Environmental Microbiology*, vol. 5, no. 7, pp. 618–627, 2003.

[11] A. A. Al-Thukair and S. Golubic, "Five new *Hyella* species from Arabian Gulf," *Algological Studies*, vol. 64, pp. 167–197, 1991.

[12] O. Koksharova and C. Wolk, "Genetic tools for cyanobacteria," *Applied Microbiology and Biotechnology*, vol. 58, no. 2, pp. 123–137, 2002.

[13] D. D. Wynn-Williams and H. G. M. Edwards, "Proximal analysis of regolith habitats and protective molecules in situ by Laser Raman Spectroscopy. Overview of terrestrial Antarctic habitats and Mars analogs," *Icarus*, vol. 144, no. 2, pp. 486–503, 2000.

[14] P. G. Becher and F. Jüttner, "Insecticidal compounds of the biofilm-forming cyanobacterium *Fischerella sp.* (ATCC 43239)," *Environmental Toxicology*, vol. 20, no. 3, pp. 363–372, 2005.

[15] M. . Pírez, G. Gonzalez-Sapienza, D. Sienra et al., "Limited analytical capacity for cyanotoxins in developing countries may hide serious environmental health problems: simple and affordable methods may be the answer," *Journal of Environmental Management*, vol. 114, pp. 63–71, 2013.

[16] S. Klitzke, C. Beusch, and J. Fastner, "Sorption of the cyanobacterial toxins cylindrospermopsin and anatoxin-a to sediments," *Water Research*, vol. 45, no. 3, pp. 1338–1346, 2011.

[17] L. C. Bowling and P. D. Baker, "Major cyanobacterial bloom in the Barwon-Darling River, Australia, in 1991, and underlying limnological conditions," *Marine and Freshwater Research*, vol. 47, no. 4, pp. 643–657, 1996.

[18] L. T. Tan, "Filamentous tropical marine cyanobacteria: a rich source of natural products for anticancer drug discovery," *Journal of Applied Phycology*, vol. 22, no. 5, pp. 659–676, 2010.

[19] F. Heidari, H. Riahi, M. Yousefzadi, and M. Asadi, "Antimicrobial activity of cyanobacteria isolated from hot spring of geno," *Middle-East Journal of Scientific Research*, vol. 12, no. 3, pp. 336–339, 2012.

[20] R. Prasanna, A. Sood, P. Jaiswal et al., "Rediscovering cyanobacteria as valuable sources of bioactive compounds (Review)," *Applied Biochemistry and Microbiology*, vol. 46, no. 2, pp. 119–134, 2010.

[21] R. B. Volk and F. H. Furkert, "Antialgal, antibacterial and antifungal activity of two metabolites produced and excreted by cyanobacteria during growth," *Microbiological Research*, vol. 161, no. 2, pp. 180–186, 2006.

[22] R.-B. Volk, "Antialgal actitiy of several cyanobacterial exometabolites," *Journal of Applied Phycology*, vol. 18, no. 2, pp. 145–151, 2006.

[23] T. K. Mao, J. Van De Water, and M. E. Gershwin, "Effects of a Spirulina-based dietary supplement on cytokine production from allergic rhinitis patients," *Journal of Medicinal Food*, vol. 8, no. 1, pp. 27–30, 2005.

[24] L. M. Colla, C. Oliveira Reinehr, C. Reichert, and J. A. V. Costa, "Production of biomass and nutraceutical compounds by Spirulina platensis under different temperature and nitrogen regimes," *Bioresource Technology*, vol. 98, no. 7, pp. 1489–1493, 2007.

[25] C. Sili, G. Torzillo, and A. Vonshak, "Arthrospira (Spirulina)," in *Ecology of Cyanobacteria II*, B. A. Whitton, Ed., pp. 677–705, Springer, Dordrecht, The Netherlands, 2012.

[26] B. Castiglioni, E. Rizzi, A. Frosini et al., "Development of a universal microarray based on the ligation detection reaction and 16S rRNA gene polymorphism to target diversity of cyanobacteria," *Applied and Environmental Microbiology*, vol. 70, no. 12, pp. 7161–7172, 2004.

[27] Y. M. Lu, W. Z. Xiang, and Y. H. Wen, "Spirulina (Arthrospira) industry in Inner Mongolia of China: current status and prospects," *Journal of Applied Phycology*, vol. 23, no. 2, pp. 265–269, 2011.

[28] N. Kumar, S. Singh, N. Patro, and I. Patro, "Evaluation of protective efficacy of *Spirulina platensis* against collagen-induced arthritis in rats," *Inflammopharmacology*, vol. 17, no. 3, pp. 181–190, 2009.

[29] O. Ciferri, "Spirulina, the edible microorganism," *Microbiological Reviews*, vol. 47, no. 4, pp. 551–578, 1983.

[30] A. Belay, "The potential of *Spirulina (Arthrospira)* as a nutritional and therapeutic supplement in health management," *Journal of the American Nutraceutical Association*, vol. 5, no. 2, pp. 27–49, 2002.

[31] J. C. M. Carvalho, R. P. Bezerra, M. C. Matsudo, and S. Sato, "Cultivation of *Arthrospira (Spirulina) platensis* by Fed-Batch Process," in *Advanced Biofuels and Bioproducts*, J. W. Lee, Ed., pp. 781–805, Springer, New York, NY, USA, 2013.

[32] S. L. Pahl, A. K. Lee, T. Kalaitzidis, P. J. Ashman, S. Sathe, and D. M. Lewis, "Harvesting, thickening and dewatering microalgae biomass," in *Algae for Biofuels and Energy*, M. A. Borowitzka and N. R. Moheimani, Eds., vol. 5, pp. 165–185, Springer, Dordrecht, The Netherlands, 2013.

[33] L. Trabelsi, H. Ben Ouda, F. Zili, N. Mazhoud, and J. Ammar, "Evaluation of *Arthrospira platensis* extracellular polymeric substances production in photoautotrophic, heterotrophic and mixotrophic conditions," *Folia Microbiologica*, vol. 58, pp. 39–45, 2013.

[34] M. A. Borowitzka, "High-value products from microalgae-their development and commercialization," *Journal of Applied Phycology*, vol. 1, pp. 1–14, 2013.

[35] N. Yoshikawa and A. Belay, "Single-laboratory validation of a method for the determination of c-phycocyanin and allophycocyanin in spirulina (Arthrospira) supplements and raw materials by spectrophotometry," *Journal of AOAC International*, vol. 91, no. 3, pp. 524–529, 2008.

[36] M. Filomena de Jesus Raposo, R. M. Santos Costa de Morais, and A. M. Miranda Bernado de Morais, "Bioactivity and applications of sulphated polysaccharides from marine microalgae," *Marine Drugs*, vol. 11, no. 1, pp. 233–252, 2013.

[37] T. G. Sotiroudis and G. T. Sotiroudis, "Health aspects of *Spirulina (Arthrospira)* microalga food supplement," *Journal of the Serbian Chemical Society*, vol. 78, no. 3, pp. 395–405, 2013.

[38] J. Villa, C. Gemma, A. Bachstetter, Y. Wang, I. Stromberg, and P. C. Bickford, "Spirulina, aging, and neurobiology," in *Spirulina in Human Nutrition and Health*, M. E. Gershwin and A. Belay, Eds., pp. 271–291, CRC Press, Taylor & Francis Group, Boca Raton, Fla, USA, 2007.

[39] Y. Ishimi, F. Sugiyama, J. Ezaki, M. Fujioka, and J. Wu, "Effects of spirulina, a blue-green alga, on bone metabolism in ovariectomized rats and hindlimb-unloaded mice," *Bioscience, Biotechnology and Biochemistry*, vol. 70, no. 2, pp. 363–368, 2006.

[40] M. A. Olvera-Novoa, L. J. Domínguez-Cen, L. Olivera-Castillo, and C. A. Martínez-Palacios, "Effect of the use of the microalga *Spirulina maxima* as fish meal replacement in diets for tilapia," *Aquaculture Research*, vol. 29, no. 10, pp. 709–715, 1998.

[41] S. F. Rabelo, A. C. Lemes, K. P. Takeuchi, M. T. Frata, J. C. Monteiro de Carvalho, and E. D. G. Danesi, "Development of cassava doughnuts enriched with *Spirulina platensis* biomass," *Brazilian Journal of Food Technology*, vol. 16, no. 1, pp. 42–51, 2013.

[42] G. Chamorro, M. Salazar, L. Favila, and H. Bourges, "Pharmacology and toxicology of the alga Spirulina," *Revista de Investigacion Clinica*, vol. 48, no. 5, pp. 389–399, 1996.

[43] K. P. Sharma, N. Upreti, S. Sharma, and S. Sharma, "Protective effect of Spirulina and tamarind fruit pulp diet supplement in fish (*Gambusia affinia* Baird & Girard) exposed to sublethal concentration of fluoride, aluminum and aluminum fluoride," *Indian Journal of Experimental Biology*, vol. 50, pp. 897–903, 2012.

[44] R. Kapoor and U. Mehta, "Supplementary effect of spirulina on hematological status of rats during pregnancy and lactation," *Plant Foods for Human Nutrition*, vol. 52, no. 4, pp. 315–324, 1998.

[45] S. A. Kedik, E. I. Yartsev, I. V. Sakaeva, E. S. Zhavoronok, and A. V. Panov, "Influence of spirulina and its components on the immune system (review)," *Russian Journal of Biopharmaceuticals*, vol. 3, no. 3, pp. 3–10, 2011.

[46] T. Hirahashi, M. Matsumoto, K. Hazeki, Y. Saeki, M. Ui, and T. Seya, "Activation of the human innate immune system by Spirulina: augmentation of interferon production and NK cytotoxicity by oral administration of hot water extract of *Spirulina platensis*," *International Immunopharmacology*, vol. 2, no. 4, pp. 423–434, 2002.

[47] M. Samir and P. S. Amrit, "A review of pharmacology and phytochemicals from Indian medicinal plants," *The Internet Journal of Alternative Medicine*, vol. 5, no. 1, pp. 1–6, 2007.

[48] J. E. Piero Estrada, P. Bermejo Bescós, and A. M. Villar del Fresno, "Antioxidant activity of different fractions of *Spirulina platensis* protean extract," *IL Farmaco*, vol. 56, no. 5-7, pp. 497–500, 2001.

[49] J. Zheng, T. Inoguchi, S. Sasaki et al., "Phycocyanin and phycocyanobilin from *Spirulina platensis* protect against diabetic nephropathy by inhibiting oxidative stress," *American Journal of Physiology*, vol. 304, no. 2, pp. R110–R120, 2013.

[50] V. B. Bhat and K. M. Madyastha, "C-Phycocyanin: a potent peroxyl radical scavenger in vivo and in vitro," *Biochemical and Biophysical Research Communications*, vol. 275, no. 1, pp. 20–25, 2000.

[51] E. M. Gorban', M. A. Orynchak, N. G. Virstiuk, L. P. Kuprash, T. M. Panteleimonova, and L. B. Sharabura, "Clinical and experimental study of spirulina efficacy in chronic diffuse liver diseases," *Likars'ka Sprava*, no. 6, pp. 89–93, 2000.

[52] N. Paniagua-Castro, G. Escalona-Cardoso, D. Hernández-Navarro, R. Pérez-Pastén, and G. Chamorro-Cevallos, "*Spirulina (Arthrospira)* protects against cadmium-induced teratogenic damage in mice," *Journal of Medicinal Food*, vol. 14, no. 4, pp. 398–404, 2011.

[53] A. Kuhad, N. Tirkey, S. Pilkhwal, and K. Chopra, "Effect of Spirulina, a blue green algae, on gentamicin-induced oxidative stress and renal dysfunction in rats," *Fundamental and Clinical Pharmacology*, vol. 20, no. 2, pp. 121–128, 2006.

[54] J. L. Blé-Castillo, A. Rodríguez-Hernández, R. Miranda-Zamora, M. A. Juárez-Oropeza, and J. C. Díaz-Zagoya, "*Arthrospira maxima* prevents the acute fatty liver induced by the administration of simvastatin, ethanol and a hypercholesterolemic diet to mice," *Life Sciences*, vol. 70, no. 22, pp. 2665–2673, 2002.

[55] S. K. Saha, M. Misbahuddin, R. Khatun, and I. R. Mamun, "Effect of hexane extract of spirulina in the removal of arsenic from isolated liver tissues of rat," *Mymensingh Medical Journal*, vol. 14, no. 2, pp. 191–195, 2005.

[56] S. Ray, K. Roy, and C. Sengupta, "In vitro evaluation of antiperoxidative potential of water extract of *Spirulina platensis* (blue green algae) on cyclophosphamide-induced lipid peroxidation," *Indian Journal of Pharmaceutical Sciences*, vol. 69, no. 2, pp. 190–196, 2007.

[57] S. M. Farooq, D. Asokan, R. Sakthivel, P. Kalaiselvi, and P. Varalakshmi, "Salubrious effect of C-phycocyanin against oxalate-mediated renal cell injury," *Clinica Chimica Acta*, vol. 348, no. 1-2, pp. 199–205, 2004.

[58] R. Rodríguez-Sánchez, R. Ortiz-Butrón, V. Blas-Valdivia, A. Hernández-García, and E. Cano-Europa, "Phycobiliproteins or C-phycocyanin of *Arthrospira (Spirulina) maxima* protect against $HgCl_2$-caused oxidative stress and renal damage," *Food Chemistry*, vol. 135, no. 4, pp. 2359–2365, 2012.

[59] G. E. El-Desoky, S. A. Bashandy, I. M. Alhazza, Z. A. Al-Othman, M. A. M. Aboul-Soud, and K. Yusuf, "Improvement of mercuric chloride-induced testis injuries and sperm quality deteriorations by *Spirulina platensis* in rats," *PLoS ONE*, vol. 8, no. 3, 2013.

[60] V. A. Ionov and M. M. Basova, "Use of blue-green microseaweed *Spirulina platensis* for the correction of lipid and hemostatic disturbances in patients with ischemic heart disease," *Voprosy Pitaniia*, vol. 72, no. 6, pp. 28–31, 2003.

[61] M. Khan, J. C. Shobha, I. K. Mohan et al., "Protective effect of Spirulina against doxorubicin-induced cardiotoxicity," *Phytotherapy Research*, vol. 19, no. 12, pp. 1030–1037, 2005.

[62] J. C. Tobon-Velasco, V. Palafox-Sanchez, L. Mendieta et al., "Antioxidant effect of *Spirulina (Arthrospira) maxima* in a neurotoxic model caused by 6-OHDA in the rat striatum," *Journal of Neural Transmission*, 2013.

[63] R. Makhlouf and I. Makhlouf, "Evaluation of the effect of Spirulina against Gamma irradiation-induced oxidative stress and tissue injury in rats," *International Journal of Applied Sciences and Engineering Research*, vol. 1, no. 2, pp. 152–164, 2012.

[64] W. L. Chu, Y. W. Lim, A. K. Radhakrishnan, and P. E. Lim, "Protective effect of aqueous extract from *Spirulina platensis* against cell death induced by free radicals," *BMC Complementary and Alternative Medicine*, vol. 10, article 53, 2010.

[65] V. A. Privalov, A. V. Lappa, O. V. Seliverstov et al., "Clinical trials of a new chlorin photosensitizer for photodynamic therapy of malignant tumors," in *Optical Methods for Tumor Treatment and Detection: Mechanisms and Techniques in Photodynamic Therapy XI*, T. J. Dougherty, Ed., vol. 4612 of *Proceedings of SPIE*, pp. 178–189, January 2002.

[66] Y. Akao, T. Ebihara, H. Masuda et al., "Enhancement of antitumor natural killer cell activation by orally administered Spirulina extract in mice," *Cancer Science*, vol. 100, no. 8, pp. 1494–1501, 2009.

[67] Y. Kawanishi, A. Tominaga, H. Okuyama et al., "Regulatory effects of *Spirulina* complex polysaccharides on growth of murine RSV-M glioma cells through Toll-like receptor-4," *Microbiology and Immunology*, vol. 57, no. 1, pp. 63–73, 2013.

[68] M. L. Parages, R. M. Rico, R. T. Abdala-Díaz, M. Chabrillón, T. G. Sotiroudis, and C. Jiménez, "Acidic polysaccharides of *Arthrospira (Spirulina)* platensis induce the synthesis of TNF-α in RAW macrophages," *Journal of Applied Phycology*, vol. 24, pp. 1537–1546, 2012.

[69] B. Li, X. Zhang, M. Gao, and X. Chu, "Effects of CD59 on antitumoral activities of phycocyanin from *Spirulina platensis*," *Biomedicine and Pharmacotherapy*, vol. 59, no. 10, pp. 551–560, 2005.

[70] J. Yasuhara-Bell and Y. Lu, "Marine compounds and their antiviral activities," *Antiviral Research*, vol. 86, no. 3, pp. 231–240, 2010.

[71] A. M. S. Mayer, A. D. Rodríguez, R. G. S. Berlinck, and M. T. Hamann, "Marine pharmacology in 2005-6: marine compounds with anthelmintic, antibacterial, anticoagulant, antifungal, anti-inflammatory, antimalarial, antiprotozoal, antituberculosis, and antiviral activities; affecting the cardiovascular, immune and nervous systems, and other miscellaneous mechanisms of action," *Biochimica et Biophysica Acta*, vol. 1790, no. 5, pp. 283–308, 2009.

[72] L. P. Blinkova, O. B. Gorobets, and A. P. Baturo, "Biological activity of Spirulina," *Zhurnal Mikrobiologii Epidemiologii i Immunobiologii*, no. 2, pp. 114–118, 2001.

[73] T. Hayashi, K. Hayashi, M. Maeda, and I. Kojima, "Calcium spirulan, an inhibitor of enveloped virus replication, from a blue-green alga *Spirulina platensis*," *Journal of Natural Products*, vol. 59, no. 1, pp. 83–87, 1996.

[74] S. Rechter, T. König, S. Auerochs et al., "Antiviral activity of Arthrospira-derived spirulan-like substances," *Antiviral Research*, vol. 72, no. 3, pp. 197–206, 2006.

[75] S. M. Abdo, M. H. Hetta, W. M. El-Senousy, R. A. Salah El Din, and G. H. Ali, "Antiviral activity of freshwater algae," *Journal of Applied Pharmaceutical Science*, vol. 2, no. 2, pp. 21–25, 2012.

[76] S. Ayehunie, A. Belay, T. W. Baba, and R. M. Ruprecht, "Inhibition of HIV-1 replication by an aqueous extract of *Spirulina platensis (Arthrospira platensis)*," *Journal of Acquired Immune Deficiency Syndromes and Human Retrovirology*, vol. 18, no. 1, pp. 7–12, 1998.

[77] Z. Khan, P. Bhadouria, and P. S. Bisen, "Nutritional and therapeutic potential of Spirulina," *Current Pharmaceutical Biotechnology*, vol. 6, no. 5, pp. 373–379, 2005.

[78] J. Balzarini, "Carbohydrate-binding agents: a potential future cornerstone for the chemotherapy of enveloped viruses?" *Antiviral Chemistry and Chemotherapy*, vol. 18, no. 1, pp. 1–11, 2007.

[79] J. Teas, J. R. Hebert, J. H. Fitton, and P. V. Zimba, "Algae-a poor man's HAART?" *Medical Hypotheses*, vol. 62, no. 4, pp. 507–510, 2004.

[80] M. Sharaf, A. Amara, A. Aboul-Enein et al., "Molecular authentication and characterization of the antiherpetic activity of the cyanobacterium *Arthrospira fusiformis*," *Die Pharmazie*, vol. 65, no. 2, pp. 132–136, 2010.

[81] A. Hernández-Corona, I. Nieves, M. Meckes, G. Chamorro, and B. L. Barron, "Antiviral activity of *Spirulina maxima* against herpes simplex virus type 2," *Antiviral Research*, vol. 56, no. 3, pp. 279–285, 2002.

[82] E. A. Shalaby, S. M. M. Shanab, and V. Singh, "Salt stress enhancement of antioxidant and antiviral efficiency of *Spirulina*

platensis," *Journal of Medicinal Plant Research*, vol. 4, no. 24, pp. 2622–2632, 2010.

[83] M. A. Quereshi, R. A. Ali, and R. Hunter, "Immuno-modulatory effects of *Spirulina platensis* supplementation in chickens," in *Proceedings of the 44th Western Poultry Disease Conference*, pp. 117–121, Sacramento, Calif, USA, 1995.

[84] F. Kokou, P. Makridis, M. Kentouri, and P. Divanach, "Antibacterial activity in microalgae cultures," *Aquaculture Research*, vol. 43, no. 10, pp. 1520–1527, 2012.

[85] H. M. Najdenski, L. G. Gigova, I. I. Iliev et al., "Antibacterial and antifungal activities of selected microalgae and cyanobacteria," *International Journal of Food Science and Technology*, 2013.

[86] D. V. L. Sarada, C. S. Kumar, and R. Rengasamy, "Purified C-phycocyanin from *Spirulina platensis* (Nordstedt) Geitler: a novel and potent agent against drug resistant bacteria," *World Journal of Microbiology and Biotechnology*, vol. 27, no. 4, pp. 779–783, 2011.

[87] A. Duda-Chodak, "Impact of water extract of Spirulina (WES) on bacteria, yeasts and molds," *ACTA Scientiarum Polonorum Technologia Alimentaria*, vol. 12, pp. 33–39, 2013.

[88] J. Sivakumar and P. Santhanam, "Antipathogenic activity of Spirulina powder," *Recent Research in Science and Technology*, vol. 3, no. 4, pp. 158–161, 2011.

[89] M. Soltani, A.-R. Khosravi, F. Asadi, and H. Shokri, "Evaluation of protective efficacy of *Spirulina platensis* in Balb/C mice with candidiasis," *Journal of Medical Mycology*, vol. 22, no. 4, pp. 329–334, 2012.

[90] M. Moraes de Souza, L. Prietto, A. C. Ribeiro, T. Denardi de Souza, and E. Badiale-Furlong, "Assessment of the antifungal activity of *Spirulina platensis* phenolic extract against *Aspergillus flavus*," *Ciencia e Agrotecnologia*, vol. 35, no. 6, pp. 1050–1058, 2011.

[91] N. Akhtar, M. M. Ahmed, N. Sarker, K. R. Mahbub, and M. A. Sarker, "Growth response of *Spirulina platensis* in papaya skin extract and antimicrobial activities of Spirulina extracts in different culture media," *Bangladesh Journal of Scientific and Industrial Research*, vol. 47, no. 2, pp. 147–152, 2012.

[92] O. B. Gorobets, L. P. Blinkova, and A. P. Baturo, "Stimulating and inhibiting effect of *Spirulina platensis* on microorganisms," *Zhurnal Mikrobiologii Epidemiologii i Immunobiologii*, no. 6, pp. 20–24, 2001.

[93] L. Varga, J. Szigeti, R. Kovács, T. Földes, and S. Buti, "Influence of a *Spirulina platensis* biomass on the microflora of fermented ABT milks during storage (R1)," *Journal of Dairy Science*, vol. 85, no. 5, pp. 1031–1038, 2002.

[94] T.-S. Vo, D.-H. Ngo, and S.-K. Kim, "Marine algae as a potential pharmaceutical source for anti-allergic therapeutics," *Process Biochemistry*, vol. 47, no. 3, pp. 386–394, 2012.

[95] N. H. Yang, E. H. Lee, and H. M. Kim, "*Spirulina platensis* inhibits anaphylactic reaction," *Life Sciences*, vol. 61, pp. 1237–1244, 1997.

[96] C. Cingi, M. Conk-Dalay, H. Cakli, and C. Bal, "The effects of spirulina on allergic rhinitis," *European Archives of Oto-Rhino-Laryngology*, vol. 265, no. 10, pp. 1219–1223, 2008.

[97] D. Remirez, A. González, N. Merino et al., "Effect of phycocyanin in zymosan-induced arthritis in mice—phycocyanin as an antiarthritic compound," *Drug Development Research*, vol. 48, no. 2, pp. 70–75, 1999.

[98] Y. Wang, C. F. Chang, J. Chou et al., "Dietary supplementation with blueberries, spinach, or spirulina reduces ischemic brain damage," *Experimental Neurology*, vol. 193, no. 1, pp. 75–84, 2005.

[99]   T. Karaca and N. Şimşek, "Effects of spirulina on the number of ovary mast cells in lead-induced toxicity in rats," *Phytotherapy Research*, vol. 21, no. 1, pp. 44–46, 2007.

[100] J. Campbell, S. E. Stevens, and D. L. Balkwill, "Accumulation of polyhy-B-hydroxybutyrate in *Spirulina platensis*," *Journal of Bacteriology*, vol. 1499, no. 1, p. 361, 1982.

[101] M. H. Jau, S. P. Yew, P. S. Y. Toh et al., "Biosynthesis and mobilization of poly(3-hydroxybutyrate) [P(3HB)] by *Spirulina platensis*," *International Journal of Biological Macromolecules*, vol. 36, no. 3, pp. 144–151, 2005.

[102] A. J. Anderson and E. A. Dawes, "Occurrence, metabolic role and industrial uses of bacterial polyhydroxyalkanoates," *Microbiological Reviews*, vol. 54, no. 4, pp. 450–474, 1990.

[103] G. Q. Chen and Q. Wu, "The application of polyhydroxyalkanoates as tissue engineering materials," *Biomaterials*, vol. 26, no. 33, pp. 6565–6578, 2005.

[104] L. Gallimberti, M. Ferri, S. D. Ferrara, F. Fadda, and G. L. Gessa, "Gamma-hydroxybutyric acid in the treatment of alcohol dependence: a double-blind study," *Alcoholism*, vol. 16, no. 4, pp. 673–676, 1992.

12

New Records of Atypical Coral Reef Habitat in the Kimberley, Australia

Z. T. Richards, M. Bryce, and C. Bryce

Aquatic Zoology, West Australian Museum, Locked Bag 49, Welshpool DC, WA 6986, Australia

Correspondence should be addressed to Z. T. Richards; zoe.richards@museum.wa.gov.au

Academic Editor: Jakov Dulčić

New surveys of the Kimberley Nearshore Bioregion are beginning to fill knowledge gaps about the region's marine biodiversity and the national and international conservation significance of this little-known tropical reef system. Here we report the recent finding of two unique coral habitats documented at Adele Island and Long Reef during the *Woodside 2009/2010 Collection Project* surveys. Firstly, we report the finding of a subtidal zone of mixed corallith and rhodolith habitat which appears on current records, to be unprecedented in Australia. Secondly, we report the discovery of an atypical Organ Pipe Coral habitat zone and provide empirical evidence that this commercially valuable species reaches an unparalleled level of benthic cover. We provide additional details about the wider hard and soft coral assemblages associated with these unique habitats; discuss the potential biological causes and consequences of them, and make recommendations to benefit their conservation.

1. Introduction

Around the globe, most tropical reef locations have been the focus of at least some scientific studies. One of the last regions of shallow-water reef remaining to be explored is the Kimberley (north-west Australia). Renowned for huge tidal exchanges up to 11 m, frequent cyclones and crocodiles, this region has experienced little reef-based research apart from a series of biodiversity surveys conducted by the Western Australian Museum in the 1990s. Despite the growing public and industrial interest in this frontier region, the diversity of the tropical Kimberley reefs remains largely unknown, even at the coarse habitat level.

Geomorphological surveys of reef development in Western Australia suggest that reefs in the Kimberley coastal bioregion are uniquely characterized by the development of Holocene accretionary veneers of coral-algal limestone (>12,000 years old) on a Proterozoic basement (2,500–543 million years old) [1]. Neighbouring reefs on the northwest shelf (i.e., within the Western Pilbara and West Coast Bioregions) have origins in the Pleistocene (25 million–12,000 years ago). Thus, contemporary reefs in the Kimberley

appear to have more recent origins than others in Western Australia; however, the extent to which the difference in age translates to compositional differences between reefs remains to be resolved.

Along the Kimberley continental edge, there is extensive development of oceanic reefs. The offshore reefs (such as Ashmore, Cartier, and Scott Reefs) have been extensively surveyed as part of National and Regional management plans [2–9] and contain a surprising diversity of marine life given their isolation, including a high diversity of hard and soft corals. Despite forming a separate and distinct bioregion to the Oceanic Shoals [10, 11], only occasional coral collecting has been conducted within the nearshore Kimberley bioregion [12].

To address this information gap, a new project began in 2008 intended to quantify the diversity, distribution, and abundance of hard and soft corals along the Kimberley coast with that of other marine fauna and flora. The Woodside Collection Project (Kimberley): 2008–2011 (henceforth, the Woodside Project) (http://www.museum.wa.gov.au/kimberley/marine-life-kimberley-region) will combine inshore and offshore data to fill various important biogeographic and taxonomic knowledge gaps. This will help define the current knowledge concerning the region's biodiversity to aid conservation managers, industry, and government in establishing the national and international conservation significance of the region.

Here we describe two unusual benthic habitats discovered during the Woodside 4 Project surveys in 2009 at Adele Island and in 2010 at Long Reef. We will also discuss their potential biological causes, consequences, and significance and make recommendations to benefit their conservation.

2. Methods

2.1. Site Description. Adele Island (15°31.3'S, 123°9.5.0'E) is situated approximately 90 kms from the coastline on the inner section of the north-west shelf. Approximately 17 km long and 9 km wide, Adele is a true reef platform with a vegetated centre. The reef is considered one of the largest and mature of the Sahul Shelf [13]. Wilson [14] describes the reef platform as, "a bioherm … [whose] top is a cap of Quaternary reefal limestone, built on an inundated rocky hill of the dissected Kimberley Basis margin over a Proterozoic basement." Prevailing westerly wind, swell, and diurnal tides, with up to 11 m range on high-water springs, have shaped the intertidal reef platform (rampart) that is heavily etched with small drainage channels and low, widely spaced ridges running parallel to the reef edge that resemble a series of long low and wide corrugations. Beyond the rampart, the reef slopes into sublittoral fore-reef and back-reef zones featuring hard corals, soft corals, hydroids, bryozoans and macroalgae, such as *Sargassum*.

Long Reef (15°56.5'S, 124°12.5''E) is situated approximately 28 kms from the coastline, also on the inner part of the NW Sahul Shelf. It is a north-south orientated elongated platform reef that is approximately 27 km long and 12 km wide, with a similar geological history to Adele Island. It was built on antecedent Proterozoic features that are remnants of topographic high points of the submerged Kimberley Basin margin [14]. Contemporary reef growth is controlled by wind, swell and extreme tidal exchanges. Long Reef features a large intertidal reef pavement zone resembling others in the inshore Kimberley such as Adele, but in contrast to Adele an island has not formed. There is, however, a central sandy spine running down the centre of the reef. The flat midlittoral reef pavement is covered with turf algae and tide pools which are fringed with small coral colonies. The reef pavement extends down to a narrow, honeycombed, and terraced fore-reef ramp that forms small drainage cascades resembling the low tide cascades of Montgomery Reef [15]. Subtidally, there is a steep fore-reef slope that descends beyond the 20 m contour and a fractured reef base with hard corals, sponges, hydroids, and other fauna and flora. The reef on the leeward side is gently sloping.

2.2. Field Surveys. Hard and soft coral biodiversity at Adele Island and Long Reef (Figure 1) were surveyed for the first time using standard benthic monitoring techniques. At Adele Island 13 stations were surveyed in October 2009

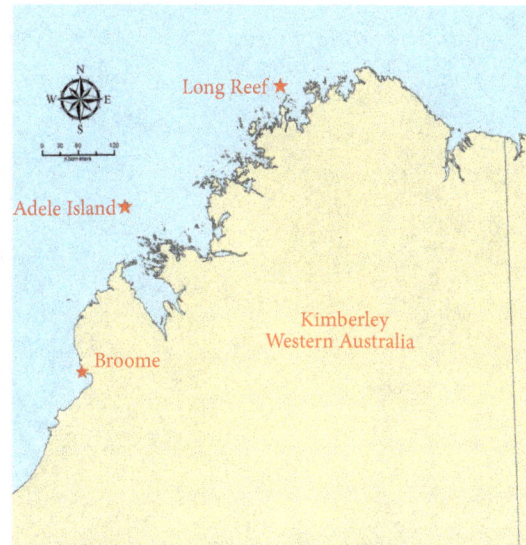

FIGURE 1: Map of the Kimberley coastline marking the location of Adele Island and Long Reef.

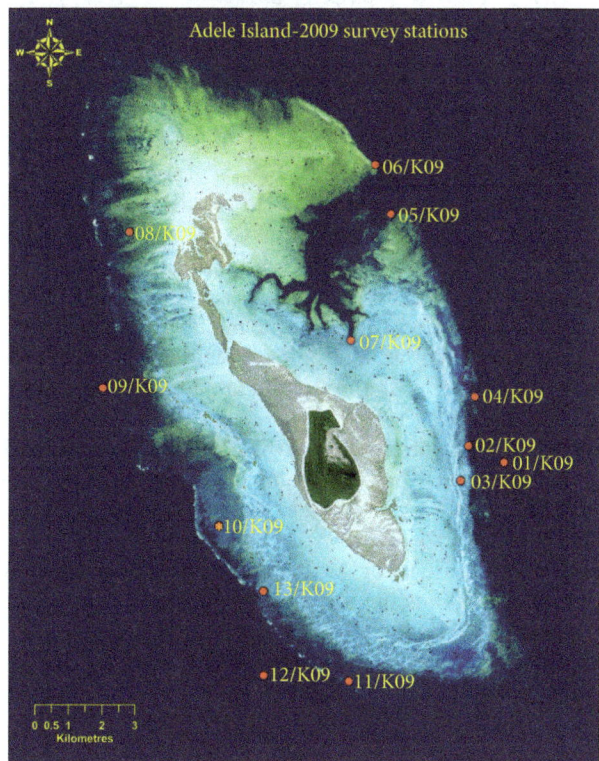

FIGURE 2: Satellite image of the extensive coral reef surrounding Adele Island with survey stations highlighted, the rollolith habitat zone is denoted by a star.

(Figure 2). Eleven stations were surveyed using SCUBA and two intertidal stations were surveyed. At Long Reef 15 stations were surveyed in October 2010 (Figure 3), nine stations using SCUBA and six stations were intertidal.

The mean species richness (+SE) of hard corals was calculated at each station from four replicated 15 m × 1 m wide

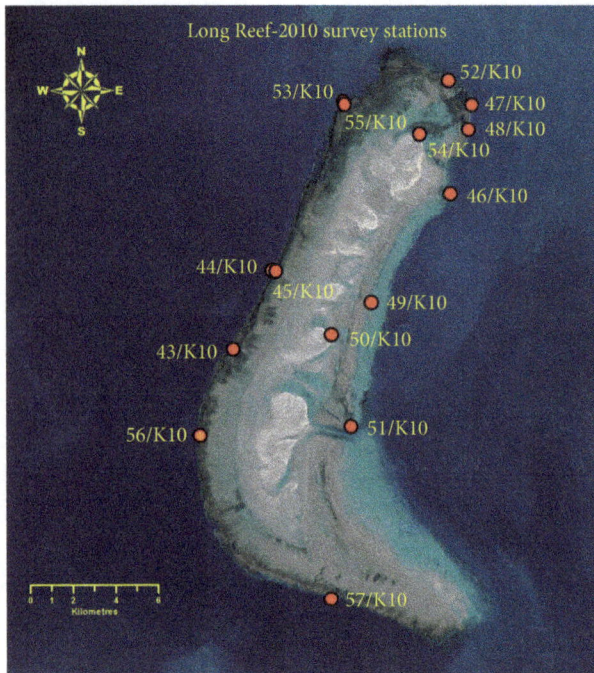

FIGURE 3: Satellite image of Long Reef with survey stations high-lighted, the *Tubipora* habitat denoted by a star.

coral biodiversity belt transects. In addition, a 20-minute rapid visual search over the station (covering approximately 1000 m^2) was conducted to detect rare and infrequent species. The total hard coral assemblage at each station was documented from pooled transect and rapid visual assessment records. For soft corals, a 60-minute rapid visual search was conducted over each station to obtain a thorough list of the soft coral assemblage.

In addition, during the spring low tide of the 24th October 2010, six 10 m long point intercept transects were conducted in the exposed fore-reef ramp to quantify the benthic coverage of *Tubipora* sp. Our intention here is to describe two unusual benthic habitats that were encountered on these expeditions in an ecological context and articulate the extent of surveying we have conducted at these inshore Kimberley locations in order to establish the prevalence and significance of the new habitats.

3. Results

3.1. Coralliths and Rhodoliths.
Intertidal and subtidal surveys of Adele Island revealed extensive development of hard and soft coral communities. Seventy-eight species of scleractinian coral were recorded in intertidal and shallow water stations (<6 m depth). The level of hard coral diversity increased to 176 species with additional surveying at subtidal sites (>14 m depth). These records indicate that approximately half of all the hard coral species currently known to occur in the offshore and inshore Kimberley bioregions are present at Adele Island [16].

Octocoral biodiversity in intertidal and shallow water stations under 6 m depth was species poor, increasing considerably with greater depth. Only four species of octocoral were recorded in habitats under 6 m depth (*Heliopora coerulea*, *Lobophytum* sp., and *Anthelia* sp., *Sansibia* sp.), whereas 28 species were recorded at a depth of 6 m to 14 m.

At station 10/K9, on the south west side of Adele Island a curious zone of benthic fauna was observed at the southern edge. This station was in <4 m depth on the midlittoral back reef. It was a high energy, current swept consolidated reef platform with many depressions leading to a labyrinth of underplatform tunnels. The depressions contained course sand and coral rubble. Fifty-five species of hermatypic coral were recorded at this station with the five most commonly encountered species on transects being *Goniastrea retiformis*, *Acropora papillare*, *Favites halicora*, *Goniastrea pectinata*, and *Porites annae*. Within the Alcyonaria only the blue coral, *Heliopora coerulea*, and one species of *Anthelia* sp. were recorded.

At the southern edge of station 10/K9, the consolidated habitat gave way to a rubble bed. This was dominated by a mixed assemblage of coralliths (free-living live coral nodules) and rhodoliths (free-living balls of coralline algae) (herein called the rhollolith zone, *sensu* Baarli, 2010—where rollolith refers to motile balls of mixed origin, i.e., coral, algae, polychaete or bryozoan). Among the coralliths, four different species of coral were identified (*Pavona venosa*, *Psammocora profundacella*, *Cyphastrea chalcidicum*, and *Millepora* sp.) (Figures 4(a), 4(b), and 4(d)). Milleporinid coralliths were the most abundant and between 3 and 5 coralliths of the other genera were observed within the section of the rollolith zone that was examined. The maximum size of the coralliths was 12 cm (greater diameter).

Rhodoliths (Figure 4(c)) dominated the assemblage in the rollolith zone and covered the substrate. The largest rhodolith recorded was 8 cm (greater diameter). Generally the rhodoliths formed as a cluster of short (>1 cm) dense branches, some of which appeared to be formed around coral rubble or shells, others appeared to be entirely of algal origin.

3.2. Long Reef: Tubipora.
Intertidal and subtidal surveys at Long Reef revealed extensive development of hard coral and soft coral communities. Preliminary systematic studies indicate 200 species of scleractinian corals occur at Long Reef, 68 of which occur in the intertidal zone. Octocoral biodiversity followed the same trend as Adele Island with low biodiversity in intertidal and shallow water stations under 6 m depth, and increasing to 29 species between 6 m and 14 m.

During a spring low tide surveys of Long Reef we encountered an atypical habitat zone dominated by the Organ Pipe Coral, *Tubipora* sp. on the reef platform bordering station 56/K10 (Figures 5(a) and 5(b)). The platform was on the south west side of Long Reef and featured many tide pools and a covering of algal turf. The platform extended down to a narrow, honeycombed fore-reef ramp that was terraced forming small drainage waterfalls. Numerous blowholes occurred along the edge of the reef; however, these were only evident for a short duration over the lowest point of the tide.

FIGURE 4: Rolloliths collected from Adele Island. (a) *Cyphastrea chalcidicum* corallith. (b) *Pavona venosa* corallith. (c) Red coralline algae rhodolith. (d) Milleporina corallith. Skeletal photos Roger Springthorpe Australian Museum.

FIGURE 5: Organ Pipe Coral at Long Reef. (a) A single colony of Organ Pipe Coral (*Tubipora* sp.). (b) The narrow fore-reef habitat on the western side of Long Reef showing the prolific development of *Tubipora* as the tide advances.

Point-intercept transects conducted to quantify the percent cover of *Tubipora* in the zone verified that *Tubipora* was the dominant benthic organism with a mean of 27.67% (±3.24 SE) cover. Turf algae (22% ± 5.49 SE); sand/rubble (18% ± 4.26 SE); and coralline algae (9% ± 2.5 SE) inhabited the majority of the remaining space. Ten other genera of scleractinian coral were recorded on benthic cover transects within this habitat (*Platygyra, Acropora, Goniastrea, Favia, Favites, Galaxea, Symphyllia, Goniopora, Seriatopora,* and *Porites*); however, their combined coverage was <20%. Six additional species of octocorallia were recorded within this *Tubipora* habitat zone (*Heliopora coerulea, Lobophytum* sp.,

Cladiella sp., *Capnella* sp., *Xenia* sp., and *Briareum* sp.), as well as a small number of sponges and hydroids.

4. Discussion

4.1. Rolloliths. Free-living spheroidal corals (i.e., coralliths) were first noted from shallow tropical reef environments in Vanuatu and Indonesia in the late 1800s [17], and palaeontological studies have dated coralliths back to the Eocene (approx. 65 mya, [18]). Throughout the 70s, numerous papers arose concerning coralliths, conferring they occur right across the globe from Bermuda to Madagasgar [19, 20]. More recently, assemblages of coralliths have been recorded in the Cook Islands [21]; Barbados [22]; the North West Indian Ocean [23]; Galapagos [24]; Mexico; and the Red Sea [25]. In Australian waters, there are currently only isolated records of coralliths. For example, *Porites lutea* has been observed to form motile balls on the depositional reef flats at Heron Island [26] and *Cyphastrea microphthalma* forms motile balls at the Cocos Islands [27].

The largest corallith encountered at Adele Island was a colony of *Pavona venosa* which was 12 cm (greatest diameter). This is small in comparison to the normal size of this species which has the potential to grow to at least 1 m (greatest diameter) when attached to the substrate. Coralliths are likely to be size restricted because if they grow too large they would become partially or completely buried. It is also worthwhile noting that no attached forms of the four corallith species were recorded at station 10/K9; however, these four coral species were encountered in low abundance at other Adele Island stations.

Rhodolith beds, however, are very widespread. They are distributed from the poles to the tropics and can accumulate to form extensive communities on a wide variety of sediments in both deep and shallow water [28, 29]. The largest beds reported are on the Brazilian Shelf spanning 23 degrees of latitude [28, 30, 31]. In Brazil, rhodoliths are so abundant they are commercially mined [32]. Numerous studies have reported that rhodolith beds are important ecologically because they support a high diversity of associated flora and infauna [28, 33–35].

In Australia, rhodolith beds are also relatively widespread, occurring off Western Australia [36] in cold-temperate waters of southern Australia [37] and in southern Queensland where they cover up to 50% of the sea-floor from 50 to 110 m depth [38]. A study conducted at Montgomery Reef in the Kimberley describes rhodoliths as being important contemporary reef builders [15]. Rhodoliths are reportedly so abundant at Montgomery Reef that they form containment banks which impound a raised lagoonal habitat, the water from which feeds the unique system of low-tide cascades. Some of the Montgomery Reef rhodoliths have a core of coral fragment, but many are constructed entirely of algae. Similarly, on South Turtle Reef in Talbot Bay (inshore Kimberley), rhodolith banks form along the reef crest and enclose small pools that retain water over the low tide [39].

While there are records of rhodoliths from the Kimberley, there are, however, no current records of coralliths (at Montgomery Reef, Talbot Bay, or any other Kimberley locations). Moreover, there are no current records of Australian coral reef habitat featuring a mixed assemblage of living coralliths (from a variety of genera) and rhodoliths. Thus, the Adele Island fore-reef rolloith zone appears to be relatively rare. Therefore, our findings provide the first record of mixed corallith and rhodolith habitat for the Kimberley coastal bioregion and also for Australia.

Further taxonomic study is needed to confirm which species of algae form the rhodolith community at Adèle Island. Nongeniculate coralline algae are considered to be challenging to identify, and at least eight genera belonging to three families of Corallinales (Hapalidiaceae, Corallinaceae,

and Sporolithaceae) contain species that commonly form rhodoliths. The rhololith bed off Western Port, Victoria, for example, is composed of four species (*Hydrolithon rupestre* (Foslie) Penrose, *Lithothamnion superpositum* Foslie, *Mesophyllum engelhartii* (Foslie) Adey, and *Neogoniolithon brassica-florida* (Harvey) Setchell and Mason) [37] so it is likely that the Kimberley rhodolith beds also contains multiple species.

The rolloliths of Adele Island most likely form in the lee of prevailing wave action, as a cumulative response to unidirectional currents and the tidal pull on surface sediments. This maintains periodic yet constant movement enabling persistent growth on all sides of the colony [21]. The frequent disturbance of bottom sediments by browsing fishes and gastropods (bioturbation) has also been hypothesized to contribute towards the unique spherical form [19, 40, 41]. Holothurians and sea urchins have also been observed to aid in keeping coralliths and rhodoliths in motion [25].

4.2. Tubipora. *Tubipora* sp. is the only known hard, calcitic, reef-building alcyonarian (soft) coral. Its skeleton consists of calcite spicules fused into upright, parallel tubes connected by transverse platforms hence its common name of Organ Pipe Coral. Organ Pipe Coral is normally sporadic to rare throughout its range which spans the Indo-Pacific from the Red Sea and East Africa to Southeast Asia, Micronesia, Australia, and the Pacific [42]. *Tubipora* has, however, been observed in large quantities near Rat Island off Gladstone in Queensland, Australia (Philip Alderslade *pers. comm.*), and at an undisclosed location in Papua New Guinea [43]. The highest published benthic coverage estimate of *Tubipora* sp. is 11.9% from one location in the Red Sea [44]. Our quantitative data showing 27.67% (+3.24 SE) coverage of *Tubipora* at this one Long Reef location suggests this is an exceptionally rare, high density population.

Tubipora was once thought to be a single ubiquitous species: *Tubipora musica* Linnaeus 1758 (Family Tubiporidae, Order Stolonifera, Class Anthozoa, Phylum Cnidaria) however there are now several nominal species of *Tubipora*. Our preliminary systematic study has raised doubt about which species of *Tubipora* is present in the Kimberley. Hence, further investigations, based on gross morphology of its skeleton and molecular sequencing, are necessary to confirm or refute prior assumptions about the species identity. In addition to the required taxonomic research, further biological research is needed to understand the age, growth rate, reproductive biology, and molecular ecology of the Long Reef *Tubipora* in order to understand why they are so prevalent at this particular location.

Shared among the various nominal species of *Tubipora* is the distinctive bright red colour of the skeleton. The red colouration is a result of organic pigments, including carotenoids, which give the orange colour to carrots and the red colour to cooked lobsters. Carotenoids occur naturally in plants but are not manufactured by animals, so it is likely the Organ Pipe Coral assimilates the pigments via heterotrophic feeding on phytoplankton. Incorporating pigment into the skeleton is likely to serve the coral in two ways: firstly, it would assist light absorption and maximize the potential for photosynthesis by symbiotic dinoflagellates; and secondly, carotenoids may help protect chlorophyll from photo damage [45].

4.3. Conservation Implications. Both hard coral and calcified macroalgae are vital components of nearshore ecosystems in tropical and temperate systems through their contributions to carbon cycling and productivity [46]; the provision of habitats and their associated biodiversity; their role as nursery areas [47, 48]; and their contribution to the development of reef structures [49]. Rhodoliths also offer the potential to serve as paleoclimate proxies [40, 50–52]. A recent study measuring carbonate deposition in two slow-growing rhodolith species ($<200\,\mu m\,y^{-1}$) permitted biweekly sampling resolution leading the authors to suggest rhodoliths are unique, globally distributed palaeothermometers, which may help refine regional climate histories during the Holocene [53].

The remote location and lack of conservation management of Adele Island and Long Reef combined with the high density of rhodoliths/coralliths and Organ Pipe Coral in the unusual habitat zones we describe here renders these marine resources highly susceptible to harvesting pressure in the future. Traditional fishers have utilized the marine resources of the Kimberley region since the early eighteenth century [54, 55], but there is currently no historical record of hard corals, soft corals, or calcifying macroalgae being targeted. There is, however, substantial and rapidly growing trade in corals [56] including Organ Pipe Coral, which is used in the curio, jewellery, and aquarium industries because of its attractive bright red skeleton [57]. In consideration of this, Organ Pipe Coral is currently listed as Near-Threatened on the IUCN red-list of threatened species [57].

Scleractinian and nonscleractinian corals are afforded some amount of protection from international trade as they are listed under Appendix I and II of CITES; however, calcifying macroalgae are not listed. At an international level, we advocate the inclusion of rhodoliths under CITES regulation and that special conditions be made to ensure the trade of coralliths is monitored. Further, more timely access to national-level CITES trade data is needed to monitor trends in the export of marine resources.

It is important that regional marine resource managers direct special attention towards these atypical Kimberley habitats because corals and calcifying macroalgae are vulnerable to changes in the environment. Specific recommendations for conserving these unique benthic habitats include the establishment of marine protected areas in the Kimberley encompassing Adele Island and Long Reef and the inclusion of these locations in the existing program of Customs surveillance. Further research into *Tubipora* and rhodolith taxonomy is also important to support conservation efforts. Ongoing adaptive management and long-term monitoring of wild coral harvest in Western Australia are needed to manage risks and to ensure Western Australia's unique natural resources are protected. The Kimberley is the new frontier for tropical reef research and given the industrial interest in the region, more information is needed to provide the basis for sound decision-making and sustainable resource management.

Acknowledgments

This research was conducted as part of the Woodside Collection (Kimberley) project. The project is a Western Australian Museum initiative and sponsored by Woodside Energy. Project partner organisations are the Australian Museum, Queensland Museum, Museum and Art Gallery of the Northern Territory, Museum Victoria, and the Herbarium of Western Australia (DEC). The second author was funded as part of the Australian node of the Census of Marine Life (CReefs Program) studying the biodiversity of Australian biota, funded by the Australian Government under auspices of the Australian Institute of Marine Science, the Great Barrier Reef Research Foundation, and BHP Billiton. This research is part of Australian Biological Resources Study (ABRS) National Taxonomic Research Grant (no. 209-05) Taxonomy of tropical Australian Octocorallia (Anthozoa: Coelenterata) primarily from the Census of Marine Life "CReefs" expeditions (J. N. A. Hooper & P. N. A. Alderslade). The authors are thankful for the help received by the project scientific staff and ship's company during survey voyages aboard the *Kimberley Quest II*. Special thanks to Dr. Barry Wilson for his useful discussions. Thanks also to Alison Sampey for the compilation of Figures 1–3.

References

[1] B. Brooke, "Geomorphology," in *Marine Biological Survey of the Southern Kimberley, Western Australia*, F. E. Wells, J. R. Hanley, and D. I. Walker, Eds., pp. 21–51, Western Australian Museum, Perth, Australia, 1995.

[2] J. K. Griffith, "The corals collected during September/October 1997 at Ashmore Reef, Timor Sea," A Report to Parks Australia, Western Australian Museum, Perth, Australia, 1997.

[3] M. Kospartov, M. Beger, D. Ceccarelli, and Z. Richards, *An Assessment of the Distribution and Abundance of Sea Cucumbers, Trochus, Giant Clams, Coral, Fish and Invasive Marine Species at Ashmore Reef National Nature Reserve and Cartier Island Marine Reserve: 2005*, UniQuest Pty Limited, DEWHA, 2006.

[4] D. Cecarelli, M. Kospartov, M. Beger, Z. Richards, and C. Birrell, *An Assessment of the Impacts of Illegal Fishing on Invertebrate Stocks at Ashmore Reef National Nature Reserve, 2006*, C & R Consulting, DEWHA, 2007.

[5] P. F. Berry, Ed., *Faunal Surveys of the Rowley Shoals, Scott Reef, and Seringapatam Reef, North-Western Australia*, vol. 25 of *Records of the Western Australian Museum Supplement*, Western Australian Museum, Perth, Australia, 1986.

[6] P. F. Berry, Ed., *Marine Faunal Surveys of Ashmore Reef and Cartier Island, North-Western Australia*, vol. 44 of *Records of the Western Australian Museum Series*, Western Australian Museum, Perth, Australia, 1993.

[7] C. Bryce, *Marine Biodiversity Survey of Mermaid Reef (Rowley Shoals), Scott and Seringapatam Reef*, vol. 77 of *Records of the Western Australian Museum Supplement*, Western Australian Museum, Perth, Australia, 2009.

[8] K. Fabricius, "A brief photo guide to the shallow-water octocorals of the Rowley Shoals, Western Australia," Report,

Department of Environment and Conservation, Government of Western Australia, 2008.

[9] Z. Richards, M. Beger, J. P. Hobbs, T. Bowling, K. Chongseng, and M. Pratchett, "Ashmore Reef National Nature Reserve and Cartier Island Marine Reserve Marine Survey 2009," ARC Centre of Excellence for Coral Reef Studies, Department of the Environment, Water, Heritage and the Arts, 2009.

[10] R. Thackway and I. D. Cresswell, *Interim Marine and Coastal Regionalisation for Australia: An Ecosystem-Based Classification for Marine and Coastal Environments*, Environment Australia, Canberra, Australia, 1998.

[11] M. Wood and D. Mills, *A Turning of the Tide: Science for Decisions in the Kimberley-Browse Marine Region*, A Western Australian Marine Science Institution (WAMSI), Perth, Australia, 2008.

[12] J. E. N. Veron and L. M. Marsh, *Hermatypic Corals of Western Australia: Records and Annotated Species List*, vol. 29 of *Records of the Western Australian Museum*, Western Australian Museum, Perth, Australia, 1988.

[13] C. Teichert and R. W. Fairbridge, "Some coral reefs of the Sahul Shelf," *Geographical Review*, vol. 28, no. 2, pp. 222–249, 1948.

[14] B. R. Wilson, *The Biogeography of the Australian North West Shelf*, Elsevier, New York, NY, USA, 2013.

[15] B. Wilson and S. Blake, "Notes on the origins and biogeomorphology of Montgomery Reef, Kimberley, Western Australia," *Journal of the Royal Society of Western Australia*, vol. 94, pp. 107–119, 2011.

[16] Z. T. Richards, A. Sampey, and L. Marsh, "Synthesis of historic marine species data for the Kimberley, Western Australia (1880s–2009): hard corals," *Records of the Western Australian Museum*. In press.

[17] M. Weber, *Introduction et description de l'Expedition*, Mongraph no. 1, Siboga-Expeditie, 1902.

[18] P. D. Taylor and D. N. Lewis, *Fossil Invertebrates*, Natural History Museum, London, UK, 2005.

[19] P. W. Glynn, "Rolling stones amongst the scleractinia: mobile coralliths in the Gulf of Panama," in *Proceedings of the 2nd International Coral Reef Symposium*, vol. 2, pp. 183–198, 1974.

[20] M. Pichon, "Free-living scleractinian coral communities in the coral reefs of Madagascar," in *Proceedings of the 2nd International Coral Reef Symposium*, vol. 2, pp. 173–181, 1974.

[21] T. P. Scoffin, D. R. Stoddart, A. W. Tudhope, and C. Woodroffe, "Rhodoliths and coralliths of Muri Lagoon, Rarotonga, Cook Islands," *Coral Reefs*, vol. 4, no. 2, pp. 71–80, 1985.

[22] J. B. Lewis, "Spherical growth in the Caribbean coral *Siderastrea radians* (Pallas) and its survival in disturbed habitats," *Coral Reefs*, vol. 7, no. 4, pp. 161–167, 1989.

[23] M. R. Claereboudt, "*Porites decasepta*: a new species of scleractinian coral (Scleractinia, Poritidae) from Oman," *Zootaxa*, no. 1188, pp. 55–62, 2006.

[24] P. W. Glynn and G. M. Wellington, *Corals and Coral Reefs of the Galapagos Islands*, University of California Press, Berkeley, Calif, USA, 1983.

[25] H. Reyes-Bonilla, R. Riosmena-Rodriguez, and M. S. Foster, "Hermatypic corals associated with rhodolith beds in the Gulf of California, México," *Pacific Science*, vol. 51, no. 3, pp. 328–337, 1997.

[26] G. Roff, "Corals on the move: morphological and reproductive strategies of reef flat coralliths," *Coral Reefs*, vol. 27, no. 2, pp. 343–344, 2008.

[27] J. E. N. Veron, *Re-Examination of the Reef Corals of Cocos (Keeling)*, vol. 14, Records of the Western Australian Museum, Western Australian Museum, Perth, Australia edition, 1990.

[28] M. S. Foster, "Rhodoliths: between rocks and soft places," *Journal of Phycology*, vol. 37, no. 5, pp. 659–667, 2001.

[29] B. Konar, R. Riosmena-Rodriguez, and K. Iken, "Rhodolith bed: a newly discovered habitat in the North Pacific Ocean," *Botanica Marina*, vol. 49, no. 4, pp. 355–359, 2006.

[30] M. Kempf, "Notes of the benthic bionomy of the N-NE Brazilian shelf," *Marine Biology*, vol. 5, no. 3, pp. 213–224, 1970.

[31] G. M. Amado-Filho, G. Maneveldt, R. C. C. Manso, B. V. Marins-Rosa, M. R. Pacheco, and S. M. P. B. Guimarães, "Structure of rhodolith beds from 4 to 55 meters deep along the southern coast of Espírito Santo State, Brazil," *Ciencias Marinas*, vol. 33, no. 4, pp. 399–410, 2007.

[32] G. B. Baarli, M. Cachao, C. M. da Silva, M. E. Johnson, J. Ledesma-Vazquez, and A. M. E. Santos, "Fossil nodules of free living biota from the upper Pleistocene Mulegé formation, Playa La Palmita, Baja California Sur, Mexico," *Publicaciones del Seminario de Paleontología de Zaragoza*, vol. 9, pp. 75–78, 2010.

[33] C. Barbera, C. Bordehore, J. A. Borg, M. Glémarec et al., "Conservation and management of northeast Atlantic and Mediterranean maerl beds," *Aquatic Conservation: Marine and Freshwater Ecosystems*, vol. 13, no. 1, pp. S65–S76, 2003.

[34] G. Hinojosa-Arango and R. Riosèmena-Rodríguez, "Influence of rhodolith-forming species and growth-form on associated fauna of rhodolith beds in the Central-West gulf of California, México," *Marine Ecology*, vol. 25, no. 2, pp. 109–127, 2004.

[35] J. Hall-Spencer, J. Kelly, and C. A. Maggs, *Assessment of Maerl Beds in the OSPAR Area and the Development of a Monitoring Program*, Department of Environment, Heritage and Local Government, Dublin, Ireland, 2008.

[36] N. Goldberg, "Age estimates and description of rhodoliths from Esperance Bay, Western Australia," *Journal of the Marine Biological Association of the United Kingdom*, vol. 86, no. 6, pp. 1291–1296, 2006.

[37] A. S. Harvey and F. L. Bird, "Community structure of a rhodolith bed from cold-temperate waters (Southern Australia)," *Australian Journal of Botany*, vol. 56, no. 5, pp. 437–450, 2008.

[38] M. Lund, P. J. Davies, and J. C. Braga, "Coralline algal nodules off Fraser Island, Eastern Australia," *Facies*, no. 42, pp. 25–34, 2000.

[39] B. Wilson, S. Blake, D. Ryan, and J. Hacker, "Reconnaissance of species-rich coral reefs in a muddy, macro-tidal enclosed embayment —Talbot Bay, Kimberley, Western Australia," *Journal of the Royal Society of Western Australia*, vol. 94, pp. 251–165, 2011.

[40] M. S. Foster, R. Riosmena-Rodríguez, D. L. Steller, and W. J. Woelkerling, "Living rhodolith beds in the Gulf of California and their implications for paleoenvironmental interpretation," *Geological Society of America Bulletin*, vol. 318, pp. 127–139, 1997.

[41] E. C. Marrack, "The relationship between water motion and living rhodolith beds in the Southwestern Gulf of California, Mexico," *Palaios*, vol. 14, no. 2, pp. 159–171, 1999.

[42] K. Fabricius and P. Alderslade, *Soft Corals and Sea Fans—A Comprehensive Guide to the Tropical Shallow Water Genera of the Central-West Pacific, the Indian Ocean and the Red Sea*, Australian Institute of Marine Science, Townsville, Australia, 2001.

[43] J. E. N. Veron, *Corals of the World*, vol. 1–3, Australian Institute of Marine Science, 2000.

[44] Y. Benayahu and Y. Loya, "Space partitioning by stony corals soft corals and benthic algae on the coral reefs of the Northern Gulf of Eilat (Red Sea)," *Helgoländer Wissenschaftliche Meeresunter-suchungen*, vol. 30, no. 1-4, pp. 362–382, 1977.

[45] J. Cvejic, S. Tambutte, S. Lotto, M. Mikov, I. Slacanin, and D. Allemand, "Determination of canthaxanthin in the red coral (*Corallium rubrum*) from Marseille by HPLC combined with UV and MS detection," *Marine Biology*, vol. 152, no. 4, pp. 855–862, 2007.

[46] S. Hetzinger, J. Halfar, B. Riegl, and L. Godinez-Orta, "Sedimentology and acoustic mapping of modern rhodolith facies on a non-tropical carbonate shelf (Gulf of California, Mexico)," *Journal of Sedimentary Research*, vol. 76, no. 3-4, pp. 670–682, 2006.

[47] N. A. Kamenos, P. G. Moore, and J. M. Hall-Spencer, "Small-scale distribution of juvenile gadoids in shallow inshore waters; what role does maerl play?" *ICES Journal of Marine Science*, vol. 61, no. 3, pp. 422–429, 2004.

[48] N. A. Kamenos, P. G. Moore, and J. M. Hall-Spencer, "Nursery-area function of maerl grounds for juvenile queen scallops *Aequipecten opercularis* and other invertebrates," *Marine Ecology Progress Series*, vol. 274, pp. 183–189, 2004.

[49] W. A. Nelson, "Calcified macroalgae critical to coastal ecosystems and vulnerable to change: a review," *Marine and Freshwater Research*, vol. 60, no. 8, pp. 787–801, 2009.

[50] A. Freiwald, R. Henrich, P. Schäfer, and H. Willkomm, "The significance of high-boreal to subarctic maerl deposits in northern Norway to reconstruct holocene climatic changes and sea level oscillations," *Facies*, vol. 25, no. 1, pp. 315–339, 1991.

[51] C. E. Cintra-Buenrostro, M. S. Foster, and K. H. Meldahl, "Response of nearshore marine assemblages to global change: a comparison of molluscan assemblages in Pleistocene and modern rhodolith beds in the Southwestern Gulf of California, México," *Palaeogeography, Palaeoclimatology, Palaeoecology*, vol. 183, no. 3-4, pp. 299–320, 2002.

[52] R. Nalin, C. S. Nelson, D. Basso, and F. Massari, "Rhodolith-bearing limestones as transgressive marker beds: fossil and modern examples from North Island, New Zealand," *Sedimentology*, vol. 55, no. 2, pp. 249–274, 2008.

[53] N. A. Kamenos, M. Cusack, and P. G. Moore, "Coralline algae are global palaeothermometers with bi-weekly resolution," *Geochimica et Cosmochimica Acta*, vol. 72, no. 3, pp. 771–779, 2008.

[54] S. Fox, "Reefs and shoals in Australia-Indonesian relations: traditional Indonesian fisherman," in *Australian in Asia: Episodes*, A. C. Milner and M. Quilty, Eds., Oxford University Press, Melbourne, Australia, 1988.

[55] N. Stacey, *Boats to burn: bajo fishing in the Australian fishing zone [Ph.D. thesis]*, Northern Territory University, Commonwealth of Australia, 2002, Ashmore Reef National Nature Reserve and Cartier Island Marine Reserve Management Plans, Environment Australia, Canberra, Australia, 1999.

[56] A. Rhyne, R. Rotjan, A. Bruckner, and M. Tlusty, "Crawling to collapse: ecologically unsound ornamental invertebrate fisheries," *PLoS ONE*, vol. 4, no. 12, Article ID e8413, 2009.

[57] D. Obura, D. Fenner, B. Hoeksema, L. Devantier, and C. Sheppard, "*Tubipora musica*," in *IUCN, 2012. IUCN Red List of Threatened Species: Version 2012. 2*, 2008, http://www.iucnredlist.org/.

13

Gametogenesis and Spawning of *Solenastrea bournoni* and *Stephanocoenia intersepta* in Southeast Florida, USA

Jenna R. Lueg,[1] **Alison L. Moulding,**[1] **Vladimir N. Kosmynin,**[2] **and David S. Gilliam**[1]

[1] *Oceanographic Center, Nova Southeastern University, 8000 N. Ocean Drive, Dania Beach, FL 33004, USA*
[2] *Bureau of Beaches and Coastal Systems, Florida Department of Environmental Protection, 3900 Commonwealth Boulevard, MS 49, Tallahassee, FL 32399, USA*

Correspondence should be addressed to David S. Gilliam, gilliam@nova.edu

Academic Editor: Baruch Rinkevich

This study constitutes the first report of the gametogenic cycle of the scleractinian corals *Solenastrea bournoni* and *Stephanocoenia intersepta*. Tissue samples were collected near Ft. Lauderdale, Florida, USA between July 2008 and November 2009 and processed for histological examination in an effort to determine reproductive mode and potential spawning times. Both *S. bournoni* and *S. intersepta* are gonochoric, broadcast spawning species. Gametogenesis of *S. bournoni* began in April or May while *S. intersepta* had a much longer oogenic cycle that began in December with spermatogenesis beginning in July. Though spawning was not observed *in situ*, spawning was inferred from the decrease of late stage gametes in histological samples. In addition, histological observations of oocyte resorption and released spermatozoa were used to corroborate spawning times. Data indicate that *S. bournoni* spawns in September while *S. intersepta* spawns after the full moon in late August or early September.

1. Introduction

Reproduction of scleractinian corals is one of the most important processes influencing their continued existence [1]. Sexual reproduction is the first step in establishing new colonies necessary for repopulating degraded reefs. It creates genetic variability essential for adapting to changing environmental conditions. Information on the reproductive biology and ecology of corals is critical for understanding their distribution, evolutionary mechanisms [2, 3], and for management and restoration of damaged areas [4].

For broadcast spawning species, synchronous maturation and release of gametes is essential for successful fertilization [5]. Coral gametogenesis and spawning are thought to be driven by a number of environmental cues including sea temperature, day length, moonlight tidal cycles, and daylight cycles [6, 7], but the exact association remains unknown [1, 7, 8]. For a number of species, temperature defines the season and month of spawning and timing tends to correlate strongly with lunar phase [5]. Sunset time can define the hour of spawning [6, 9, 10] and tidal fluctuations may decrease water movement and lead to increased fertilization

opportunities [11]. It is very likely that coral gametogenesis and spawning are not prompted by a single environmental cue, but rely on more than one environmental signal [12].

In the last few decades, there have been numerous studies on scleractinian coral reproduction [4, 7, 13–16]. About 60% of the approximate 60 known Caribbean species have been investigated [3]. Species of particular interest in southeast Florida include *Solenastrea bournoni* and *Stephanocoenia intersepta* because of their local abundance. *Stephanocoenia intersepta* and *Solenastrea bournoni* are common species found offshore in all three counties (Miami-Dade, Broward, and Palm Beach) and on all reef types (nearshore ridge complex, Inner reef, Middle reef, and Outer reef) within the southeast Florida region [17, 18]. These species are not as common in the rest of the Caribbean where *S. intersepta* generally comprises less than 5% of the coral population in areas such as the US Virgin Islands, Venezuela, Turks and Caicos, Mexico, Belize, and Bahamas and *S. bournoni* is even less abundant [19]. There is no information in the published literature on reproductive mode or spawning time of *S. bournoni*. *Stephanocoenia intersepta* has been reported to be a gonochoric species that has been observed broadcast

TABLE 1: Criteria for differentiating oocyte and spermary developmental stage based on Szmant-Froelich et al. [27].

Stage	Oocytes	Spermaries
I	Nucleus large, located in or adjacent to mesoglea	Small cluster of ≤10 cells surrounded by mesoglea
II	Accumulation of cytoplasm around nucleus, located in mesoglea	Larger cluster of ≥10 cells
III	Increased amount of cytoplasm, but no vitelline membrane	Cells closer together, central lumen developed
IV	Larger, elongated and presence of vitelline membrane	Cells very small, undergoing meiosis, tails in lumen

spawning in Bonaire 3–7 nights after the August full moon [16] and in the Gulf of Mexico from 6 to 11 nights after the August or September full moon between the hours of 19:30 and 20:00 [9, 20]. Females of this species have been observed releasing a high percentage of fertilized eggs [20, 21] and it has been proposed that eggs are fertilized in the tentacles just prior to release [21]. Cycles of oogenesis and spermatogenesis have not been previously described for these two species. The purpose of this study was to document the gametogenic cycle of *S. bournoni* and *S. intersepta* and to determine their spawning times in southeast Florida.

2. Materials and Methods

2.1. Collection and Histological Processing. Solenastrea bournoni and *S. intersepta* were sampled twice per month from July 2008 through July 2009 and weekly in August and September 2009. During each collection date, five colonies of *S. bournoni* and *S. intersepta* were sampled. Additionally, five *S. bournoni* samples were collected twice per week in October and November 2009. Colonies were sampled in southeast Florida near Ft. Lauderdale (26°03′N to 26°19′N) in 4 to 18 m depth. Tissue samples measuring approximately 6–12 cm^2 in size were removed using a hammer and chisel or a 2.5 cm steel core. All samples were taken from the middle of the colony to avoid less fecund edges [22–24]. Only colonies that had a diameter of over 10 cm were sampled to ensure sexual maturity [14, 24–26] and each colony was sampled only once. Epoxy was placed on the exposed skeleton where tissue was removed to minimize potential settlement of algae and boring organisms. Collection methods caused minimal damage to *S. bournoni* colonies since tissue completely regrew over the sample area within 6 months. However, *S. intersepta* had much slower regeneration rates and showed minimal regrowth over the sampled areas during the study period. Collected samples were fixed immediately in 10% aqueous zinc-buffered formalin (Z-fix) for 24 hours. Tissue samples were decalcified using a buffered 10% hydrochloric acid solution, dehydrated using a series of ethanol and xylene, infiltrated with paraffin, and embedded in paraffin blocks.

Serial sections of 5 μm were cut with a microtome at three depths of the tissue and placed on slides. The first serial sections were taken at a location just below the actinopharynx, and subsequent sections were taken approximately 0.15–0.3 mm and 0.45–0.6 mm below the actinopharynx. Two replicate slides were made at each of the three depths, yielding 6 slides for each sample. Three of these slides, one at each depth, were stained with Heidenhain's

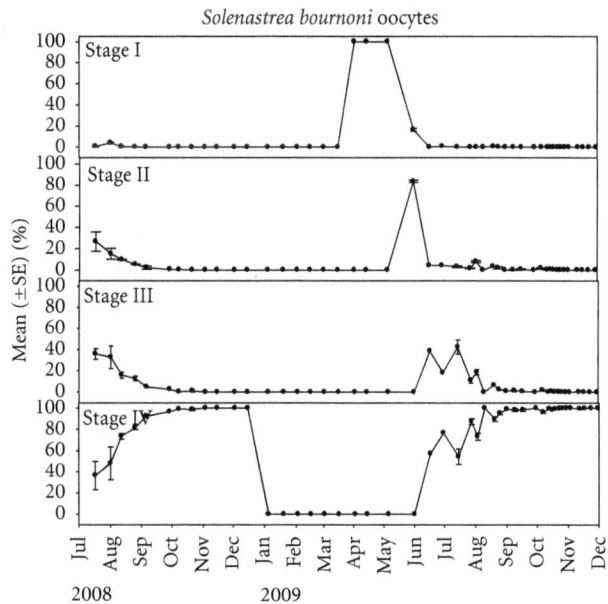

FIGURE 1: *Solenastrea bournoni* mean (±SE) percentage of oocytes per colony quantified through histological examination.

azocarmine aniline blue to highlight reproductive structures. The rest were stained with Harris' hematoxylin and eosine (H&E) in order to confirm the presence of released spermatozoa in which the nuclei stain dark purple. All staining characteristics described are with Heidenhain's azocarmine aniline blue unless otherwise noted.

2.2. Quantification of Gametes and Fecundity. Gamete developmental stages were determined from the slide that contained the most gametes which was usually the deepest slide. Criteria determining gamete developmental stage (Table 1) were modified from Szmant-Froelich et al. [27]. Using the slide with the most gametes, the percentage per colony of each gamete stage was calculated per sampling date using the total number of gametes in 7–10 polyps per colony. Resorption and released spermatozoa percentages per sampling date were calculated based on the number of mesenteries of 7–10 polyps per colony that contained resorbed oocytes or spermatozoa released from spermaries. Polyp fecundity was calculated for every female colony sampled during the month prior to spawning. Fecundity was calculated by multiplying the mean number of mature (stage IV) oocytes in cross sections of ten polyps per colony and the mean number of mature oocytes in longitudinal sections of 10 mesenteries, yielding mean stage IV oocytes

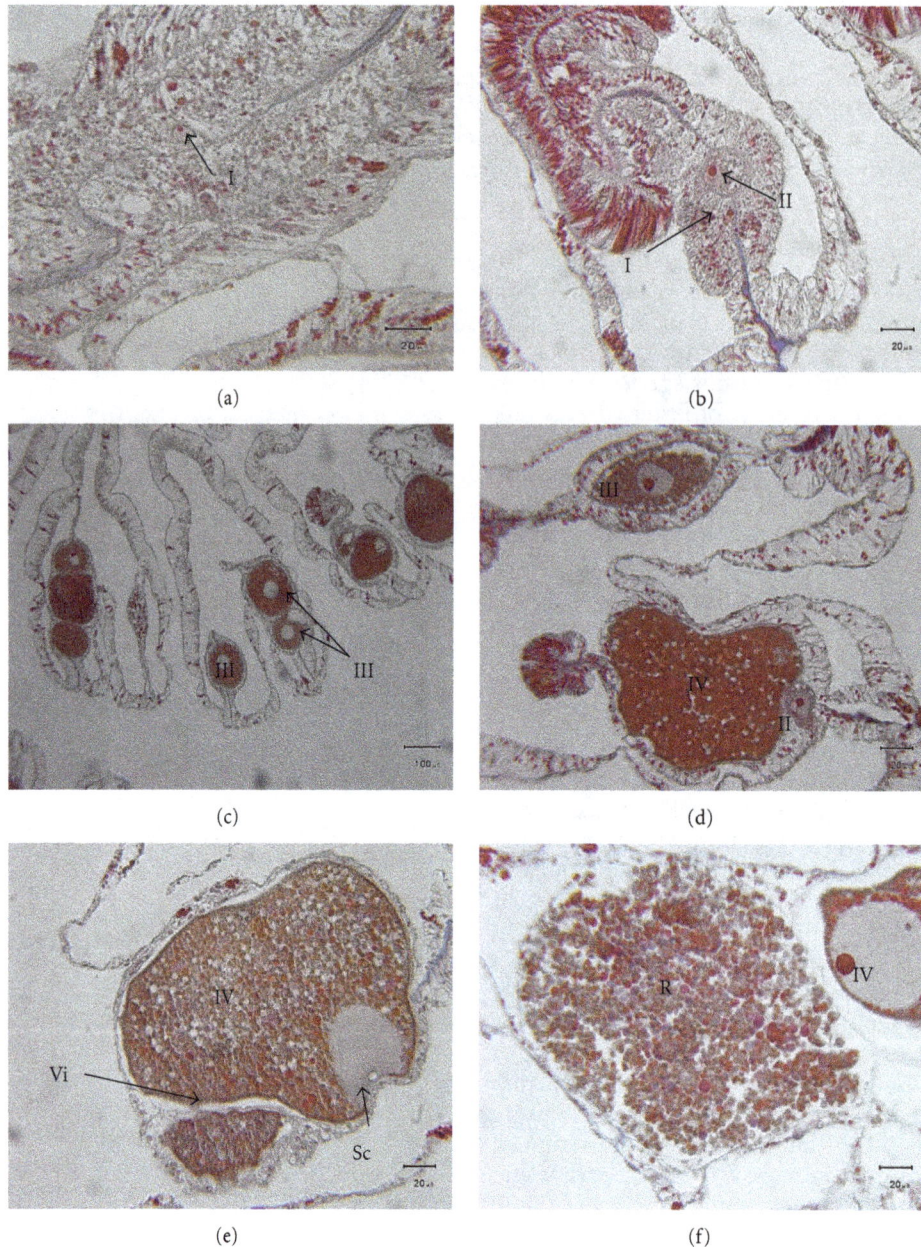

FIGURE 2: Photomicrographs of *Solenastrea bournoni* oocyte stages I–IV (defined by Table 1). (a) Stage I oocyte; (b) stages I and II oocytes; (c) stage III oocytes; (d) stages II, III and IV; (e) stage IV oocytes, Vi = vitelline membrane, Sc = scalloping of nucleus; (f) stage IV oocyte adjacent to oocyte resorption (R). All scale bars 20 μm except for (c) which is 100 μm.

per polyp. For description of developmental stages, mean oocyte diameters were measured from multiple colonies over several sampling dates when abundance was highest. Mean abundance of stage IV oocytes per colony was calculated and compared to environmental factors including moon phase and temperature. Chi-squared tests were used to determine sex ratio significance.

2.3. Environmental Data. Temperature loggers were deployed at 10 meters depth where most coral samples were collected. Regression analyses were preformed to

examine the relationship between oocyte abundance and temperature. Stage IV oocyte abundance was plotted with moon phase to determine if spawning times could be linked to lunar cycles.

3. Results

3.1. Solenastrea bournoni. Solenastrea bournoni is a gonochoric species with a 1 : 1 sex ratio ($\chi^2 = 0.157$, df $= 1$, $P = 0.692$). A total of 112 females and 118 males were sampled, and seven colonies did not have gametes and

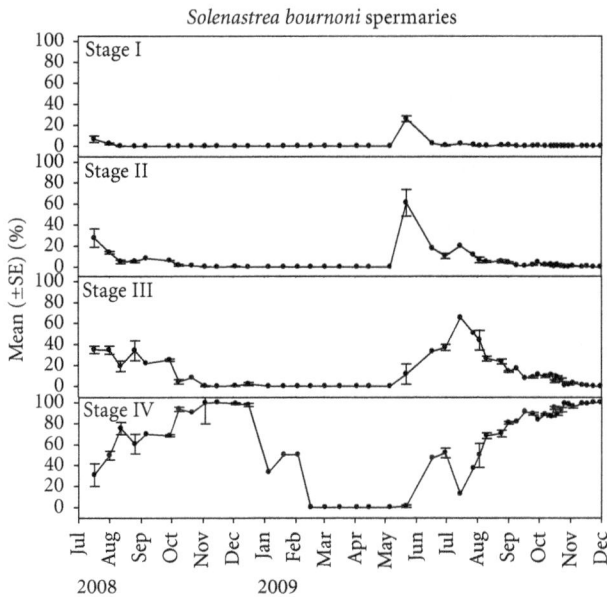

FIGURE 3: *Solenastrea bournoni* mean (±SE) percentage spermary stages per colony quantified through histological examination.

could not be determined to be male or female. Out of 237 colonies sampled, there were two male colonies that contained a single mature oocyte in the same mesentery that contained spermatocytes and one male colony that contained two oocytes in separate polyps surrounded by mature spermatocytes. No larvae were seen in any colony, and late stage oocytes did not contain zooxanthellae. Fecundity was calculated as 166.34 ± 42.00 (SE) oocytes per polyp ($n = 8$ colonies) in 2008 and 305.33 ± 78.48 oocytes per polyp ($n = 10$) in 2009.

3.1.1. Gametogenesis.

Oogenesis in *Solenastrea bournoni* began in April, and stage I oocyte abundance peaked from April to early May (Figure 1). Mean diameter of stage I oocytes observed during their peak abundance was $25.24 \pm 0.98 \mu m$ and ranged in size from 16 to $30 \mu m$ ($n = 26$ oocytes). Early stage I oocytes were located in the gastrodermal layer and migrated into the mesoglea (Figures 2(a) and 2(b)). They were characterized by large nuclei in relation to the amount of cytoplasm. Nuclei stained light pink to grey with a magenta nucleolus, and the cytoplasm stained light grey.

Stage II oocytes were first observed in early June (Figure 1) but may have developed earlier (no females were collected in late May). Highest percentages per colony were seen in early June. Mean diameter of randomly chosen stage II oocytes from June samples was $57.32 \pm 2.83 \mu m$ and ranged from 35 to $98 \mu m$ ($n = 30$). These larger oocytes were located in the mesoglea and stained light grey with a light pink nucleus and magenta nucleolus (Figure 2(b)). The size of cytoplasm was greater than the nucleus.

Stage III oocytes began to appear in mid-June, and abundance peaked in mid-July 2008 and in early August 2009

(Figure 1). Stage III oocytes measured during this period were larger than stage II with a mean diameter of $102.12 \pm 3.65 \mu m$ and a range of $69–149 \mu m$ ($n = 30$). Stage III oocytes stained a darker pink in color and had a centrally located nucleus (Figures 2(c) and 2(d)).

Stage IV oocytes developed by mid-June, peaked in August, and continued to comprise a high percentage of total oocytes through the end of sampling in late November (Figure 1). Mean diameter of stage IV oocytes was $225.59 \pm 10.18 \mu m$ and ranged from 153 to $341 \mu m$ ($n = 30$). Stage IV oocytes were characterized by their large size and dark vitelline membrane (Figure 2(d)). Depending on the section, the nucleus had usually migrated to one side, and the vitelline membrane sometimes had a scalloped appearance (Figure 2(e)). Stage IV oocytes also had an increased amount of lipid vacuoles that stained orange, pink, magenta, and light purple.

Resorption of oocytes was observed throughout most of the year in 2009 although in small percentages. In early resorption (Figure 2(f)), the vitelline membrane broke apart, and lipid vacuoles were clumped together. Lipid vacuoles slowly dispersed into the gastrodermis and degraded. Resorption after February was seen in less than 1% of all mesenteries examined. In July, there was a low occurrence of resorption (4.46% of mesenteries in all colonies sampled that month), and in early October one colony showed resorption of oocytes in 20% of its mesenteries. Higher levels of resorption were observed through the end of sampling in late November.

Spermatogenesis began rapidly in May with all stages appearing in the same two-week period (Figure 3). Stage I and stage II spermaries stained light pink and were differentiated by the number of cells (Figures 4(a) and 4(d)). Stage I spermaries were characterized by fewer than 10 small, loosely clustered cells located in the mesoglea, and stage II spermaries contained more than 10 cells. Both stage I and stage II spermaries had the highest percentages in May (Figure 3).

Stage III spermaries peaked in mid-July (Figure 3) and were comprised of larger and darker cells that were located closer together around a centrally located lumen (Figures 4(b) and 4(d)). The highest percentages of stage IV spermaries were in October, and they were observed through February (Figure 3). The magenta cells were very small, and spermatozoa tails were lined up in the lumen (Figures 4(c) and 4(d)).

Low percentages of released spermatozoa were observed in the gastrodermis during many of the sampling dates (Figures 4(d) and 4(e)). Small spikes in percentage of released spermatozoa occurred in mid-August and October 2008 followed by the highest percentage observed in November. In 2009, released spermatozoa were seen in variable percentages between colonies starting in January. Percentage of released spermatozoa decreased greatly in May, but they were still observed in small percentages until July. Then percentages increased at the end of September, continuing through November to near 100%. In stains of hematoxylin and eosine (H&E), the heads of the spermatozoa stained dark purple shortly after being released (Figure 4). Months later,

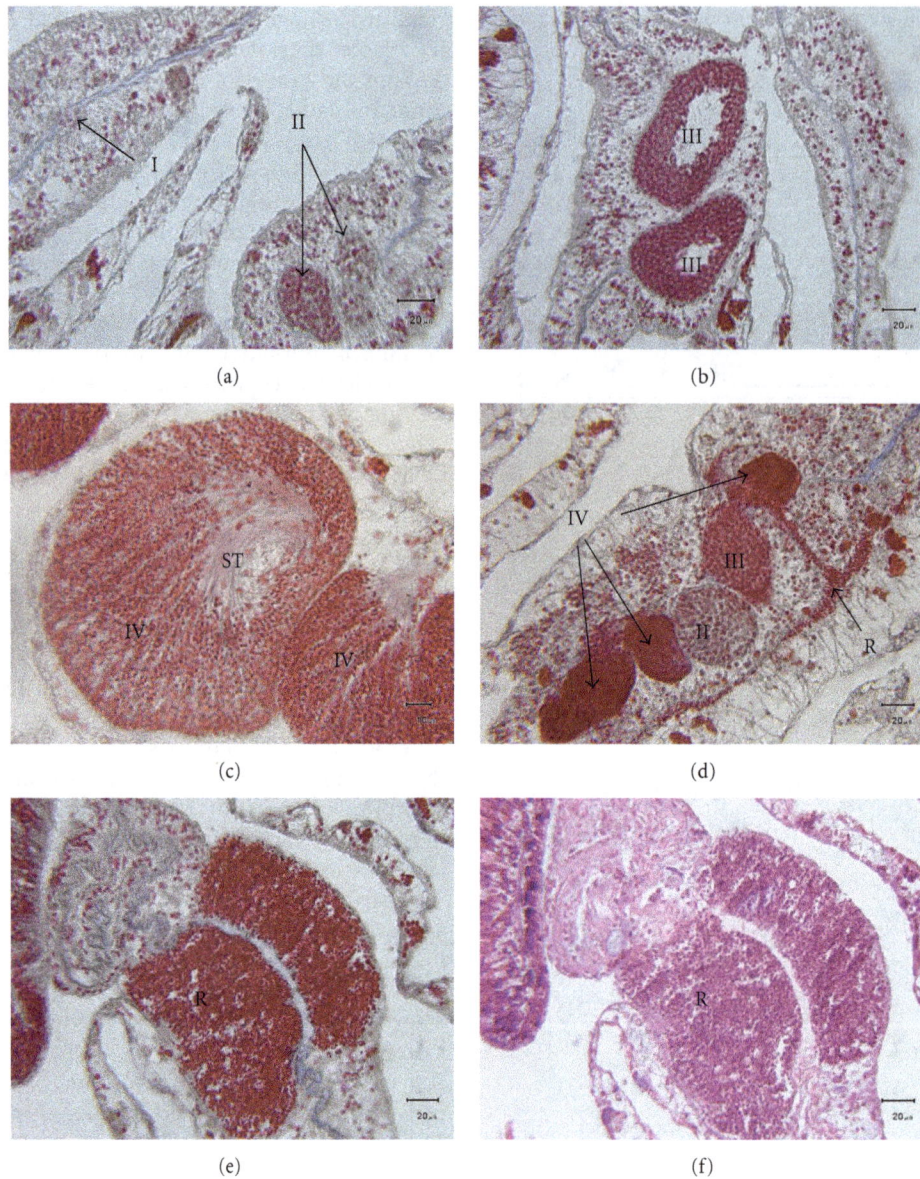

FIGURE 4: Photomicrographs of *Solenastrea bournoni* spermary stages I–IV (defined by Table 1). (a) Stages I and II spermaries; (b) stage III spermaries; (c) stage IV spermaries, ST = spermatozoa tails lined up; (d) stages II, III, and IV, R = spermatozoa being released into gastrovascular cavity; (e) released spermatozoa (R) stained with Heidenhain's azocarmine aniline blue; (f) released spermatozoa (R) stained with H&E. All scale bars 20 μm except for (c) which is 10 μm.

residual spermatozoa did not stain purple, indicating they had degraded and were no longer viable.

3.1.2. Lunar Periodicity. Mean abundance of stage IV oocytes per colony was plotted with lunar phase (Figure 5). Peak abundances of stage IV oocytes occurred in late August and early September and experienced the largest decrease between the full and new moon of September. Smaller decreases in abundance were also observed between the full and new moons of October though few female colonies were sampled before the October 2008 full moon and after the October 2009 full moon.

3.1.3. Temperature. Mean stage IV oocyte abundance tracked mean daily water temperatures (Figure 6) with oocyte abundance decreasing and increasing in unison with temperature. In 2009, peaks in oocyte abundance in July, September, and August occurred after peaks in temperature. To test this correlation, a simple linear regression was performed plotting all stages of oocyte abundance against temperature (Figure 7). A significant relationship was found between oocyte abundance and temperature ($R^2 = 0.67$, $P < 0.001$), so the same analysis was performed for all stages of spermary abundance (Figure 8). A significant relationship was also found between spermary abundance and temperature ($R^2 = 0.65$, $P < 0.001$).

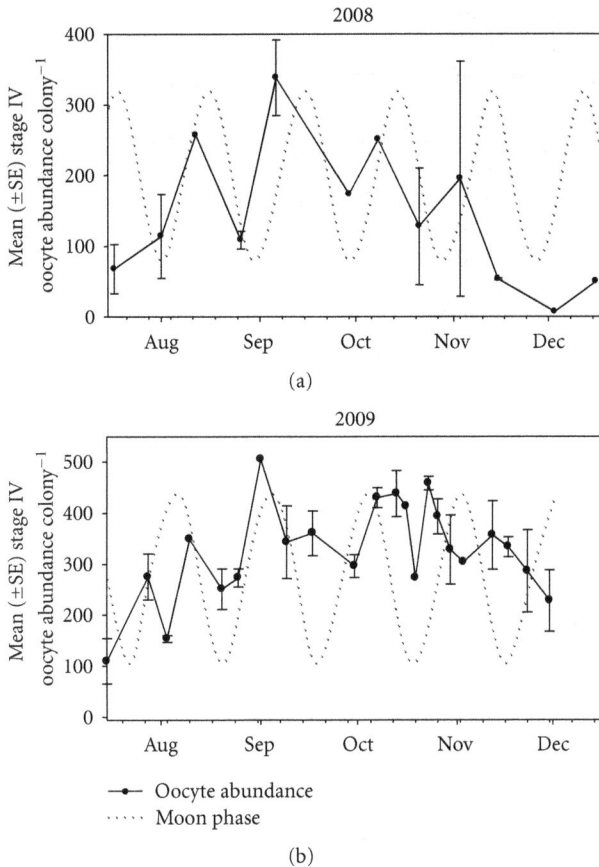

(a)

(b)

FIGURE 5: *Solenastrea bournoni* mean ($\pm$SE) abundance of stage IV oocytes per colony at each sampling date with moon phases for years 2008 and 2009; full moons represented by peaks in moon phase line.

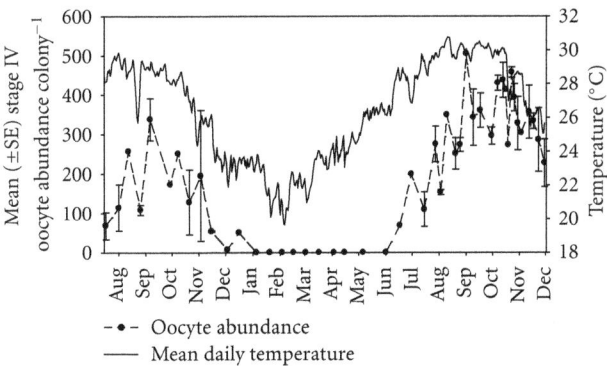

FIGURE 6: *Solenastrea bournoni* mean ($\pm$SE) abundance of stage IV oocytes per colony at each sampling date with temperature for sampling dates 2008 through 2009.

3.1.4. Predicted Spawning Times. Results indicate that *Solenastrea bournoni* spawned after the full moon of September. In 2008, the greatest decrease in mean oocyte abundance was observed between 6 and 29 September, from 338 to 173 oocytes per colony. The full moon occurred on September 15. Oocyte resorption was first observed in October in

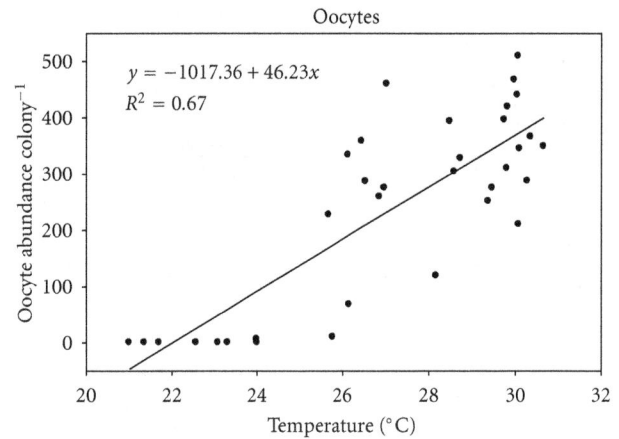

FIGURE 7: Simple regression of temperature and *Solenastrea bournoni* oocyte abundance, including all stages per colony at each sample date. $R^2 = 0.67$, $P < 0.001$.

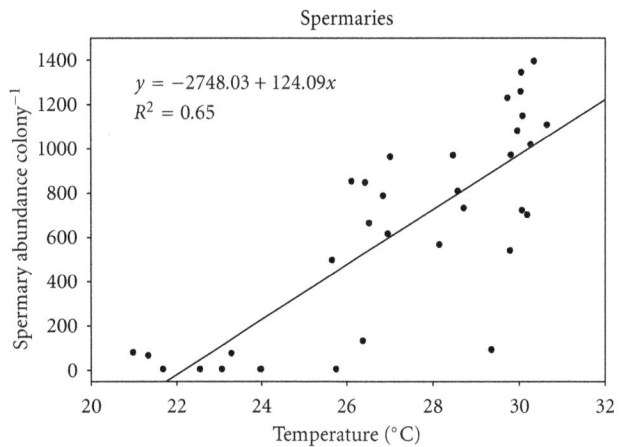

FIGURE 8: Simple regression of temperature and *Solenastrea bournoni* spermary abundance, including all stages per colony at each sample date. $R^2 = 0.65$, $P < 0.001$.

low percentages, and percentages increased in November. Released spermatozoa were observed in August but were not present again until October. A greater amount of oocyte resorption and released spermatozoa in October indicates that the main spawning event took place before October. In 2009, the highest peak in mean abundance of stage IV oocytes was on 9 September with 505 oocytes per colony. A full moon occurred on September 4, and mean abundance of stage IV oocytes decreased after 9 September. Oocyte resorption also started in October 2009. The amount of released spermatozoa was moderate on 17 September, but increased through December of 2009. Unfortunately, the fact that only one of the five colonies sampled after the potential spawning time in 2008 was female and only one of the five colonies sampled before the potential spawning in 2009 was female precludes statistical analysis of significant decreases in oocyte abundance to support the proposed spawning period. However a similar decrease in mature oocyte abundance after the full moon of September in both years, coupled

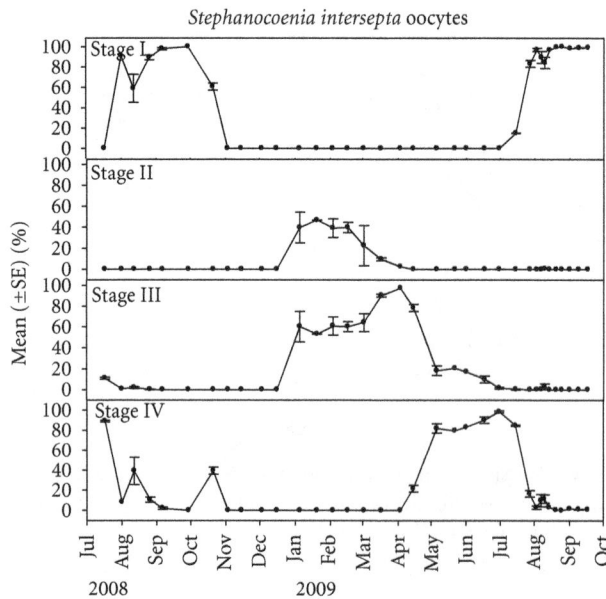

FIGURE 9: *Stephanocoenia intersepta* mean (±SE) percentages of oocyte per colony quantified through histological examination.

with observed oocyte resorption and increases of released spermatozoa in October, indicates the main spawning event likely occurred after the full moon of September.

3.2. Stephanocoenia intersepta. No cases of hermaphroditism were observed in 181 colonies sampled over 15 months. Sixty-five colonies were found to contain no gametes. No sex could be assigned to 31 of these 65 colonies because 11 had no gametes during the reproduction period and 20 were sampled during the winter nonreproduction period. The remaining 34, sampled from January to June, were assumed to be males that had not yet developed gametes since spermatogenesis was not observed until July. Sex ratio was $1:1$ (78 females, 72 males, $\chi^2 = 0.24$, df = 1, $P > 0.01$). Female fecundity of *S. intersepta* was 64.51 ± 12.04 oocytes per polyp in 2008 ($n = 8$) and 92.67 ± 9.27 oocytes per polyp in 2009 ($n = 19$).

3.2.1. Gametogenesis. Oocytes were observed in samples of *S. intersepta* from December through October while spermatocytes were present from late July through October. Resorption of oocytes was present for 2 to 3 months in September to early November. Released spermatozoa were present in September and October, but by mid-November, no gametes or remnant gametes were seen.

Stage I oocytes were first observed in mid-December in a longitudinal section of one colony but were not seen in any cross sections which were used to quantify gamete stages. Stage I oocytes peaked in percentage in January and were observed through August 2009 in low percentages (Figure 9). Stage I oocytes randomly chosen during January through August had a mean diameter of $15.81 \pm 0.84\,\mu m$ and ranged from 9.9 to $28.86\,\mu m$ ($n = 30$ oocytes). These oocytes were found in the gastrodermis adjacent to mesoglea or in the

mesoglea (Figure 10(a)). They had large nuclei in relation to the amount of cytoplasm, and the nuclei stained pink with a magenta nucleolus. If cytoplasm was seen, it was light pink or grey in color.

Stage I oocytes developed into stage II oocytes as early as December but were seen only in a longitudinal section of one colony. The highest percentage of Stage II oocytes was observed in April (Figure 9). Stage II oocytes randomly chosen from all dates observed had a mean diameter of $47.81 \pm 2.09\,\mu m$, and they ranged in size from 31.33 to $63.71\,\mu m$ ($n = 30$). Nuclei stained grey or light pink, and cytoplasm stained grey. Stage II oocytes were larger with the amount of cytoplasm greater than the nucleus, and they resided in the mesoglea (Figure 10(b)).

Stage III oocytes developed in April, and highest percentages were present in late June (Figure 9). The mean diameter of Stage III oocytes was $140.62 \pm 5.69\,\mu m$ and ranged from 79.86 to $206.3\,\mu m$ ($n = 32$). Cytoplasm stained light pink, and oocytes usually had a centrally located nucleus (Figure 10(c)).

Stage IV oocytes first developed in mid-July, peaked in percentage in August, and declined after September (Figure 9). Stage IV oocytes randomly chosen from July through September had a mean diameter of $318.84 \pm 13.78\,\mu m$ and ranged from 184 to $435.2\,\mu m$ ($n = 30$). The lipid vacuoles in the cytoplasm were more defined than they were in stage III and stained pink (Figure 10(d)). The vitelline membrane stained blue to purple and appeared scalloped in late stage IV oocytes (Figure 10(d)).

In 2008, resorption of *S. intersepta* oocytes occurred from early September through late October with no signs of resorption in November. In 2009, oocyte resorption was seen in low percentages in late July through early September, but in mid-September, oocyte resorption was at 100% in all mesenteries.

Stage I spermaries formed in late July, and the highest percentage was seen in August (Figure 11). Stage I spermaries were characterized by a small cluster of less than 10 loosely packed cells that stained light pink and were found in the mesoglea (Figure 12(a)). Stage II spermaries were observed in late July to early August with the highest percentages occurring in September 2008 and late August 2009 (Figure 11). Stage II spermaries stained light pink and consisted of more than 10 loosely packed cells (Figure 12(b)). Stage III spermaries were first observed in late July to early August, and peak percentages occurred in August (Figure 11). Stage III spermaries were larger with more compact, darker staining cells and had a centrally located lumen (Figure 12(c)). Stage IV spermaries were first observed in August, and the highest percentages occurred in early August 2008 and early September 2009 (Figure 11). Spermatocytes of stage IV spermaries stained dark magenta with pink staining tails (Figure 12(d)).

Released spermatozoa appeared in late September 2008 and were at their highest percentage in late October (Figure 11). In 2009, spermatozoa were observed outside of the mesoglea in early September, and in mid-September (Figures 12(e) and 12(f)) they were observed in almost at 60% of all mesenteries counted.

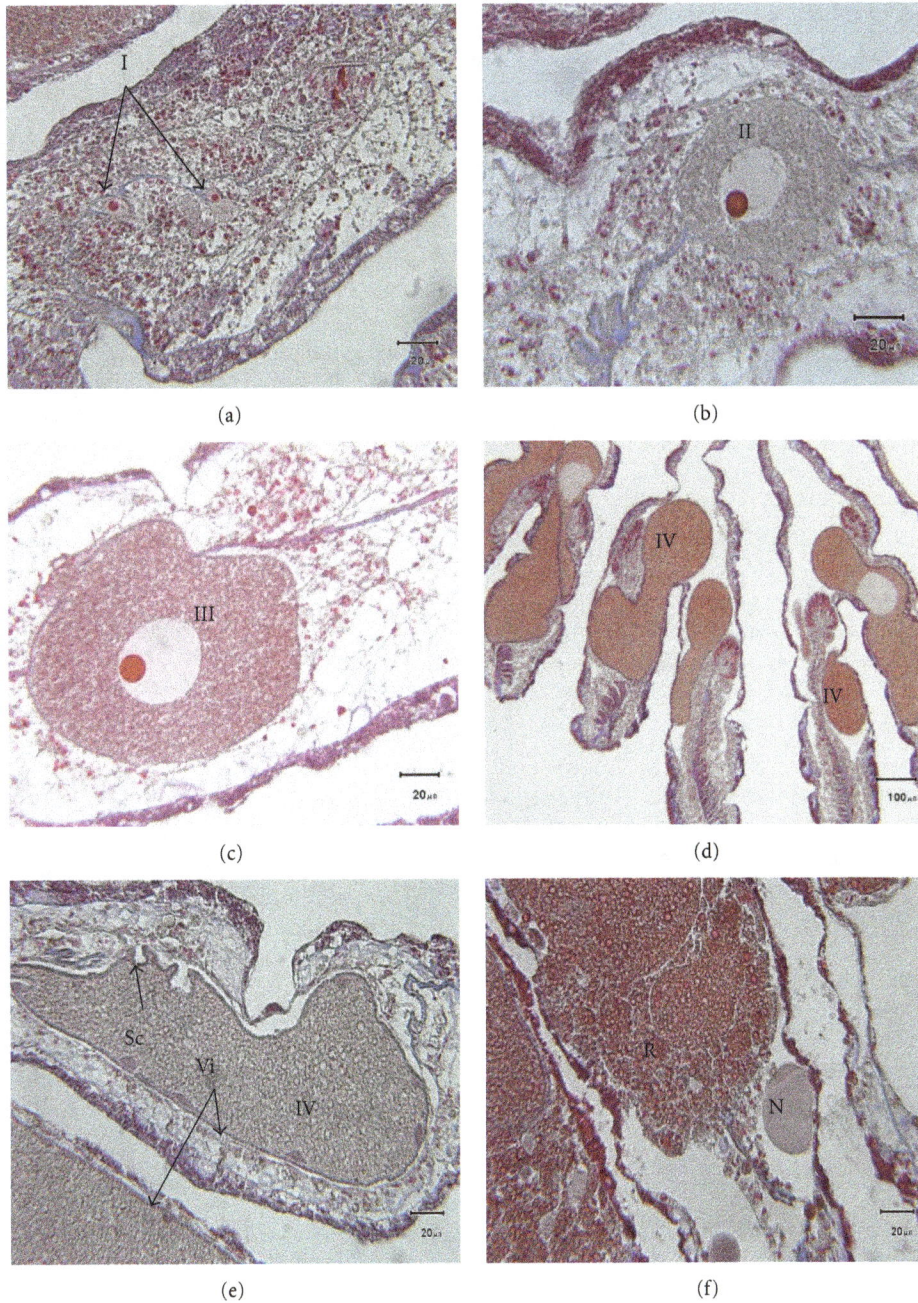

FIGURE 10: Photomicrographs of *Stephanocoenia intersepta* oocyte stages I–IV (defined by Table 1). (a) Stage I oocytes; (b) stage II oocyte; (c) stage III oocyte; (d) stage IV oocytes; (e) stage IV oocyte, Vi = vitelline membrane, Sc = scalloping of vitilline membrane; (f) oocyte resorption (R), N = Nucleus. All scale bars 20 μm except (d) which is 100 μm.

3.2.2. Lunar Periodicity. *Stephanocoenia intersepta* mean abundance of stage IV oocytes per colony on each sampling date was plotted with lunar phase (Figure 13). In 2008, there was a decrease in abundance between 26 August and 6 September which occurred after the full moon of 16 August. A decrease in stage IV oocytes occurred after 9 September 2009 following the full moon on 4 September.

3.2.3. Temperature. Oogenesis began during a time of cooler water temperatures in December. Abundance of all

spermary stages was plotted against temperature, and a linear regression analysis indicated a very low correlation that was not significant ($R^2 = 0.09$, $P = 0.17$). However maximum abundance of late stage gametes occurred during the warmest time of the year.

3.2.4. Predicted Spawning Times. In 2008, highest abundance of stage IV oocytes was 10 days after the full moon of 16 August. Results of a Mann-Whitney rank-sum test indicated a significant difference ($P < 0.013$) in abundance of stage

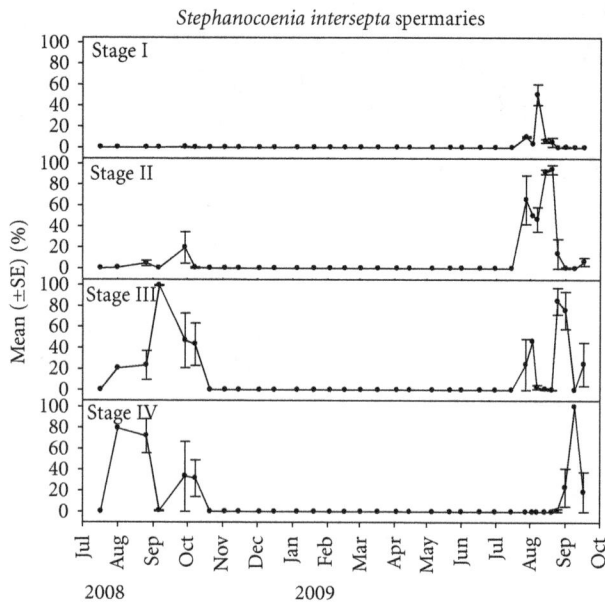

FIGURE 11: *Stephanocoenia intersepta* mean (±SE) percentages of spermaries per colony quantified through histological examination.

IV oocytes between colonies collected on 26 August and 6 September 2008, supporting the idea that *S. intersepta* spawned at least 10 days after the August 2008 full moon. In 2009, the greatest decrease in abundance of stage IV oocytes occurred between 9 September and 17 September. Unfortunately, only one female colony was sampled on 17 September which was the last sampling date. This colony showed oocyte resorption in 100% of all mesenteries examined. The four male colonies sampled on this date released spermatozoa in a mean of 60% of examined mesenteries, indicating a spawning event likely occurred before this date. These data indicate that *S. intersepta* spawned between 9 and 17 September 2009, following the full moon of 4 September.

4. Discussion

Gametogenic cycles of *S. bournoni* and *S. intersepta* differed greatly even though they likely spawned during similar times of the year. Most Caribbean broadcast spawning coral species studied to date begin gametogenesis a few months after spawning [24]. *Stephanocoenia intersepta* followed this pattern while *S. bournoni* did not. Gametogenesis of *S. bournoni* is similar to the Caribbean gonochoric broadcast spawning species *Oculina varicosa* which starts gametogenesis in early summer (*S. bournoni* starts in spring) and spawns in late summer and fall [8] as *S. bournoni* does. No reports of other Caribbean broadcast spawning species that begin gametogenesis in spring were found, indicating that this is an unusual occurrence.

Gametogenesis was more rapid in *S. bournoni* than in *S. intersepta*. Oogenesis in *S. bournoni* took less than 3 months (from early April to late June) to produce fully developed oocytes, but oocytes continued to develop for three more months through mid-October. The oogenic cycle

of *S. intersepta* began in December, and oocytes slowly matured over 9 months. Spermaries developed rapidly for both species. No spermaries were seen in *S. bournoni* in early May. However, all stages were observed only 12 days later. *Stephanocoenia intersepta* spermaries took only one month to mature after spermatogenesis began in mid-to late July.

Temperature has been thought to play a potential role in inducing gametogenesis [28], and many Caribbean broadcast spawning species release their gametes during the highest annual water temperatures [20, 24]. Gametogenesis in *S. bournoni* was significantly correlated with temperature, and increases and decreases of *S. bournoni* oocyte abundance followed the temperature regime. These data suggest that the trend of warming or the change in temperature may trigger *S. bournoni* to begin gametogenesis. *Stephanocoenia intersepta* gametogenesis did not show a relationship with increasing temperature indicating that there may be other exogenous factors such as moonlight [26] or day length that cue gamete production. However, similar to other Caribbean corals, maximum mature gamete abundance coincided with the warmest time of the year suggesting that temperature may play a role in gamete maturation [29].

Both species are gonochoric, but three colonies of *S. bournoni* were found to contain individual cosexual polyps that functioned predominantly as male but contained one mature oocyte per polyp. This sexual system is called andromonoecious when male and cosexual polyps are found on the same colony [30]. When male, female, and cosexual polyps are found on separate colonies, this sexual system is called polygamodioecious, and species that have this pattern may be sequential cosexuals with overlap of alternating cycles of male and female function leading to cosexual polyps [30]. However, because such a small proportion of the colonies sampled (3 out of 237 colonies) exhibited this pattern, a small number of polyps (one or two) in the tissue sample displayed this pattern, and only one oocyte occurred in each of the four observed cosexual polyps; it is unlikely that this species is sequential cosexual. Thus, *S. bournoni* can be described as stable gonochoric since it is not unusual to find a low amount of hermaphroditism among gonochoric species [28].

Solenastrea bournoni female fecundity was greater than *S. intersepta* in terms of the number of oocytes per polyp. The number of stage IV oocytes varied between species likely due to differences in oocyte and polyp morphology [31]. The polyps of *S. bournoni* were larger than *S. intersepta*, creating space for a greater number of oocytes. Diameter of *S. intersepta* stage IV oocytes was larger than *S. bournoni* possibly due to a longer period of oogenesis. Fecundity of *S. bournoni* varied between years as has been reported in other species [23] and may be a result of the small number of female colonies sampled prior to spawning. Fecundity of *S. intersepta* was not as variable between years.

The most widely accepted cue for coral spawning is lunar phase [7, 13, 32], and corals have blue-sensitive photoreceptors that have the ability to detect moonlight [33]. Although spawning was not observed directly, histological evidence indicated that *S. bournoni* is a broadcast spawner. No planulae were observed in any of the sampled colonies, and varying stages of gamete development followed by a

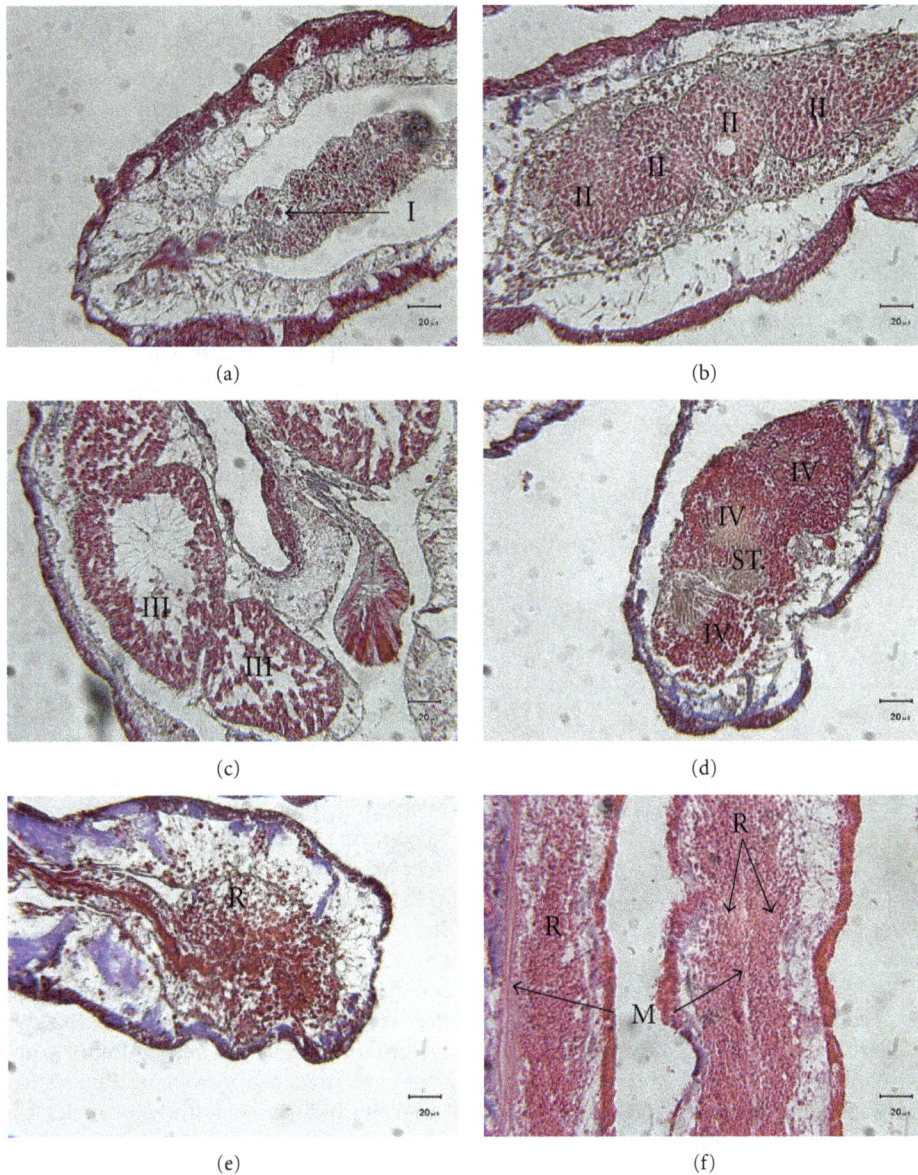

FIGURE 12: Photomicrographs of *Stephanocoenia intersepta* spermary stages I–IV (defined by Table 1). (a) Stage I spermary; (b) stage II spermaries; (c) stage III spermaries; (d) stage IV spermaries, ST = spermatozoa tails lined up; (e) released spermatozoa (R) stained with Heidenhain's azocarmine aniline blue; (f) released spermatozoa stained with H&E, M = mesoglea. All scale bars 20 μm.

sharp decrease in the abundance of mature gametes indicated spawning activity [24, 34, 35]. Based on results of this study, *S. bournoni* appears to spawn after the full moon of September. Although the main spawning event is thought to occur in September, it is possible that this species may also spawn in August or October due to the presence of mature gametes during these months and the decreases in abundance of mature oocytes. There was a smaller secondary decrease in abundance of mature oocytes after the full moon of October, but small samples sizes before and after the full moon limited interpretation of the data. It is not unusual for a coral species to spawn over more than one lunar period [14–16, 20, 36] especially if gametogenesis is asynchronous [8]. With its short gametogenic cycle and extended periods

with mature gametes, it may be possible for *S. bournoni* to spawn over multiple months. Since gametogenesis is closely tied to temperature in this species, spawning may occur sooner or over multiple months in areas closer to the equator. However, we do not have data to confirm these suggestions.

It is likely that *S. intersepta* spawned 5–10 days after the full moon in August or September depending on how early the full moon occurred. Histological evidence suggested that spawning by *S. intersepta* occurred at least 10 days after the 16 August 2008 full moon and at least 5 days after the 4 September 2009 full moon. Spawning could not have been successful if eggs were released after the 6 August 2009 full moon because there were no mature spermaries until late August. Reports from Bonaire indicate that *S. intersepta*

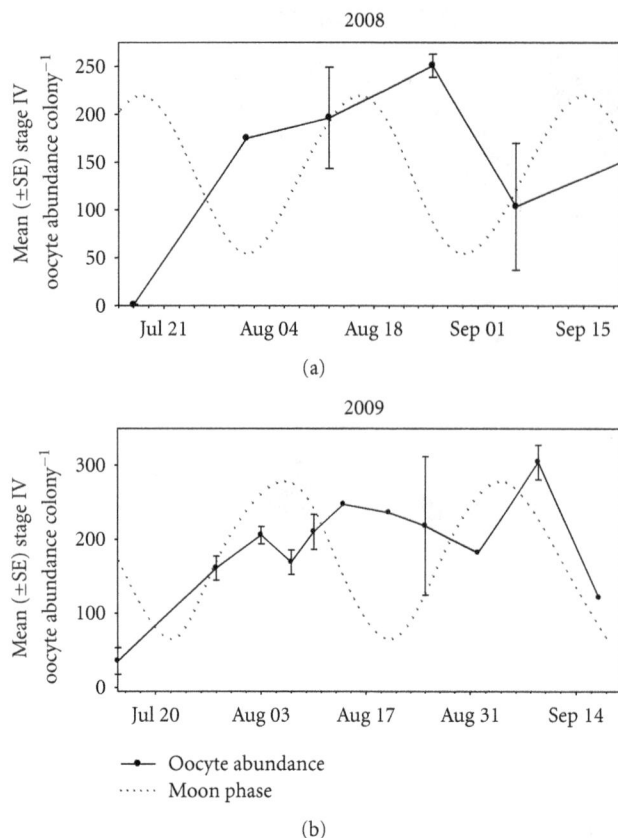

FIGURE 13: *Stephanocoenia intersepta* mean (±SE) abundance of stage IV oocytes per colony at each sampling date plotted with moon phases for years 2008 and 2009; full moons represented by peaks in moon phase line.

spawns between 3 to 7 nights after the August full moon [16]. At the Flower Garden Banks in the Gulf of Mexico, a region with a difference of only 157 km in latitude from sample sites in southeast Florida, *S. intersepta* spawns 7 to 9 days after the August or September full moon according to Vize et al. [9] and 6 to 10 days after the August full moon as reported by Hagman et al. [20]. Thus, the results of the current study agree with other published records of spawning times from the Caribbean and Gulf of Mexico.

Resorption of oocytes and released spermatozoa may help to confirm spawning times. Resorption of oocytes is not fully understood, but it is thought that by breaking down the large amount of lipid vesicles in oocytes, energy can be absorbed back into the coral [7]. Since oocytes are energetically costly, it would be beneficial to use any excess energy from oocytes that were not released during spawning. Assuming that resorption of oocytes and residual spermatozoa occurs after a spawning event, presence of resorbed gametes may provide additional evidence of when a spawning event occurred. Resorption in *S. bournoni* began in mid-November 2008, indicating that spawning occurred prior to this date. Increased percentages of released spermatozoa in October indicate that spawning may have occurred in September. Resorption and release of spermatozoa in 2009 occurred through much of the year in *S. bournoni*. Oocytes

were broken down over a long period of time, and residual spermatozoa that were not viable, as indicated by a lack of purple staining with H&E, persisted for a very long time. High amounts of resorption occurred in November 2009, and high percentages of released spermatozoa were observed in mid-September, again supporting the idea that *S. bournoni* spawned in September. In most coral reproduction studies, samples are not taken year round, and no other examples of this long-term breakdown of gametes were found in the literature. Oocyte resorption over long periods suggests that *S. bournoni* may not immediately require the energy gained from oocyte resorption. Given *S. bournoni*'s delayed onset of gametogenesis until spring, perhaps this extra energy is not needed until production of gametes commences. A slow breakdown may provide energy over a long period of time during winter months when metabolism rates may be slower with cooler temperatures.

Resorption of *S. intersepta* oocytes began in August and occurred in high percentages in September 2008. This observation, coupled with the appearance of a high percentage of released spermatozoa beginning at the end of September, suggests that a spawning event happened before this time and supports the idea of spawning after the August 2008 full moon. All residual oocytes and spermatocytes were gone by early November. In 2009, oocyte resorption was seen in low amounts (less than 2%) from July through early September, but on 17 September, the one female colony sampled showed resorption in 100% of mesenteries. Low percentages of released spermatozoa were observed in September until the 17th when released spermatozoa were at 60%. This evidence supports that spawning of *S. intersepta* occurred between 9 and 17 September 2009. After spawning, *S. intersepta* quickly broke down leftover gametes. *Stephanocoenia intersepta* may need this energy to support gametogenesis which began in December. The role of resorption in coral energetics and its timing relative to gametogenesis needs further study.

In summary, this study provides the first description of the gametogenic cycles of *S. bournoni* and *S. intersepta*. Spawning of *S. bournoni* is thought to occur in southeast Florida between 5 to 14 days after the full moon of September with a possible secondary spawning event occurring after the full moon of October. *Stephanocoenia intersepta* is thought to spawn 2 to 12 days after the late August or early September full moon which is consistent with reports from the Gulf of Mexico and Caribbean. More intense sampling during these potential spawning windows is needed to more precisely pinpoint the exact time of spawning, but this study provides a good foundation for further investigation.

Acknowledgments

This research was supported in part by the Florida Department of Environmental Protection and the Hillsboro Inlet District. Coral samples were collected under the Florida Fish and Wildlife Conservation Commission Special Activity License 08RP-1050. Special thanks to Adam T. St. Gelais, and D. Abigail Renegar for their support in laboratory techniques. Field assistance was provided by Elizabeth A. Larson, Stephanie J. Bush, Vanessa I. P. Brinkhuis, Allison

S. Brownlee, Paola Espitia-Hecht, Daniel P. Fahy, Mauricio Lopez, Jennifer Mellein, and Zach Ostroff.

References

[1] J. R. Guest, A. H. Baird, K. E. Clifton, and A. J. Heyward, "From molecules to moonbeams: spawning synchrony in coral reef organisms," *Invertebrate Reproduction and Development*, vol. 51, no. 3, pp. 145–149, 2009.

[2] M. J. A. Vermeij, E. Sampayo, K. Bröker, and R. P. M. Bak, "The reproductive biology of closely related coral species: gametogenesis in *Madracis* from the southern Caribbean," *Coral Reefs*, vol. 23, no. 2, pp. 206–214, 2004.

[3] A. H. Baird, J. R. Guest, and B. L. Willis, "Systematic and biogeographical patterns in the reproductive biology of scleractinian corals," *Annual Review of Ecology, Evolution, and Systematics*, vol. 40, pp. 551–571, 2009.

[4] E. Weil and W. L. Vargas, "Comparative aspects of sexual reproduction in the caribbean coral genus *Diploria* (Scleractinia: Faviidae)," *Marine Biology*, vol. 157, no. 2, pp. 413–426, 2010.

[5] V. J. Harriott, "Reproductive ecology of four scleratinian species at Lizard Island, Great Barrier Reef," *Coral Reefs*, vol. 2, no. 1, pp. 9–18, 1983.

[6] R. L. Harrison, R. C. Babcock, G. D. Bull, J. K. Oliver, C. C. Wallace, and B. L. Willis, "Mass spawning in tropical reef corals," *Science*, vol. 223, no. 4641, pp. 1186–1189, 1984.

[7] P. L. Harrison and C. C. Wallace, "Reproduction, dispersal and recruitment of scleractinian corals," in *Coral Reefs*, Z. Dubinsky, Ed., pp. 133–207, Elsevier, Amsterdam, The Netherlands, 1990.

[8] S. Brooke and C. M. Young, "Reproductive ecology of a deep-water scleractinian coral, *Oculina varicosa*, from the southeast Florida shelf," *Continental Shelf Research*, vol. 23, no. 9, pp. 847–858, 2003.

[9] P. D. Vize, J. A. Embesi, M. Nickell, D. P. Brown, and D. K. Hagman, "Tight temporal consistency of coral mass spawning at the Flower Garden Banks, Gulf of Mexico, from 1997–2003," *Gulf of Mexico Science*, vol. 23, no. 1, pp. 107–114, 2005.

[10] R. Van Woesik, F. Lacharmoise, and S. Köksal, "Annual cycles of solar insolation predict spawning times of Caribbean corals," *Ecology Letters*, vol. 9, no. 4, pp. 390–398, 2006.

[11] R. C. Babcock, G. D. Bull, P. L. Harrison et al., "Synchronous spawnings of 105 scleractinian coral species on the Great Barrier Reef," *Marine Biology*, vol. 90, no. 3, pp. 379–394, 1986.

[12] A. Szmant-Froelich, M. Reutter, and L. Riggs, "Sexual reproduction of *Favia fragum* (Esper): lunar patterns of gametogenesis, embryogenesis and planulation in Puerto Rico," *Bulletin of Marine Science*, vol. 37, pp. 880–892, 1985.

[13] Y. H. Fadlallah, "Sexual reproduction, development and larval biology in scleractinian corals—a review," *Coral Reefs*, vol. 2, no. 3, pp. 129–150, 1983.

[14] A. M. Szmant, "Reproductive ecology of Caribbean reef corals," *Coral Reefs*, vol. 5, no. 1, pp. 43–53, 1986.

[15] R. H. Richmond and C. L. Hunter, "Reproduction and recruitment of corals: comparisons among the Caribbean, the Tropical Pacific, and the Red Sea," *Marine Ecology Progress Series*, vol. 60, pp. 185–203, 1990.

[16] M. De Graaf, G. J. Geertjes, and J. J. Videler, "Observations on spawning of scleractinian corals and other invertebrates on the reefs of Bonaire (Netherlands Antilles, Caribbean)," *Bulletin of Marine Science*, vol. 64, no. 1, pp. 189–194, 1999.

[17] D. S. Gilliam, "Southeast Florida coral reef evaluation and monitoring project 2010 year 8 final report," Florida DEP Report #RM085, Miami Beach, Fla, USA, 2011.

[18] D. S. Gilliam, R. E. Dodge, R. E. Speiler, C. J. Walton, K. Kilfoyle, and C. Miller, Marine biological monitoring in Broward County, Florida: year 9 Annual Report. Prepared for the Broward County Board of County Commissioners, Broward County Natural Resources Planning and Management Division. pp. 114, 2011.

[19] J. C. Lang, "Status of coral reefs in the western Atlantic: results of initial surveys, Atlantic and Gulf Rapid Reef Assessment (AGRRA) program," Atoll Research Bulletin Technical Report 496, 2003.

[20] D. K. Hagman, S. R. Gittings, and K. J. P. Deslarzes, "Timing, species participation, and environmental factors influencing annual mass spawning at the Flower Garden Banks (northwest Gulf of Mexico)," *Gulf of Mexico Science*, vol. 16, no. 2, pp. 170–179, 1998.

[21] M. J. A. Vermeij, K. L. Barott, A. E. Johnson, and K. L. Marhaver, "Release of eggs from tentacles in a Caribbean coral," *Coral Reefs*, vol. 29, no. 2, pp. 411–411, 2010.

[22] E. A. Chornesky and E. C. Peters, "Sexual reproduction and colony growth in the scleractinian coral *Porites asteroides*," *Biological Bulletin*, vol. 172, pp. 161–177, 1987.

[23] B. Rinkevich and Y. Loya, "Variability in the pattern of sexual reproduction of the coral Stylophora pistillata at Eilat, Red Sea: a long-term study," *Biological Bulletin*, vol. 173, pp. 335–344, 1987.

[24] A. M. Szmant, "Sexual reproduction by the Caribbean reef corals Montastrea annularis and *M. cavernosa*," *Marine Ecology Progress Series*, vol. 74, no. 1, pp. 13–25, 1991.

[25] B. L. Kojis and N. J. Quinn, "1981Aspects of sexual reproduction and larval development in the shallow water hermatypic coral, *Goniastrea australensis* (Edwards and Haime, 1857)," *Bulletin of Marine Science*, vol. 31, pp. 558–573, 1981.

[26] P. L. Jokiel, R. Y. Ito, and P. M. Liu, "Night irradiance and synchronization of lunar release of planula larvae in the reef coral *Pocillopora damicornis*," *Marine Biology*, vol. 88, no. 2, pp. 167–174, 1985.

[27] A. Szmant-Froelich, P. Yevich, and M. E. Q. Pilson, "Gametogenesis and early development of the temperate coral *Astrangia danae* (anthozoa: scleractinia)," *Biological Bulletin*, vol. 158, pp. 257–269, 1980.

[28] A. C. Giese and J. S. Pearse, "Introduction: general principles," in *Reproduction of Marine Invertebrates*, A. C. Giese and J. S. Pearse, Eds., pp. 1–49, Academic Press, New York, NY, USA, 1974.

[29] K. Soong, "Sexual reproductive patterns of shallow-water reef corals in Panama," *Bulletin of Marine Science*, vol. 49, no. 3, pp. 832–846, 1991.

[30] J. R. Guest, A. H. Baird, B. P. L. Goh, and L. M. Chou, "Sexual systems in scleractinian corals: an unusual pattern in the reef-building species Diploastrea heliopora," *Coral Reefs*, vol. 31, no. 2, pp. 1–9, 2012.

[31] V. R. Hall and T. P. Hughes, "Reproductive strategies of modular organisms: comparative studies of reef-building corals," *Ecology*, vol. 77, no. 3, pp. 950–963, 1996.

[32] S. C. Wyers, H. S. Barnes, and S. R. Smith, "Spawning of hermatypic corals in Bermuda: a pilot study," *Hydrobiologia*, vol. 216-217, no. 1, pp. 109–116, 1991.

[33] M. Y. Gorbunov and P. G. Falkowski, "Photoreceptors in the cnidarian hosts allow symbiotic corals to sense blue moonlight," *Limnology and Oceanography*, vol. 47, no. 1, pp. 309–315, 2002.

[34] P. W. Glynn, S. B. Colley, C. M. Eakin et al., "Reef coral reproduction in the eastern Pacific: Costa Rica, Panama, and Galapagos Islands (Ecuador) 2. Poritidae," *Marine Biology*, vol. 118, no. 2, pp. 191–208, 1994.

[35] R. G. Waller, P. A. Tyler, and J. D. Gage, "Reproductive ecology of the deep-sea scleractinian coral *Fungiacyathus marenzelleri* (Vaughan, 1906) in the northeast Atlantic Ocean," *Coral Reefs*, vol. 21, no. 4, pp. 325–331, 2002.

[36] B. Vargas-Ángel, S. B. Colley, S. M. Hoke, and J. D. Thomas, "The reproductive seasonality and gametogenic cycle of *Acropora cervicornis* off Broward County, Florida, USA," *Coral Reefs*, vol. 25, no. 1, pp. 110–122, 2006.

14

Raphides in the Uncalcified Siphonous Green Seaweed, *Codium minus* (Schmidt) P. C. Silva

Jeffrey S. Prince

Dauer Electron Microscopy Laboratory, Department of Biology, The University of Miami, P.O. Box 249118, Coral Gables, FL 33124, USA

Correspondence should be addressed to Jeffrey S. Prince, jeffprince@miami.edu

Academic Editor: Wen-Xiong Wang

The vacuole of utricles, the outermost cell layer of the siphonous green seaweed, *Codium minus*, had numerous single needles and needle bundles. The crystals composing each needle appeared arranged in a twisted configuration, both ends were pointed, and each needle was contained in a matrix or membrane; bundles of needles appeared enclosed by a matrix. Chemical and electron diffraction analysis indicated that the needles consisted of calcium oxalate. This is the first paper on terrestrial plant-like raphides in an alga.

1. Introduction

Bundles of acicular crystals of calcium oxalate formed in specialized cells, idioblasts, are termed raphides in embryophytes [1–3]. Only one bundle of needles occurs per idioblast; the needles are long; needle length from several key pacific economic plants had a mean minimum length of 43 μm [4]. Each needle in a bundle is enclosed in a membrane, the crystal chamber [5–7], and bundles are enclosed by a water-soluble organic matrix, termed the vacuolar matrix [5, 8]. Raphides can burst through mature idioblasts due to swelling of the large amounts of mucilage contained in the cell [9]. Single needles do not occur in terrestrial plants but only as a bundle of needles [9, 10]. Individual needles, raphide, are found as microfossils in soils [4].

Abundant evidence supports the role of these needles in deterring vertebrate and invertebrate herbivory [1, 3, 7, 11, 12]. Raphides abrade the mouth and digestive tract of terrestrial herbivores causing edema; in addition, grooves in the needles may inject noxious plant metabolites and bacteria into these grazers [2].

Defense against herbivory in the marine environment generally involves calcification of the outer surface of green, red, and brown seaweeds [13]. This hard outer surface of calcium carbonate not only increases the difficulty of getting to the soft inner tissue, but consumption of the hard outer matrix can also alter the digestive pH, a deterrent for several herbivores [1, 13]. No alga has been found to have raphides. Where needles (acicular crystals) have been found, they appear singly, not in bundles of needles; they are surrounded by a vacuolar membrane or crystal chamber but not both, are generally very small, or reside individually in the cytoplasm and not the cell vacuole [14–18]. Single needles also occur in the cell vacuole of both lightly calcified parts of otherwise heavily calcified seaweeds [19–21].

We provide the first report of calcium oxalate crystals with all the characteristics that typify raphides of terrestrial plants. Large numbers of single needles and raphides (bundles of them) were found in the cell vacuole of an uncalcified green seaweed. The crystals in each acicular needle appeared twisted; individual needles were enclosed in a matrix or membrane, while bundles of needles were surrounded by a matrix. Single needles and raphides were located within the central cell vacuole and against the peripheral cytoplasmic layer.

2. Materials and Methods

Codium minus was collected on 13 July, 2003 from the Izu Peninsula, Honshu, Japan, fixed in copious 4% neutral formalin in sea water for several days, loosely wrapped in paper toweling saturated with fixative, placed in Ziploc bags,

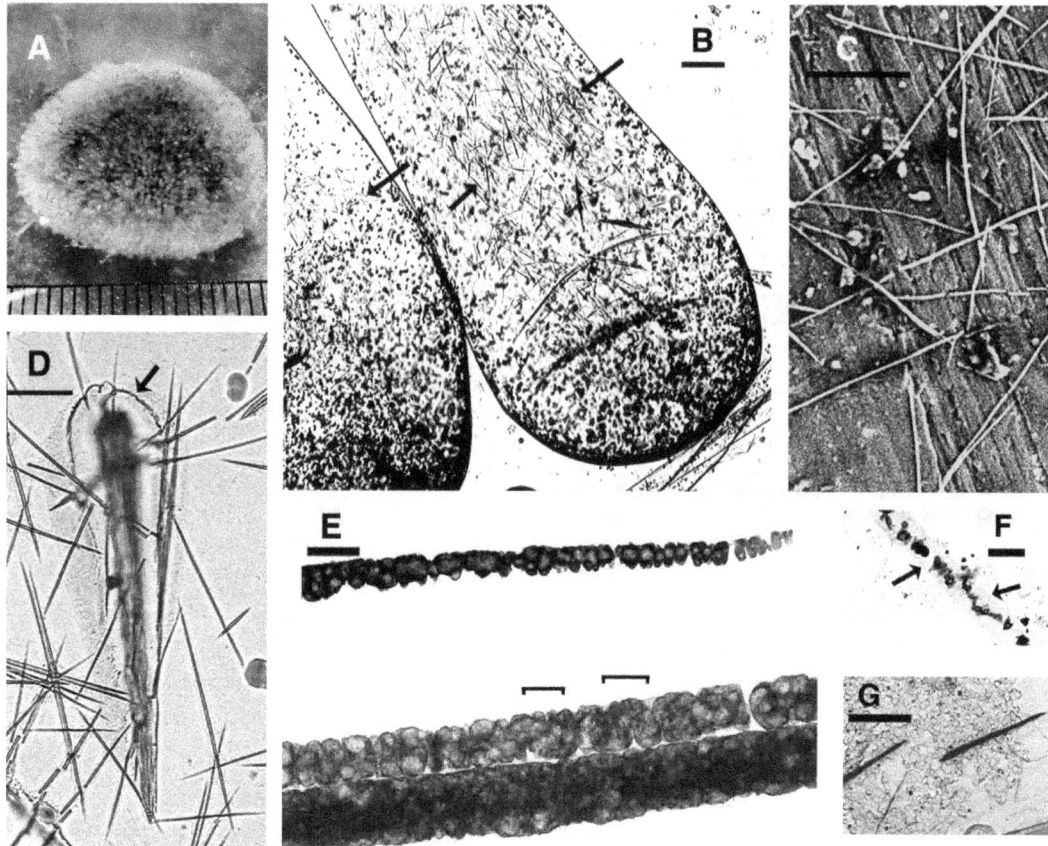

FIGURE 1: Calcium oxalate needles and bundles of needles in *Codium minus*: (A) *C. minus*; (B) utricle with needle bundles and individual needles (arrows). (C) SEM of individual needles. (D) Individual needles and a needle bundle, the latter contained within a matrix or membrane (arrow); (E) and (F) TEM of needle tips showing twisting of crystal (E, brackets) and an apparent membrane about individual needles (arrows); (G). Needle bundle staining intensely for calcium oxalate (Yasue [22]). Scale bars: (A) ruler in mm; (B) 100 μm; (C) and (D) 20 μm; (E) and (F) 0.5 μm; (G) 50 μm.

and shipped to the USA. Digital light micrographs of whole utricles were taken with an Olympus BX60. For scanning electron microscopy (SEM), utricles were rinsed in distilled water, and their contents were isolated onto stubs, air dried, and then visualized uncoated with a Jeol 5600LV scanning electron microscope.

For elemental analysis, samples with a similar preparation as above were analyzed uncoated using a Link/Oxford ISIS 300 (EDS) system ancillary to a Philips/FEI XL30 ESEM-FEG scanning electron microscope (SEM) operated in the environmental mode (1 to 10 torr.). Needles were analyzed so as to minimize beam interaction with the stub (an aluminum alloy containing copper, Figure 2).

For transmission electron microscopy (TEM), utricles were first rinsed in distilled water, and their contents were squeezed onto formvar-coated grids and examined, unstained with a Philips 300 electron microscope at 60 kV. Electron diffraction ring patterns were also obtained with the Philips 300 at 80 kV and 0 tilt.

Solubilities of the needles were tested by exposing whole specimens to 1.0 N hydrochloric acid, 5.25% sodium hypochlorite (commercial bleach), or 67% aqueous acetic acid [16, 19, 22]. The Yasue [22] method for calcium oxalate

localization was done according to that described by Pueschel [16] using dithiooxamide (Sigma Chemical) which results in calcium oxalate inclusions staining densely black.

3. Results

The thallus of *Codium minus*, a pulvinate Codiaceae, had a mean diameter of 30 mm ($\pm$8.1, 21; SD, N; Figure 1(A)). Needles and raphides (needle bundles) were found in all terminal utricles in such numbers that they caused a glistening hue to the utricle under low magnification (Figure 1(B)). They were absent from the medullary filaments that compose the mass of the central portion of the thallus. Both ends of the needles were drawn out into a point. Sizes ranged from 31 to 72 μm long (55.1 $\pm$ 0.7; 12 mean, SD, N) by 0.5 to 1.3 μm (0.7 $\pm$ 0.3; 12 mean, SD, N) wide (Figures 1(B), 1(C), and 1(D)). All needles in needle bundles, raphides, were oriented in the same direction, forming a large compound needle (Figures 1(D) and 1(G)). Optical section found that individual needles and bundles were located throughout the large central vacuole of the utricle as well as against the cytoplasm lining the cell wall. Light microscopy of raphides found a membrane or matrix surrounding the

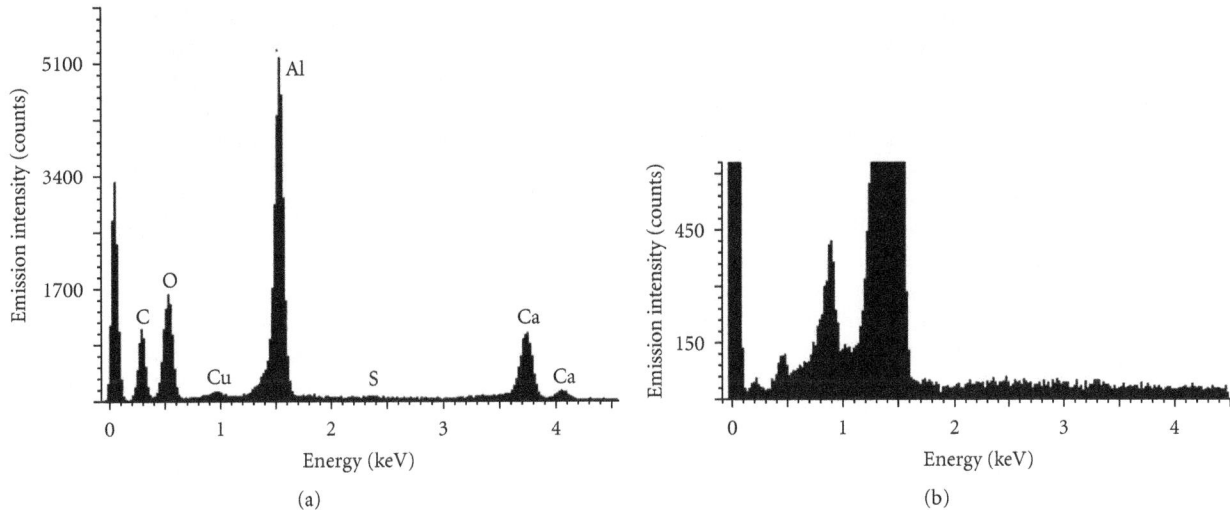

FIGURE 2: Energy dispersive X-ray spectrum (EDS) of a needle (a) and stub (b). Emission intensity (counts) is plotted versus energy (keV) characteristic for various elements (Al: aluminum; C: carbon; Ca: calcium; Cu: copper; O: oxygen; S: sulfur).

entire structure (Figure 1(D)). TEM showed a membrane-like matrix surrounding the tips of individual needles (Figure 1(F)); this membrane became obscure in the needle body. Crystalline globular material that composed the needle appeared to be arranged in a helical pattern (Figure 1(E), brackets). Under the SEM, however, the needles appeared smooth and uniform, and no globular substructure was apparent (Figure 1(C)).

Chemical tests found that needles failed to dissolve after 2 hrs in cold or warm (70°C) acetic acid (67%) or Clorox (5.25%). Hydrochloric acid (1 N) dissolved the needles in 5 to 10 min but without release of any gas. The Yasue [16, 22] method for localization of calcium oxalate found both single and compound needles outside of the utricles stained densely (Figure 1(G)), at times with a precipitate about them; needles within utricles did not stain.

EDS-SEM of the needles found strong signals for calcium, while signals for copper and aluminum appeared to originate from the stub (Figure 2). Electron diffraction patterns for the needles appeared to more closely match, those for calcium oxalate than those for calcium carbonate (Table 1).

4. Discussion

This is the first paper on embryophyte-type raphides (bundles of acicular needles) in an alga. Raphides apparently dissociated within the vacuole of the utricle forming quantities of single needles. Single needles and needle bundles were found in the vacuole of utricles of Codium minus in such numbers that these cells glisten under low magnification. EDS-SEM found strong signals for calcium, while electron diffraction analysis suggested that the needles consisted of calcium oxalate. This was further supported by various chemical tests. Needles were found only in the utricles (the

TABLE 1: D-spacing from electron diffraction of needles from Codium minus and for calcium oxalate [19] and standards (JCPDS, 1986) for calcium carbonate as calcite (24–27C; 5–586).

Needles	Oxalate	CaCO$_3$	CaCO$_3$
2.767	2.75	2.83	2.845
2.434	2.4	2.495	2.495
1.695	1.69	1.6259	1.626

outer most, green layer of Codium species) and were not found in the internal portion of the plant composed of colorless intertwined medullary filaments. But in this group of seaweeds, there are no cross-walls; thus, utricles and medullary filaments share a common cytoplasm [23].

Raphides in embryophytes are formed in idioblasts, specialized cells that form crystals of various size, morphology and composition [3]. Idioblasts that form raphides are located in all portions of terrestrial plants and have an enlarged nucleus, a dense cytoplasm, and a small to absent vacuole [4, 7]. A small vacuole appears and is the site for crystal nucleation; the vacuole membrane and cell itself markedly expand during growth of the crystal [6]. The vacuole membrane, vacuolar matrix, appears granular at maturity under the scanning electron microscope and is apparently water soluble [5]. Idioblasts that give rise to raphides produce a single bundle of multiple, parallel-aligned, needle-like crystals [5, 11]. In some cases, needles within a bundle can be disorganized [3]. Each needle of a bundle, raphide, is contained within an additional membrane, the crystal chamber, which apparently dictates both the shape and precipitation of the crystal itself [6]. Raphides at maturity are often contained in very large cells, the raphides consuming most of the idioblast, while mucilage frequently fills the remaining available space [9, 10]. Swelling of the mucilage causes the cell to burst ejecting the raphides to the outside of the plant [9]. A single needle, raphide, furthermore, refers to needles outside the plant or microfossils found in soils [3, 4].

The definition of raphides as described above for embryophytes, therefore, requires the presence of several traits: needles developing in a specialized cell (idioblast) with an enlarged nucleus and dense cytoplasm; needles occurring in bundles; large needle size, one bundle per cell; a crystal chamber about each needle; a vacuolar matrix about a bundle of needles. As mentioned above, no paper on raphides in marine or fresh water algae meets these requirements including those found in *C. minus*, the current paper. But the needles in *C. minus* do meet the following traits for raphides: they have a mean length greater than the minimum mean length of raphides in many pacific economic plants [4]; they are composed of calcium oxalate; the needles occur in bundles, each bundle surrounded by a matrix (vacuolar matrix?); a crystal chamber apparently surrounds each needle. But raphides in embryophytes occur singly per cell, while each utricle has several needle bundles. But the whole thallus of *Codium minus* is a single cell as crosswalls are absent. The traits of *C. minus*, that do not meet the definition of terrestrial raphides are, therefore, several bundles per cell, the cell also being multinucleate. Is there a connection between production of needle bundles in the multinucleate, siphonous alga, *C. minus*, and evolution of raphides in land plants?

Dawes [24] and Leliaert and Coppejans [25] found abundant calcium oxalate needles (35–55 μm in length) in the vacuole of all cells in the endemic Australian coenocytic green alga, *Apjohnia laetevirens* Harvey. Calcium oxalate crystals have been recorded for other noncalcified algae. Pueschel [16] found small, cruciate crystals in the peripheral cytoplasm but not within the vacuole of the fresh water green alga, *Spirogyra*. *Antithamnion*, a marine red alga, had single needles up to 30 μm in length in the peripheral cytoplasm [15, 18]. The length of the needles in *Antithamnion*, though shorter than those in *C. minus*, suggested that they may adversely affect herbivores [18], but the bipyramidal morphology of crystals (apparently surrounded by an organic matrix = the vacuolar membrane?) in *Chaetomorpha*, a marine green seaweed, probably negates their role in grazer defense [26]. Other papers [14, 15, 17] also describe needles, in some cases more than 100 μm long. All of these papers describe, however, only single needles, not bundles of them and, in addition, may describe the presence of a crystal chamber or vacuolar matrix but not both.

The acicular shape of raphides appears to be a critical component of plant defense against herbivory [7, 11]. These acicular crystals may also bear barbs or grooves [27]. Arnott and Webb [11] found that raphides in grape (*Vitis*) are twinned (indicated by one end being forked, the opposite pointed), the twinning occurs along the long axis of the raphide, and the crystal lattice rotates around the twin axis. This rotation of the crystal lattice structure is thought to enhance crystal stability and growth [11]. The calcium oxalate needles described for *C. minus* share similarities to raphides described for grape. (1) Needles in *C. minus* and those in grape have a similar range in length, but those in

C. minus were approximately half as wide. Raphides in grape, as mentioned above, are twinned crystals, and the width of one twin is approximately, that of the nontwinned needle in *C. minus* (both ends of which were pointed). (2) The crystals composing the needle appear to be twisted (twisting in *C. minus*, however, appeared to occur at a greater frequency than that in grape). (3) Each needle in *C. minus*, as mentioned above, is enclosed in a crystal chamber, a bundle of needles in a vacuolar matrix.

The crystal chamber surrounding each needle in a bundle appears to share many similarities with the silicalemma of diatoms [23]. Both are single-membrane sacks, each apparently provides an internal surface for crystal nucleation and deposition, and both determine the mature crystal morphology [23]. The silicalemma deposits the highly characteristic, taxonomically important, silicon frustule of diatoms, while the crystal chamber appears to be involved with calcium deposition as either carbonate or oxalate in the shape of a needle or more elaborate shape.

Raphides within noncalcified green seaweeds is one possible mechanism for defense against herbivory. Another is the lack of the production of a chemical attractant for grazers. Prince and Leblanc [28] found that *Codium fragile* did not produce an attractant for *Strongylocentrotus droebachiensis*, the green sea urchin. Green sea urchins would graze on *C. fragile* only if they happen to come across it. On the other hand, the brown kelp, *Laminaria saccharina*, produces a strong attractant for green sea urchins resulting in urchin barrens where once kelp beds were dominant [28].

Acknowledgments

The author would like to thank Dr. Cynthia Trowbridge for collecting *Codium minus*, Dr. Matt Lynn who provided the electron diffraction patterns and elemental analysis of the needles, and Dr. Barbara Whitlock for her comments on the paper.

References

[1] M. E. Hanley, B. B. Lamont, M. M. Fairbanks, and C. M. Rafferty, "Plant structural traits and their role in antiherbivore defence," *Perspectives in Plant Ecology, Evolution and Systematics*, vol. 8, no. 4, pp. 157–178, 2007.

[2] S. Lev-Yadun and M. Halpern, "External and internal spines in plants insert pathogenic microorganisms into herbivore's tissue for defense," in *Microbial Ecology Research Trends*, T. Van Dijk, Ed., pp. 155–168, Nova Science Pubs., Inc., New York, NY, USA, 2008.

[3] G. G. Coté, "Diversity and distribution of idioblasts producing calcium oxalate crystals in Dieffenbachia seguine (Araceae)," *American Journal of Botany*, vol. 96, no. 7, pp. 1245–1254, 2009.

[4] A. Crowther, "Morphometric analysis of calcium oxalate raphides and assessment of their taxonomic value for archeological microfossils," in *Archeological Science Beneath the Microscope: Studies in Residue and Ancient DNA Analysis in Honor of Thomas H. Loy*, H. Haslam, G. Robertson, A. Crowther, S. Nugent, and L. Kirkwood, Eds., pp. 102–210, Australian National University Press, Canberra, Australia, 2009.

[5] M. A. Webb, J. M. Cavaletto, N. C. Carpita, L. E. Lopez, and H. J. Arnott, "The intravacuolar organic matrix associated with calcium oxalate crystals in leaves of Vitis," *Plant Journal*, vol. 7, no. 4, pp. 633–648, 1995.

[6] C. J. Prychid, R. S. Jabaily, and P. J. Rudall, "Cellular ultrastructure and crystal development in Amorphophallus (Araceae)," *Annals of Botany*, vol. 101, no. 7, pp. 983–995, 2008.

[7] V. R. Franceschi and P. A. Nakata, "Calcium oxalate in plants: formation and function," *Annual Review of Plant Biology*, vol. 56, pp. 41–71, 2005.

[8] V. R. Franceschi and H. T. Horner, "Calcium oxalate crystals in plants," *The Botanical Review*, vol. 46, no. 4, pp. 361–427, 1980.

[9] K. Esau, *Plant Anatomy*, John Wiley & Sons, New York, NY, USA, 1965.

[10] C. R. Metcalf and L. Chalk, *Anatomy of Dicotyledons, Leaves, Stem, and Wood in Relation to Taxonomy with Notes on Economic Uses*, vol. 1-2, Claredon Press, Oxford, UK, 1950.

[11] H. J. Arnott and M. A. Webb, "Twinned raphides of calcium oxalate in grape (Vitis): implications for crystal stability and function," *International Journal of Plant Sciences*, vol. 161, no. 1, pp. 133–142, 2000.

[12] J. M. Kingsbury, *Poisonous Plants of the United States and Canada*, Prentice-Hall Inc., Englewood Cliffs, NJ, USA, 1964.

[13] M. E. Hay, Q. E. Kappel, and W. Fenical, "Synergisms in plant defenses against herbivores: interactions of chemistry, calcification, and plant quality," *Ecology*, vol. 75, no. 6, pp. 1714–1726, 1994.

[14] D. Menzel, "Fine structure of vacuolar inclusions in the siphonous green alga *Chlorodesmis fastigiata* (Udoteaceae, Caulerpales) and their contribution to plug formation," *Phycologia*, vol. 26, no. 2, pp. 205–221, 1987.

[15] C. M. Pueschel, "Calcium oxalate crystals in the red alga *Antithamnion kylinii* (Ceramiales): cytoplasmic and limited to indeterminate axes," *Protoplasma*, vol. 189, no. 1-2, pp. 73–80, 1995.

[16] C. M. Pueschel, "Calcium oxalate crystals in the green alga *Spirogyra hatillensis* (Zygnematales, Chlorophyta)," *International Journal of Plant Sciences*, vol. 162, no. 6, pp. 1337–1345, 2001.

[17] C. M. Pueschel, "Calcium oxalate crystals in the green alga *Spirogyra hatillensis* (Zygnematales, Chlorophyta)," *Journal of Phycology*, vol. 32, s. 3, pp. 55–56, 2002.

[18] C. M. Pueschel and J. A. West, "Effects of ambient calcium concentration on the deposition of calcium oxalate crystals in *Antithamnion* (Ceramiales, Rhodophyta)," *Phycologia*, vol. 46, no. 4, pp. 371–379, 2007.

[19] E. I. Friedmann, W. C. Roth, J. B. Turner, and R. S. Mcewen, "Calcium oxalate crystals in the aragonite-producing green alga *Penicillus* and related genera," *Science*, vol. 177, no. 4052, pp. 891–893, 1972.

[20] J. B. Turner and E. I. Friedmann, "Fine structure of capitular filaments in the coenocytic green alga *Penicillus*," *Journal of Phycology*, vol. 10, no. 2, pp. 125–134, 1974.

[21] L. Bohm and D. Futterer, "Algal calcification in some Codiaceae (Chlorophyta): ultrastructure and location of skeletal deposits," *Journal of Phycology*, vol. 14, no. 4, pp. 486–493, 1978.

[22] T. Yasue, "Histochemical identification of calcium oxalate," *Acta Histochemistry and Cytochemistry*, vol. 2, no. 3, pp. 83–95, 1969.

[23] R. E. Lee, *Phycology*, Cambridge University Press, New York, NY, USA, 1980.

[24] C. J. Dawes, "A study of the ultrastructure of a green alga, *Apjohnia laetevirens* Harvey with emphasis on the cell wall structure," *Phycologia*, vol. 8, no. 2, pp. 77–84, 1969.

[25] F. Leliaert and E. Coppejans, "Crystalline cell inclusions: a new diagnostic character in the Cladophorophyceae (Chlorophyta)," *Phycologia*, vol. 43, no. 2, pp. 189–203, 2004.

[26] C. M. Pueschel and J. A. West, "Cellular localization of calcium oxalate crystals in Chaetomorpha coliformis (Cladophorales; Chlorophyta): evidence of vacuolar differentiation," *Phycologia*, vol. 50, no. 4, pp. 430–435, 2011.

[27] W. S. Sakai, S. S. Shiroma, and M. A. Nagao, "Study of Raphide Microstructure in Relation to Irritation," *Scanning Electron Microscopy*, pt. 2, pp. 979–986, 1984.

[28] J. S. Prince and W. G. Leblanc, "Comparative feeding preference of *Strongylocentrotus droebachinesis* (Echinoidea) for the invasive green seaweed *Codium fragile* ssp. *tomemtosoides* (Chlorophyceae) and four other seaweeds," *Marine Biology*, vol. 113, no. 1, pp. 159–163, 1992.

15

Larval Diel Vertical Migration of the Marine Gastropod *Kelletia kelletii* (Forbes, 1850)

Melissa R. Romero,[1] Kimberly M. Walker,[1] Carmen J. Cortez,[2]
Yareli Sanchez,[3] Kimberly J. Nelson,[1] Daisha C. Ortega,[1] Serra L. Smick,[1]
William J. Hoese,[1] and Danielle C. Zacherl[1]

[1] *Department of Biological Science, California State University Fullerton, P.O. Box 6850, Fullerton, CA 92834-6850, USA*
[2] *Graduate Group in Ecology, University of California Davis, One Shields Avenue, Davis, CA 95616, USA*
[3] *Global Change Research Group, Department of Biology, San Diego State University, 5500 Campanile Drive, San Diego, CA 92182, USA*

Correspondence should be addressed to Danielle C. Zacherl, dzacherl@fullerton.edu

Academic Editor: Susumu Ohtsuka

Documenting larval behavior is critical for building an understanding of larval dispersal dynamics and resultant population connectivity. Nocturnal diel vertical migration (DVM), a daily migration towards the surface of the water column at night and downward during the day, can profoundly influence dispersal outcomes. Via laboratory experiments we investigated whether marine gastropod *Kelletia kelletii* larvae undergo nocturnal DVM and whether the behavior was influenced by the presence of light, ontogeny, and laboratory culturing column height. Larvae exhibited a daily migration pattern consistent with nocturnal diel vertical migration with lower average vertical positioning (ZCM) during day-time hours and higher vertical positioning at night-time hours. ZCM patterns varied throughout ontogeny; larvae became more demersal as they approached competency. There was no effect of column height on larval ZCM. DVM behavior persisted in the absence of light, indicating a possible endogenous rhythm. Findings from field plankton tows corroborated laboratory nocturnal DVM findings; significantly more *K. kelletii* were found in surface waters at midnight compared to at noon. Unraveling the timing of and the cues initiating DVM behavior in *K. kelletii* larvae can help build predictive models of dispersal outcomes for this emerging fishery species.

1. Introduction

In open coast marine habitats, multiple factors influence larval dispersal destinations, including abiotic factors, such as current speed and direction, and biotic factors, such as timing of larval release and pelagic larval duration (PLD). Additionally, larvae of marine species living within estuaries, on coastal shelves, or near oceanic islands have been hypothesized to use vertical positioning behaviors that, when coupled with stratified countercurrents, promote retention or return to suitable settlement habitat [1–3]. Documenting this form of larval behavior is a critical component to building an understanding of larval dispersal dynamics [3–6].

Diel vertical migration (DVM) has been implicated in both modeling [7, 8] and empirical studies [9] as a positioning behavior that can affect dispersal outcomes. In nocturnal DVM, larvae migrate toward the surface of the water column at night, and then downward in the water column during the day. Reverse DVM shows the opposite pattern, but it is also thought to influence dispersal outcomes (e.g., [10]). These vertical migratory behaviors, or sensory capabilities that are consistent with vertical movement behaviors [11], have been observed in a wide variety of taxa, including arthropods [12], mollusks [10, 13–15], fish [16], and others [17].

Diel vertical migration is primarily controlled by light, with other factors acting as modifiers of this behavior [18, 19]. Light plays a multifaceted role in DVM because it may (1) signal organisms to start or stop swimming (i.e., photokinesis) [20], (2) provide cues for the direction of swimming (i.e., positive or negative phototaxis) [21], and (3) entrain endogenous rhythms so that behavior persists in the absence of light [22]. In addition to light, other cues such as gravity, temperature, oxygen, salinity, pressure, and chemicals from phytoplankton and predators may influence DVM [18, 23–26].

While light may initiate, signal the direction of, and entrain DVM behavior, the behavior itself varies throughout ontogeny. This variation may be associated with size-dependent predation risk where younger, smaller individuals alter their DVM patterns in the presence of predators, while older, larger individuals, who are less vulnerable, do not alter their behavior [27]. Alternately, younger oyster (*Crassostrea virginica*) larvae remain evenly distributed throughout the water column while older larvae rise during the flood tide and sink during the ebb tide [28]. The latter behavior enhances retention within an estuary, enhances up-estuary transport, and provides opportunity for the larvae to sample the substrate for suitable habitat. A similar pattern has been observed in scallop larvae, *Placopecten magellanicus* [13], in the open sea and blue crab megalopae, *Callinectes sapidus* [29], in estuaries. Thus, complex larval positioning behaviors that vary across a day or through ontogeny or both can impact how and where larvae disperse [3].

There is growing interest in the dispersal dynamics of Kellet's whelk, *Kelletia kelletii*, a large predatory neogastropod inhabiting rocky reefs and kelp forests along the coast of California, USA, and Baja, Mexico. *Kelletia kelletii* is slow-growing, slow to mature, and aggregates seasonally for mating; these traits make this recently targeted fishery species vulnerable to overexploitation. The *K. kelletii* fishery has experienced a rapid increase in landings since 1995 [30], prompting the California Department of Fish and Game to designate the species as an "emerging fishery" (CA Regulatory Notice Register 2011 43-Z; Craig Shuman, personal communication). This species has a pelagic larval duration (PLD) of at least 5.5 weeks (M. Romero and D. Zacherl, unpublished data), making long-distance dispersal a possibility, though even species with long PLD are capable of retention or very short-distance dispersal [31]. Recent molecular work on *K. kelletii* based on microsatellite markers suggests broad exchange of larvae among populations (i.e., global F_{ST} = 0.00138) [32]. However, such broad exchange might be occurring over temporal scales of decades to centuries [33] and may not reflect year-to-year exchange likely to be more relevant for fishery management. Knowledge of the larval behavior of this species would facilitate development of oceanographic models of dispersal that incorporate this behavior, against which the molecular results could be compared.

Using a series of controlled laboratory experiments and field plankton tows, we explored whether *Kelletia kelletii* larvae exhibit nocturnal DVM and what factors (including light, ontogeny, and culture column artifacts) influence their DVM behavior. The following specific research questions were addressed. (1) Do larvae exhibit nocturnal DVM in laboratory cultures and, if so, does the pattern of DVM change as a function of ontogeny? (2) Does light cue DVM ascent or descent behaviors? (3) Does culturing column height affect DVM behavior? (4) Do larvae in the field exhibit distributions that are consistent with nocturnal DVM behavior?

2. Methods

2.1. Study Organism. *Kelletia kelletii* is a large predatory buccinid gastropod commonly found in subtidal kelp forests, rocky reefs and cobble-sand interfaces at depths ranging from 2 to 70 m [34] from Isla Asunción, Baja California, Mexico [35], to Monterey, CA, USA [36]. Rosenthal [34] reported onset of sexual maturity at c. 60 mm in shell length (defined as maximum shell length from the tip of the spire to the tip of the siphonal canal). Kellet's whelks reproduce annually, with egg-laying restricted to late spring and summer. The females deposit masses of egg capsules on benthic hard substrate in which larvae develop for c. 30–34 days. The hatched larvae are pelagic [34]. Laboratory culturing studies resulted in successful metamorphosis of 33% of larvae (n = 10) from weeks 5.5 through 9 in the presence of live rock dominated by *Petaloconchus montereyensis* (prey species of *K. kelletii*), as well as 100% of larvae exposed to high concentrations of KCl in weeks 8 and 9; these pilot results suggest a planktonic duration of at least 5.5–9.0 weeks, though the competency window after 9 weeks remains untested (M. Romero and D. Zacherl, unpublished data).

2.2. Laboratory: DVM and Effects of Ontogeny. To determine whether *K. kelletii* larvae undergo nocturnal DVM and if behavior is affected by ontogeny, we cultured replicate batches (n = 5) of newly-hatched larvae at 15°C during August and September, 2005. Egg masses laid by *K. kelletii* were hand-collected from McAbee beach, Monterey Bay, California (N36°37.09′ W121°53.82′), via SCUBA at depths of 15–21 meters in August 2005, and transported in coolers to CSU Fullerton. To control for genetic differences only larvae hatching from a single egg cluster were used for this experiment. Seawater (33.2 ppt) used in all experiments was collected from Scripps Institution of Oceanography (La Jolla, California), filtered to 0.2 μm (FSW = filtered sea water), and transported to CSU Fullerton. Egg masses were placed in 4 L clear glass culture jars with lids containing 3 L FSW at 15°C in a temperature-controlled growth chamber illuminated by 6 GE 35 watt high output cool white fluorescent linear lamps (F24T12-CW) with a 12 : 12 light : dark cycle. Every other day the egg masses were removed, culture jars were washed by vigorously rinsing them three times with deionized water and twice with ultrapure water (resistivity > 18.0 MΩ), and the egg masses were returned to jars with fresh FSW. Egg masses were maintained in this way until larvae hatched.

Within 15 hours of hatching, 100 sibling larvae each were placed in replicate cultures (n = 5) in 1000 ml glass jars with 800 ml of FSW for a final water column size of ~8 cm diameter (d) × 15 cm height (h). Larvae were reared in the same growth chambers with conditions as described above. Dead or fouled larvae were removed from cultures daily, and the remaining larvae were transferred daily to clean jars containing fresh FSW. After every water change, the total number of larvae in each replicate jar was recorded. We fed larvae a phytoplankton mixture of *Isochrysis galbana* (9,000 cells mL^{-1}) and *Pavlova lutheri* (9,000 cells mL^{-1}) every other day following water changes.

DVM behavior was quantified one day after hatching (i.e., week 1) and once a week through week 5, which approaches the minimum time to competency (Romero and Zacherl, unpublished data). The first observation was made at least 12 hours after water changes and feeding, and occurred four times over a 24 hr period at 0600, 1200, 1800, and 2400 hrs. Observations at 0600 and 1800 hrs occurred an hour after the light source turned on and off, respectively. Initial observations were carried out 20 hours after replicate jars were established. In order to simplify our data into one dichotomous measure, we recorded the number of demersal larvae within each jar. Demersal larvae were defined as any larvae present within 2.5 cm of the bottom of each jar. Unlike subsequent experiments (see below), detailed observations were not recorded on the vertical positions of the remaining larvae, though qualitative observations were recorded.

A two-way ANOVA was used to evaluate the factors time-of-day, ontogeny and time-of-day X ontogeny interactions on the percent of demersal larvae, with replicate as a blocking factor to take into account the repeated measures design. To more intuitively depict DVM behavior graphically (in Figure 1) we converted percent demersal values to percent nondemersal values so that higher values correspond to a higher percentage of larvae in the water columns.

2.3. Laboratory: Effects of Light and Culture Column Height on DVM.

To test whether light and culture column height influenced DVM behavior, we cultured replicate batches ($n = 4$) at 15°C during June 2007. Larvae originated from multiple egg masses that were collected at Palos Verdes, CA, USA (N33°42.67′ W118°14.66′), were allowed to hatch in the laboratory at CSU Fullerton, and were cultured together under conditions described above until they were 7-8 days old (hereafter referred to as "week 2 larvae"), except that the light:dark cycle was 16:8 to correspond to field conditions during that time of year. One hundred week 2 larvae each were placed in replicate acrylic culture columns ($n = 4$) under two different photoperiod treatments (ambient and dark) and two different water column heights, 9.5 d × 125 h cm (tall) and 9.5 d × 15 h cm (short), filled with 10.1 L and 1.22 L FSW, respectively.

Culture columns were placed in temperature-controlled walk-in incubators (15-16°C). One incubator imitated a natural photoperiod (ambient treatment) with 16 hours of light, from 0500 to 2100 h, followed by 8 hours of dark (16:8), while the second experienced 24 h of dark (dark treatment, 0:24). Four tall and 4 short columns were randomly placed in each temperature-controlled incubator and were supported by a custom yoke system (resembling a large wooden test tube rack) that enabled us to view the columns from all angles including the bottom. Columns were visually divided into 5 cm increments, with the top and bottom 5 cm sections further divided into 2.5 cm increments. Helical 26 W fluorescent bulbs were placed approximately 7.6 cm away from the top of each ambient light treatment column. Columns were isolated from one another and from the influence of adjacent lights by cardboard dividers and blackout cloth that completely surrounded each column. We placed larvae in columns at 2000 h and larval vertical

positions were recorded every 4 h in a 24-hour period starting at 0700 h the next morning. During counting, larvae in both treatments were illuminated with red LED headlamps. *Kelletia kelletii* larvae do not respond to red light (Walker, Zacherl and Hoese unpublished data), as with many other larval invertebrates (e.g., [37]).

For each photoperiod X column height treatment, we calculated the depth center of mass (ZCM) as in Tremblay and Sinclair [38]. ZCM = $\Sigma p_i z_i$ where p_i = proportion of larvae and z_i = distance from top for each depth interval. We then tested the effects of column height, photoperiod and time on ZCM using a three-way full-factorial ANOVA treating replicates as a blocking factor to account for our repeated measures design.

Finally, many factors differed between experiments carried out in 2005 versus 2007, including photoperiod, collection locations for egg masses, time of year, observation intervals, presence of phytoplankton in culture vessels (present in 2005 and absent in 2007), and size, composition of culture vessels (glass versus acrylic), and segmentation of culture vessels into increments. In order to qualitatively compare the results of the two sets of experiments we calculated % demersal larvae in our 9.5 d × 15 h cm (short) culture columns, as in the summer 2005 experiment (see above). Again, to more clearly depict DVM behavior graphically, percent demersal values were converted to percent nondemersal. Larvae of the same age (week 2) were compared to one another.

2.4. Field: Larvae in Surface Plankton Tows.

To test whether *K. kelletii* larvae in the field exhibit a distribution consistent with nocturnal DVM behavior, with higher concentrations of larvae at the surface at night compared to day, surface plankton tows ($n = 5$) were conducted at 2400 h and 1200 h. Horizontal tows were conducted perpendicular to the shore using a 0.5-meter diameter plankton net with 333 μm mesh size near the coast of Palos Verdes, CA, USA on July 10, 2007. All tows started between the following two coordinates: N33°43.462′ W118°21.173′ and N33°43.087′ W118°20.106′. The net was maintained at the surface to 1 m below the surface with floats attached to the metal ring of the plankton net while vessel speed was maintained at an average of 2 knots. After eight minutes the net was pulled out of the water vertically and the sample sprayed down with seawater into a cod-end bucket. Using a General Oceanics Inc. mechanical flow meter (Model 2030), we calculated the volume of water sampled per tow as 88.10 ± 1.65 m³ (SE). Tows were conducted on a clear day and overcast night with surface water temperature averaging 17.2°C. All samples collected were chilled and taken to California State University Fullerton for sorting.

Samples were sorted using dissecting microscopes and all *K. kelletii* larvae were isolated by visual inspection and counted. In order to positively identify *K. kelletii* larvae we compared our plankton tow samples to reference samples of *K. kelletii* larvae collected from Santa Barbara, CA, and cultured in the laboratory through settlement. The count of *K. kelletii* from each tow was converted into larvae/m³ seawater. Data were log transformed ($\log X + 1$) due to

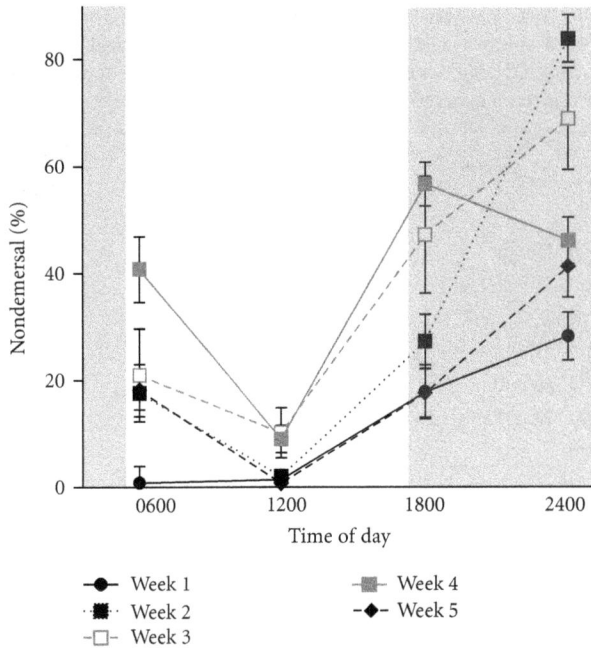

FIGURE 1: Percentage of *Kelletia kelletii* larvae that were nondemersal throughout a 24-hour period as a function of ontogeny (weeks 1–5) in 8 d × 15 h cm culturing columns ($n = 5$). Grey shading indicates lights off. Error bars represent ± 1 SE.

TABLE 1: Two-way ANOVA testing for effect of time-of-day and ontogeny on *Kelletia kelletii* demersal behavior. DF: degrees of freedom, SS: sum of squares, replicate was treated as random. Bold results indicate significance.

Source	DF	SS	F-ratio	Prob > F
Time-of-day (TOD)	3	32400.91	62.15	**<0.0001**
Ontogeny	4	10562.96	15.20	**<0.0001**
TOD*Ontogeny	12	10183.84	4.88	**<0.0001**
Replicate	1	317.52	1.83	0.18
Error	79	13728.48		
Total	99	67193.71		

heteroscedasticity and a Student's t-test was used to compare concentrations of larvae at 1200 h versus 2400 h.

2.5. Light Measurements in the Laboratory and Field. In order to ensure light intensities in laboratory experiments were comparable to light intensities experienced by larvae in the field, we completed profiles of photosynthetically active radiation (PAR, μmol s^{-1} m^{-2}) as a function of depth in laboratory columns and in the field (at 1200 h in partial sun and 1300 h in full sun) off the coast of Palos Verdes, CA (13 July 2007, 33°43′291″N, 118°20′703″W) with an LI-192 underwater quantum sensor and a LI-COR data logger. In the field, PAR was measured in replicate samples ($n = 2$) every 1 m from the surface to 5 m depth, and then every 5 m to 20 m depth. In the laboratory, PAR was measured in the tall columns at multiple depths (1, 25, 50, 75, 100 cm) to ensure that a gradient of light intensity was achieved with depth. In the short columns, we were only able to take measurements at 1 cm below the surface due to size constraints imposed by the sensor.

3. Results

3.1. DVM and Effects of Ontogeny. Kelletia kelletii larvae exhibited a daily migration pattern consistent with nocturnal diel vertical migration throughout the five-week period examined (Figure 1). Larvae were found up in the water column at midnight (almost always at the surface based upon qualitative observations made during the experimental period), and demersal during the day, with greater than 80% of larvae demersal at noon throughout their ontogeny. During the observation periods 1 hour after lights were

turned on (0600 hr) and off (1800 hr), intermediate proportions of larvae were nondemersal, with larvae scattered throughout the water column and at the surface. There were a higher proportion of demersal larvae during weeks 1 and 5 relative to other weeks, with greater than 60% of larvae being demersal at week 5 regardless of time of day. Last, during week 4, the highest percentage of nondemersal larvae shifted from midnight (2400 hr) to 1800 hr whereas the highest percentage of nondemersal larvae in other weeks peaked at midnight (time-of-day by ontogeny interaction, 2-way ANOVA, $P < 0.0001$, Table 1).

3.2. Effects of Light and Culture Column Height on DVM. In both tall and short culture columns, *K. kelletii* larvae exhibited a daily migration pattern consistent with nocturnal diel vertical migration (Figure 2) in both ambient and dark photoperiod treatments, with average larval vertical positioning higher in the water column (i.e., nearer the surface) during night-time hours and lower in the water column during day-time hours (e.g., compare 0300 to 1500 hours in Figure 2, both panels). Larvae in all treatments began their upward migration before 2100 h, while still exposed to light, and migrated downward between 0300 and 0700 h. Larvae in ambient photoperiod treatments in both tall and short columns, had significantly lower average depth center of mass (ZCM) than those in the dark, but only when the larvae were exposed to light (three-way ANOVA, light by time-of-day interaction, $P = 0.01$, Table 2, Figure 2). Post hoc Tukey comparisons ($P < 0.05$) revealed significant differences in ZCM between ambient and dark photoperiods at 0700, 1100 and 1500 h. Column height had no significant effect on ZCM (Table 2).

3.3. DVM in the Field. There were significantly more *K. kelletii* at the surface of the water column at 2400 h, with 0.32 ± 0.08 per m^3, compared to 1200 h, with 0.03 ± 0.02 per m^3 (t-test, $P = 0.018$, Figure 3).

3.4. Light Measurements. In the field, light measurements ranged from 40–1950 PAR at depths from 20 to 0.01 meters below the surface (Table 3). Light intensity diminished with depth; the decline was best described by a logarithmic function $y = -244.1 \ln(x) + 848.68$, with $R^2 = 0.99$. Light intensity in the laboratory columns ranged from 60–298 PAR; the decline in light intensity was best described by the

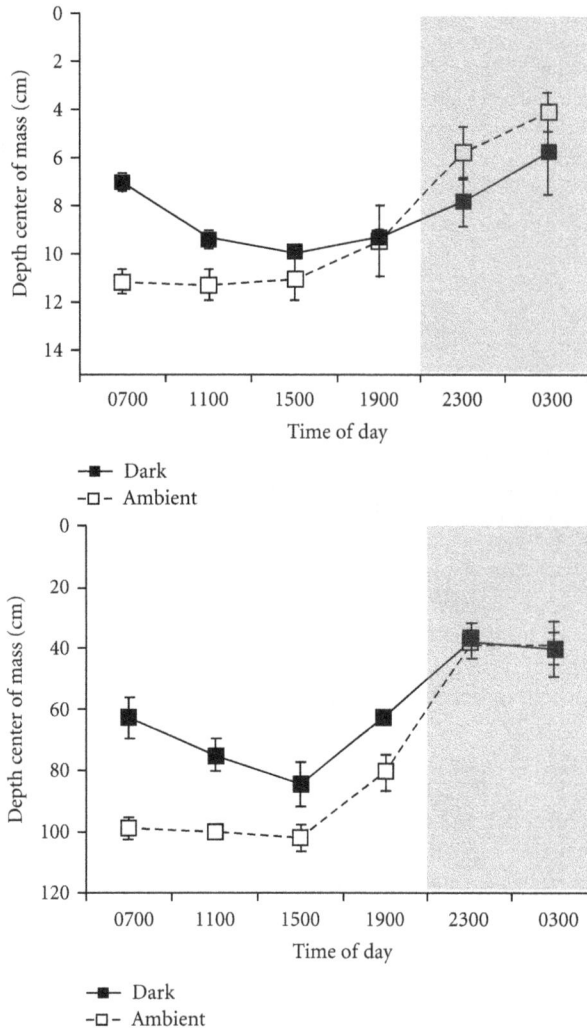

FIGURE 2: Depth center of mass for *Kelletia kelletii* larvae throughout a 24 hr period in 9.5 d × 15 h cm columns ("short," top panel) and 9.5 d × 125 h cm columns ("tall," bottom panel) in treatments (*n* = 4) of ambient photoperiod (16 : 8 h, open squares) or dark only (0 : 24 h, black squares). Grey shading indicates lights off in the ambient photoperiod. Error bars represent ± 1 SE.

exponential function, $y = 302.3e^{-1.548x}$ with $R^2 = 0.98$. At the surface of the laboratory columns in the 2007 trials, light intensity was approximately equivalent to that measured in the field at the 10 m depth. For the 2005 trials, PAR at the surface of laboratory columns was equivalent to PAR in the field at 15–20 m depths.

4. Discussion

Both laboratory and field-generated data were consistent with the hypothesis that *Kelletia kelletii* larvae exhibited a classic nocturnal diel vertical migration distribution, with larvae migrating to the surface at night and downward during the day. This pattern has been observed in other veligers, including queen conch, *Strombus gigas* [14] and scallops, *Placopecten magellanicus* [38]. Understanding the

TABLE 2: Three-way ANOVA testing for effect of light, time-of-day and column height on *Kelletia kelletii* diel vertical migratory behavior. DF: degrees of freedom, SS: sum of squares, Rep: replicate. Rep was treated as random. Bold results indicate significance.

Source	DF	SS	*F*-ratio	Prob > *F*
Light	1	0.36	1.80	0.20
Height	1	0.04	0.20	0.66
Light*Height	1	0.24	1.21	0.29
Rep (Light, Height)	12	2.39	3.28	**0.001**
Time-of-day (TOD)	5	11.09	36.59	**<0.0001**
Light*TOD	5	0.99	3.28	**0.01**
TOD*Height	5	0.57	1.89	0.11
Light*TOD*Height	5	0.11	0.37	0.87
Error	60	3.64		
Total	95	19.43		

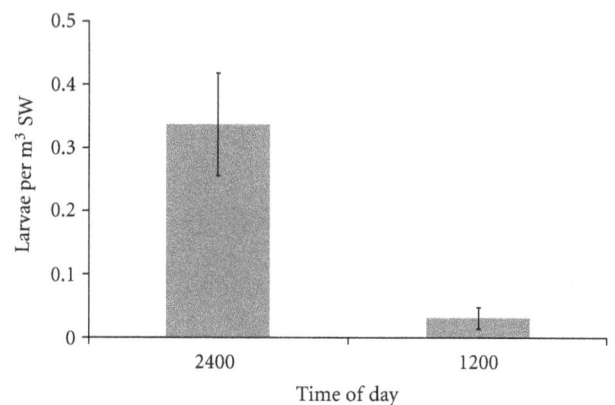

FIGURE 3: Number of *Kelletia kelletii* larvae per m^3 from replicate (*n* = 5) surface plankton tows conducted at 2400 and 1200 h off Palos Verdes, CA, in June 2007. Error bars represent ± 1 SE.

functional significance of *K. kelletii*'s vertical migration is beyond the scope of this study, though one probable scenario is that the larvae migrate to the surface waters at night to feed on phytoplankton, and return to depth during daylight hours to avoid visual predators or to avoid exposure to high levels of UV radiation [39].

Kelletia kelletii larvae became more demersal as they approached competency—by the time they were 5 weeks old, greater than 60% of all of the larvae were demersal, regardless of time of day. It is unlikely that this ontogenetic shift in vertical distribution is due to a decrease in photopositivity with age, as has been demonstrated in queen conch [14] and invertebrate larvae in general [40], since *K. kelletii* larvae did not exhibit a pattern consistent with photopositivity early in development. Indeed, their vertical positioning during week 1 was most similar to that observed during week 5, with nearly 100% of larvae located in the lowest 2.5 cm of the water column during daytime exposure to light (Figure 1). Their behavior is, however, consistent with a general trend among marine larvae that are approaching competence and preparing to settle into adult habitat—they

TABLE 3: Photosynthetically active radiation (PAR) measurements in the field during partial-sun to full-sun conditions and in laboratory culture columns. Treatments refer to tall (9.5 d × 125 h cm) and short (9.5 d × 15 h cm) column dimensions. Columns in the 2005 lab treatment measured 8 d × 15 h cm. PAR units are μmol s^{-1}m^{-2}.

Location	Year/treatment	Depth (m)	Range PAR	Avg. PAR ± 1SE (n)
Field	2007	0.01	1898–1950	1924 ± 26(2)
Field	2007	1	860–1050	955 ± 95(2)
Field	2007	5	390–650	520 ± 130 (2)
Field	2007	10	175–335	255 ± 80(2)
Field	2007	15	75–198	137 ± 62(2)
Field	2007	20	40–116	78 ± 38(2)
Lab	2007/tall	0.01	287–292	289 ± 2(3)
Lab	2007/tall	0.25	231–234	233 ± 1(3)
Lab	2007/tall	0.50	150–154	152 ± 1(3)
Lab	2007/tall	0.75	88–93	90 ± 2(3)
Lab	2007/tall	1	60–64	62 ± 1(3)
Lab	2007/short	0.01	242–298	270 ± 28(2)
Lab	2005	0.01	51–144	87 ± 12(10)

are thought to use the additional time near-bottom to sample suitable habitat for settlement [40, 41].

Light, however, had an effect on vertical distributions of our larvae. Throughout ontogeny larvae were demersal when the lights were on and nondemersal with the lights off. In addition, over our 24 hour tracking, average larval vertical position shifted upwards when lights were off. The exact mechanism of how light influenced larval movement is more difficult to characterize. Downward larval movement in response to light could be the result of negative phototaxis, positive geotaxis in response to light cues, or simply negative photokinesis because larvae are negatively buoyant. A horizontal trough with light illumination from the side could be used to isolate the factors from one another [41, 42].

Some observed behaviors in our experiments also suggested the presence of an endogenous rhythm. Larvae in our ambient treatment started migrating upward before the lights went out (Figure 2). Further, larvae in the dark treatment migrated upward during night-time hours and downward during day-time hours, even though no light cues were present. This downward larval migration in the dark treatment was less extreme during day-time hours than in the ambient treatment (Figure 2), suggesting that the presence of light in the ambient treatment induced a behavioral response. An endogenous sunset ascent is part of a theoretical model of DVM behavior in the calanoid copepod, *Calanopia americana* [43], and this ascent may be part of daily activity patterns or may be driven by hunger [44–47]. Downward migration in our study may be cued by a combination of endogenous inactivity, negative phototaxis to high light levels, or both [43]. Because we did not have a sunrise or sunset condition where light level changed gradually, *K. kelletii* larvae in this experiment could not have used relative rates of irradiance change to cue downward migration [12].

The findings of nocturnal DVM behavior in *K. kelletii* were consistent across years (Figure 4) despite different experimental conditions in the laboratory, and these findings were corroborated by our field study of larval positioning.

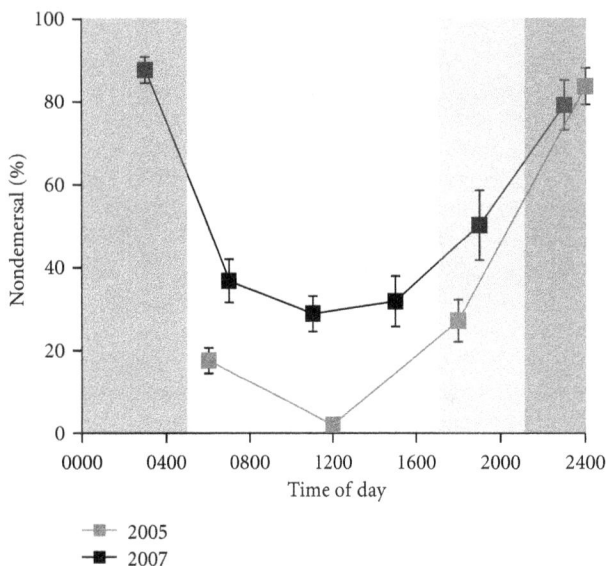

FIGURE 4: Comparison of vertical migration behavior in 2005 trials versus 2007 trials using larvae of similar ages showing the percentage of nondemersal larvae as a function of time of day. Data collected in 2007 were reanalyzed by converting to % nondemersal in order to make the datasets comparable. Shading indicates timing of light cycles. Light grey shading indicates darkness in 2005. Dark grey shading indicates darkness for both years. Error bars represent ± 1 SE.

In the laboratory, larvae were consistently found higher up in the water column during night-time hours and lower in the water column during day-time hours even with significant differences in the experimental setups in summer 2005 versus 2007 (see Table 4 for a summary of differences). Much has been discussed about experimental artifacts associated with behavior studies in the laboratory (e.g., [18]). Our culture columns provided a gradient of

TABLE 4: Comparison of experimental and control variables in day-night laboratory trials in jars in 2005 versus short columns in 2007.

Variable	2005	2007
Environment	Incubator	T-controlled walk-in incubator
Light cycle	12:12	16:8
Egg collection site	Monterey, CA	Palos Verdes, CA
Egg source	Single female	Multiple females
Data collection interval	6 hrs	4 hrs
Dark phase starting time	1700	2100
PAR range	51–144	242–298
Response factor	% demersal	ZCM
Column dimensions	8 d × 15 h cm	9.5 d × 15 h cm
Phytoplankton present	yes	no

light at realistic PAR intensities relative to field conditions (Table 3). However, light attenuated in our tall laboratory culture columns more quickly than might be expected in the field. Light intensity should attenuate 55–60% in the top 1 meter of the water column; ours attenuated 78–79.5%. The explanation for this stronger attenuation than expected is unclear; we did have a cable attached to our light sensor and it was difficult to keep the sensor vertical, which may have led to some error in our measures. Since our light source was not spectrally matched with natural sunlight, it is possible that some wavelengths that attenuate more quickly than others (e.g., red) may have made up a larger proportion of total light from our light source at the surface. In addition, the wavelengths of light experienced by our laboratory larvae at a particular intensity were not matched spectrally with field conditions. For example at PAR = 60 in the field (at a depth of approximately 20 m), most wavelengths would be blue, compared to the full spectrum of wavelengths potentially experienced by larvae at 60 PAR in the laboratory. Our angular light distribution was also unnatural; while we provided a directional light source, we did not measure whether reflected light in the culture columns affected larval behavior. We also did not mimic other field conditions known to alter larval behavior, such as presence of a thermocline [13], currents [48], and other factors (reviewed in [49] and references therein), which could each potentially override the effects of light and endogenous cues. Future studies should aim to tease out the relative importance of these additional factors on larval behavior. Despite these potential artifacts, our field plankton tows did corroborate the overall pattern of nocturnal DVM witnessed in the laboratory.

Our field measures of larval positioning at the surface at night are consistent with our conclusions that larvae do vertically migrate and that light acts, at least partially, as a cue for larval positioning. However, we do not have evidence that the larvae in the field are demersal during the daytime, as our laboratory results suggest. Based upon our field plankton tows, we only know that K. kelletii larvae are not at the surface during the day; we do not know at what depth they are found. Certainly their lower vertical positioning in the laboratory in the presence of light is evident relative to dark treatments (Figure 2, bottom panel). At their average ZCM

in the presence of light (approximately 100 cm depth) in the lab, the larvae were exposed to PAR ranging from 60–64, which is equivalent to PAR intensity experienced in the field at approximately 20 m depth on a sunny day. If larval response to light is indeed negatively phototactic or positively geotactic in the presence of light then we predict that Kellet's whelk larvae in the field would descend to a depth of at least 20 meters during midday.

The effective management of any emerging fishery requires some understanding of the connectivity among populations [50] so that important sources of the next generation can be identified [51]. Given the complexity involved with directly tracking larvae of any species from their birth location to their settlement site, many investigators have turned to high-resolution coupled biophysical models to generate a preliminary understanding of larval dispersal trajectories and subsequent connectivity among source populations (e.g., [52]). Knowledge about larval behavior is a critical component of these modeling efforts [3, 53–55]. The data presented here begin to shed insight into K. kelletii's larval behavior, suggesting that they undergo extensive daily changes in their vertical positioning possibly on the order of tens of meters. Subsequent studies should emphasize field sampling at multiple depths in the presence of varied flow and thermocline conditions to corroborate the full extent of the daily vertical shifts. Given K. kelletii's recently designated status as an emerging fishery, coupled with the knowledge that diel vertical migratory behavior can profoundly affect the dispersal outcomes of larvae, we call for focused attention on this vulnerable fishery.

Acknowledgments

The authors gratefully acknowledge funding from NSF-OCE Grant no. 0351860 to D. C. Zacherl, and NSF-UMEB Grant no. 0602922 to W. J. Hoese. Thanks for institutional support from Cabrillo Marine Aquarium, CSU Fullerton Department of Biological Science, USC Wrigley Institute for Environmental Studies, and irreplaceable assistance from Sean Walker (statistics), Meredith Raith, John Luong, Andres Carrillo, Ray Munson and Munson Engineering. Thanks to Richard Forward and James Welch for illuminating the authors' understanding of larval behavioral responses. Last, thanks to reviewers for improving the quality of this paper.

References

[1] S. G. Morgan and J. L. Fisher, "Larval behavior regulates nearshore retention and offshore migration in an upwelling shadow and along the open coast," *Marine Ecology Progress Series*, vol. 404, pp. 109–126, 2010.

[2] T. W. Cronin and R. B. Forward, "Tidal vertical migration: an endogenous rhythm in estuarine crab larvae," *Science*, vol. 205, no. 4410, pp. 1020–1022, 1979.

[3] C. B. Paris, L. M. Chérubin, and R. K. Cowen, "Surfing, spinning, or diving from reef to reef: effects on population connectivity," *Marine Ecology Progress Series*, vol. 347, pp. 285–300, 2007.

[4] A. M. Knights, T. P. Crowe, and G. Burnell, "Mechanisms of larval transport: vertical distribution of bivalve larvae varies with tidal conditions," *Marine Ecology Progress Series*, vol. 326, pp. 167–174, 2006.

[5] E. W. North, Z. Schlag, R. R. Hood et al., "Vertical swimming behavior influences the dispersal of simulated oyster larvae in a coupled particle-tracking and hydrodynamic model of Chesapeake Bay," *Marine Ecology Progress Series*, vol. 359, pp. 99–115, 2008.

[6] A. C. Hardy and E. R. Gunther, *The Plankton of the South Georgia Whaling Grounds and Adjacent Waters, 1926-1927*, Cambridge University Press, London, UK, 1935.

[7] M. Marta-Almeida, J. Dubert, A. Peliz, and H. Queiroga, "Influence of vertical migration pattern on retention of crab larvae in a seasonal upwelling system," *Marine Ecology Progress Series*, vol. 307, pp. 1–19, 2006.

[8] C. M. Aiken, S. A. Navarrete, and J. L. Pelegrí, "Potential changes in larval dispersal and alongshore connectivity on the central Chilean coast due to an altered wind climate," *Journal of Geophysical Research*, vol. 116, Article ID G04026, 2011.

[9] M. M. Criales, J. A. Browder, C. N. K. Mooers, M. B. Robblee, H. Cardenas, and T. L. Jackson, "Cross-shelf transport of pink shrimp larvae: interactions of tidal currents, larval vertical migrations and internal tides," *Marine Ecology Progress Series*, vol. 345, pp. 167–184, 2007.

[10] E. Poulin, A. T. Palma, G. Leiva et al., "Avoiding offshore transport of competent larvae during upwelling events: the case of the gastropod *Concholepas concholepas* in Central Chile," *Limnology and Oceanography*, vol. 47, no. 4, pp. 1248–1255, 2002.

[11] M. J. Kingsford, J. M. Leis, A. Shanks, K. C. Lindeman, S. G. Morgan, and J. Pineda, "Sensory environments, larval abilities and local self-recruitment," *Bulletin of Marine Science*, vol. 70, no. 1, pp. 309–340, 2002.

[12] J. H. Cohen and R. B. Forward, "Diel vertical migration of the marine copepod *Calanopia americana*. I. Twilight DVM and its relationship to the diel light cycle," *Marine Biology*, vol. 147, no. 2, pp. 387–398, 2005.

[13] S. M. Gallager, J. L. Manuel, D. A. Manning, and R. O'Dor, "Ontogenetic changes in the vertical distribution of giant scallop larvae, *Placopecten magellanicus*, in 9-m deep mesocosms as a function of light, food, and temperature stratification," *Marine Biology*, vol. 124, no. 4, pp. 679–692, 1996.

[14] P. J. Barile, A. W. Stoner, and C. M. Young, "Phototaxis and vertical migration of the queen conch (*Strombus gigas* Linne) veliger larvae," *Journal of Experimental Marine Biology and Ecology*, vol. 183, no. 2, pp. 147–162, 1994.

[15] J. L. Manuel, S. M. Gallager, C. M. Pearce, D. A. Manning, and R. K. O'Dor, "Veligers from different populations of sea scallop *Placopecten magellanicus* have different vertical migration patterns," *Marine Ecology Progress Series*, vol. 142, no. 1–3, pp. 147–163, 1996.

[16] T. P. Hurst, D. W. Cooper, J. S. Scheingross, E. M. Seale, B. J. Laurel, and M. L. Spencer, "Effects of ontogeny, temperature, and light on vertical movements of larval pacific cod (*Gadus macrocephalus*)," *Fisheries Oceanography*, vol. 18, no. 5, pp. 301–311, 2009.

[17] E. D. Garland, C. A. Zimmer, and S. J. Lentz, "Larval distributions in inner-shelf waters: the roles of wind-driven cross-shelf currents and diel vertical migrations," *Limnology and Oceanography*, vol. 47, no. 3, pp. 803–817, 2002.

[18] R. B. Forward, "Diel vertical migration: zooplankton photobiology and behavior," *Oceanography and Marine Biology*, vol. 26, pp. 361–393, 1988.

[19] J. H. Cohen and R. B. Forward, "Zooplankton diel vertical migration—a review of proximate control," *Oceanography and Marine Biology*, vol. 47, pp. 77–110, 2009.

[20] B. Diehn, M. Feinleib, W. Haupt, E. Hildebrand, F. Lenci, and W. Nultsch, "Terminology of behavioral responses of motile organisms," *Photochemistry and Photobiology*, vol. 26, pp. 559–560, 1977.

[21] J. A. Rudjakov, "The possible causes of diel vertical migrations of planktonic animals," *Marine Biology*, vol. 6, no. 2, pp. 98–105, 1970.

[22] J. T. Enright and W. M. Hamner, "Vertical diurnal migration and endogenous rhythmicity," *Science*, vol. 157, no. 3791, pp. 937–941, 1967.

[23] S. Lass and P. Spaak, "Chemically induced anti-predator defences in plankton: a review," *Hydrobiologia*, vol. 491, pp. 221–239, 2003.

[24] D. Rittschof and J. H. Cohen, "Crustacean peptide and peptide-like pheromones and kairomones," *Peptides*, vol. 25, no. 9, pp. 1503–1516, 2004.

[25] A. Pires and R. M. Woollacott, "A direct and active influence of gravity on the behavior of a marine invertebrate larva," *Science*, vol. 220, no. 4598, pp. 731–733, 1983.

[26] M. Huntley and E. R. Brooks, "Effects of age and food availability on diel vertical migration of *Calanus pacificus*," *Marine Biology*, vol. 71, no. 1, pp. 23–31, 1982.

[27] W. E. Neill, "Population variation in the ontogeny of predator-induced vertical migration of copepods," *Nature*, vol. 356, no. 6364, pp. 54–57, 1992.

[28] M. Carriker, "Ecological observations on the distribution of oyster larvae in New Jersey estuaries," *Ecological Monographs*, vol. 21, pp. 19–38, 1951.

[29] R. A. Tankersley, J. M. Welch, and R. B. Forward, "Settlement times of blue crab (*Callinectes sapidus*) megalopae during flood-tide transport," *Marine Biology*, vol. 141, no. 5, pp. 863–875, 2002.

[30] K. Barsky, D. Haas, M. Heisdorf et al., "Review of selected California fisheries for 2008: coastal pelagic finfish, market squid, ocean salmon, groundfish, California spiny lobster, spot prawn, white seabass, kelp bass, thresher shark, skates and rays, Kellet's whelk, and sea cucumber," *CalCOFI Reports*, vol. 50, pp. 14–42.

[31] S. E. Swearer, J. S. Shima, M. E. Hellberg et al., "Evidence of self-recruitment in demersal marine populations," *Bulletin of Marine Science*, vol. 70, supplement, no. 1, pp. 251–271, 2002.

[32] C. White, K. A. Selkoe, J. Watson, D. A. Siegel, D. C. Zacherl, and R. J. Toonen, "Ocean currents help explain population

genetic structure," *Proceedings of the Royal Society B*, vol. 277, no. 1688, pp. 1685–1694, 2010.

[33] S. R. Palumbi, "Population genetics, demographic connectivity, and the design of marine reserves," *Ecological Applications*, vol. 13, no. 1, pp. S146–S158, 2003.

[34] R. J. Rosenthal, "Observations on the reproductive biology of the Kellet's whelk, *Kelletia kelletii* (gastropoda: Neptuneidae)," *The Veliger*, vol. 12, no. 3, pp. 319–324, 1970.

[35] J. H. McLean, *Marine Shells of Southern California*, vol. 24 of *Science Series*, Natural History Museum of Los Angeles County, Los Angeles, Calif, USA, 1978.

[36] T. J. Herrlinger, "Range extension of *Kelletia kelletii*," *The Veliger*, vol. 24, no. 1, p. 78, 1981.

[37] J. H. Cohen and R. B. Forward, "Spectral sensitivity of vertically migrating marine copepods," *Biological Bulletin*, vol. 203, no. 3, pp. 307–314, 2002.

[38] M. J. Tremblay and M. Sinclair, "Diel vertical migration of sea scallop larvae *Placopecten magellanicus* in a shallow embayment," *Marine Ecology Progress Series*, vol. 67, no. 1, pp. 19–25, 1990.

[39] J. Ringelberg and E. Van Gool, "On the combined analysis of proximate and ultimate aspects in diel vertical migration (DVM) research," *Hydrobiologia*, vol. 491, pp. 85–90, 2003.

[40] G. Thorson, "Light as an ecological factor in the dispersal and settlement of larvae of marine bottom invertebrates," *Ophelia*, vol. 1, no. 1, pp. 167–208, 1964.

[41] D. A. McCarthy, R. B. Forward, and C. M. Young, "Ontogeny of phototaxis and geotaxis during larval development of the sabellariid polychaete *Phragmatopoma lapidosa*," *Marine Ecology Progress Series*, vol. 241, pp. 215–220, 2002.

[42] B. L. Bayne, "The responses of the larvae of *Mytilus edulis* L. to light and to gravity," *OIKOS*, vol. 15, no. 1, pp. 162–174, 1964.

[43] J. H. Cohen and R. B. Forward, "Diel vertical migration of the marine copepod *Calanopia americana* . II. Proximate role of exogenous light cues and endogenous rhythms," *Marine Biology*, vol. 147, no. 2, pp. 399–410, 2005.

[44] D. E. Stearns, "Copepod grazing behavior in simulated natural light and its relation to nocturnal feeding," *Marine Ecology Progress Series*, vol. 30, pp. 65–76, 1986.

[45] M. Pagano, R. Gaudy, D. Thibault, and F. Lochet, "Vertical migrations and feeding rhythms of mesozooplanktonic organisms in the Rhône River plume area (north-west Mediterranean Sea)," *Estuarine, Coastal and Shelf Science*, vol. 37, no. 3, pp. 251–269, 1993.

[46] G. C. Harding, W. P. Vass, B. T. Hargrave, and S. Pearre Jr., "Diel vertical movements and feeding activity of zooplankton in St Georges Bay, NS, using net tows and a newly developed passive trap," *Canadian Journal of Fisheries and Aquatic Sciences*, vol. 43, no. 5, pp. 952–967, 1986.

[47] S. Pearre Jr., "Eat and run? The hunger/satiation hypothesis in vertical migration: history, evidence and consequences," *Biological Reviews of the Cambridge Philosophical Society*, vol. 78, no. 1, pp. 1–79, 2003.

[48] J. R. Pawlik, C. A. Butman, and V. R. Starczak, "Hydrodynamic facilitation of gregarious settlement of a reef-building tube worm," *Science*, vol. 251, no. 4992, pp. 421–424, 1991.

[49] C. M. Young, "Behavior and locomotion during the dispersal phase of larval life," in *Ecology of Marine Invertebrate Larvae*, L. McEdward, Ed., CRC Press, Boca Raton, Fla, USA, 1995.

[50] R. R. Strathmann, T. P. Hughes, A. M. Kuris et al., "Evolution of local recruitment and its consequences for marine populations," *Bulletin of Marine Science*, vol. 70, no. 1, pp. 377–396, 2002.

[51] J. R. Watson, D. A. Siegel, B. E. Kendall, S. Mitarai, A. Rassweiller, and S. D. Gaines, "Identifying critical regions in small-world marine metapopulations," *Proceedings of the National Academy of Sciences of the United States of America*, vol. 108, no. 43, pp. E907–E913, 2011.

[52] R. K. Cowen, C. B. Paris, and A. Srinivasan, "Scaling of connectivity in marine populations," *Science*, vol. 311, no. 5760, pp. 522–527, 2006.

[53] C. DiBacco, D. Sutton, and L. McConnico, "Vertical migration behavior and horizontal distribution of brachyuran larvae in a low-inflow estuary: implications for bay-ocean exchange," *Marine Ecology Progress Series*, vol. 217, pp. 191–206, 2001.

[54] R. K. Cowen and S. Sponaugle, "Larval dispersal and marine population connectivity," *Annual Review of Marine Science*, vol. 1, pp. 443–466, 2009.

[55] J. M. Leis, "Behaviour as input for modelling dispersal of fish larvae: behaviour, biogeography, hydrodynamics, ontogeny, physiology and phylogeny meet hydrography," *Marine Ecology Progress Series*, vol. 347, pp. 185–193, 2007.

16

Sponge Farming Trials: Survival, Attachment, and Growth of Two Indo-Pacific Sponges, *Neopetrosia* sp. and *Stylissa massa*

Karin Schiefenhövel and Andreas Kunzmann

Department of WG Ecophysiology, Leibniz Center for Tropical Marine Ecology, Fahrenheitstraße 6, 28359 Bremen, Germany

Correspondence should be addressed to Andreas Kunzmann, andreas.kunzmann@zmt-bremen.de

Academic Editor: Wen-Xiong Wang

Sponges, an important part of the reef ecosystem, are of commercial value for public aquaria, pharmacology and chemistry. With the growing demand for sponges, natural resources are at risk of being overexploited. Growing of sponges in artificial or semi natural farms is an alternative. In this study different farming methods were tested on two Indo-Pacific sponge species, *Neopetrosia* sp. and *Stylissa massa*. Survival, growth and attachment ability were observed with different substrates (suspended ropes, coral boulders and artificial substrate), two types of aquaria with different water volume and two different field sites in Indonesia. The two species responded differently to their individual locations and environmental stresses. Survival, growth and attachment rates of *Neopetrosia* sp. at the field site are depending on the cultivation method, we found highest volume increment (27–35%) for a horizontal line in the field. Whereas the volume increase for *S. massa* did not show any differences for the different transplantation methods, *Neopetrosia* sp. generally showed higher rates than *S. massa*. Further aquaria experiments, for example, on nutrient supply, should be tested to receive more detailed data about sponges, particularly because almost all fragments of both species showed a decline or steady state in mean length.

1. Introduction

As filtering organisms, sponges play an important role in reef ecosystems. Like other benthic suspension feeders, they are responsible for a large share of the energy flow from the pelagic to the benthic system [1, 2]. Sponges actively move water through their body, an advantage that enables sponges to inhabit different types of habitats also at different depths [1]. Sponges were used successfully for reef rehabilitation because of their ability to clean water by filtering small particles like detritus, algae, and bacteria, and they also hold rubble and corals together [3, 4]. Sponges are also commercially important for farming (bath sponge) and ornamental trade and as a new resource of chemicals for pharmacology [5–9].

Sponges are very effective active filter feeders. The volume of water passing through a sponge can be enormous; a sponge with 10 cm in length and four cm in diameter can filter 80 L water in 24 h [10]. Due to that sponge farms are tested for their potential as biofilters near fish farms or land-based sewage water [9].

Their ability to reproduce from small pieces makes sponges very attractive for commercial farming [11, 12]. Sponges with symbiotic relationships are of high interest for public aquaria because they are more colourful, less expensive to keep, and easier to sell than asymbiotic species [11]. Some sponge species have symbionts supplying the host with nutrients and need less additional feeding [13]. In the last 20 years numerous bioactive metabolites synthesized by symbiotic microorganisms in sponges have been extracted and identified [9]. Bioactive substances in some sponges have potential uses in medicines, bactericides, pesticides, cosmetics, fungicides and antifoulings [11, 14]. Aldisine alkaloids, discovered in *S. massa*, were shown to be a potent inhibitor of mitogen-activated protein kinase kinase-1 that can inhibit the growth of certain tumor cells [15]. Peptides from a sponge in Papua New Guinea (*Neopetrosia* sp.) can inhibit amoeboid invasion by human tumor cells [16].

Sponges usually only produce trace amounts of the bio-active compound; therefore large-scale production from aquaculture is necessary to avoid overexploitation of the native stock [6, 17]. Several studies have examined the best growing methods for certain sponge species. MacMillan [12] and Corriero et al. [5] suggest attaching sponge fragments to horizontal lines, while Page et al. [18] tried mesh arrays and Duckworth and Battershill [19] favoured the use of mesh bags after they tried to grow different New Zealand sponge species (*Latrunculia* sp. nov, *Polymastia croceus,* and *Raspailia agminata*) on ropes and in mesh bags.

This study compares growth, survival, and attachment rates of two Indo-Pacific sponge species, *Neopetrosia* sp. and *S. massa,* on three different substrates: horizontal rope (as suggested by MacMillan [12]), live rock, and artificial substrate (cement plate) placed on a metal grid. The experiment was carried out in Indonesia for five months (November 2002 to March 2003) to determine a suitable method and site for culturing selected species, with the hope of using it for reef rehabilitation and/or for export trade to public aquaria. Because little is known about most sponge species and farming methods are often species specific [20], this study will also contribute to a better understanding of sponge survival, attachment, and growth performance of the selected species.

2. Material and Methods

2.1. Sponge Species. Neopetrosia sp. (de Laubenfels 1949) is a hard blue sponge with a rough porous surface [21] (Figure 1). It has a branching growth form with more than one osculum per branch. It can grow from 10 to 20 cm in length and two to four cm in diameter [21].

Stylissa massa (Carter 1881) is a soft yellow sponge, ranging in size from 7 to 20 cm in length and 5 to 11 cm in diameter [21] (Figure 1). It has an irregular but vertically extending growth form, which narrows to the attachment surface and several oscula are distributed irregularly over the sponge body.

2.2. Sampling and Study Sites

Pulau Pari—Sampling and Study Site 1. Pulau Pari is an island located in the Thousand Island reef complex north of Jakarta, the capital of Indonesia. The sampling and study site (5°51′55″S, 106°36′18″E) was located in the sandy lagoon of Western Pulau Pari, at a depth of 2.5 m and about 200 m from the main island (Figure 2). 24 donor sponges of *S. massa* with a length of 10 to 20 cm were collected in the lagoon and on the reef flat at a depth between 0.5 and two meters during low tide. These specimens were growing in the sandy bottom of the transition zone between sea grass and the outer reef as well as on coral boulders in the sandy lagoon. At this site survival, growth and attachment only of *S. massa* fragments on horizontal ropes and live rock were tested (Table 1).

Teluk Pegametan—Sampling and Study Site 2. Bali is one of the bigger islands of Indonesia (140 by 80 km) and located eight degrees south of the equator. The sampling and study

(a)

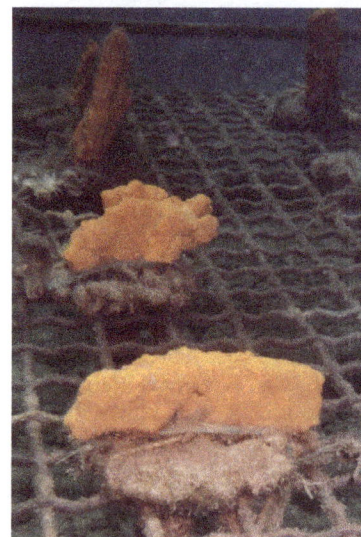

(b)

Figure 1: *Neopetrosia* sp. (a) and *Stylissa massa* (b) fragments attached to the artificial substrate and placed on a metal grid at Teluk Pegametan.

site (8°08′12″S, 114°36′E) was located in the Bay Teluk Pegametan in the north of Bali, on one of the sandy reef flats at a depth of 2.5 m at low tide and about 750 m from the mainland (Figure 3). 50 donor sponges of *Neopetrosia* sp. and 41 of *S. massa* with a length of 10 to 20 cm were collected at the same depth as in Pulau Pari. *S. massa* was found growing on coral boulders while *Neopetrosia* sp. was more often found on live corals.

Two types of aquaria were used, a glass tank and a cement tank. The glass tanks, located at CV. Dinar Goris, Northern Bali, were 80 cm in length, 40 cm high, and 25 cm deep. The cement tank was 2400 cm in length, 70 cm high, and 140 cm deep with a water depth of 40 cm. Both types of aquaria were supplied daily with seawater from the Bay Teluk Pegametan, five km from the facility. The seawater was filtered through a

TABLE 1: List of application methods of fragments at the different study sites associated with the number of transplants and the days of culture.

Species	Place	Field/aquaria	Attachment method	Cutting method	No. of transplants	Days in culture
Neopetrosia sp.	Field	TP	HL (t/m/b)	HC	36 (12/12/12)	108
Neopetrosia sp.	Field	TP	LR	HC	12	108
Neopetrosia sp.	Field	TP	AS	HC	12	83
Neopetrosia sp.	Aquaria	GT	HL	HC	24	107
Neopetrosia sp.	Aquaria	GT	LR	HC	24	107
Neopetrosia sp.	Aquaria	CT	AS	HC	12	45
S. massa	Field	PP	HL (t/m/b)	HC/VC	18/18 (6/6/6)	131
S. massa	Field	PP	LR	HC/VC	6/6	131
S. massa	Field	TP	HL (t/m/b)	HC/VC	18/18 (6/6/6)	108
S. massa	Field	TP	LR	HC/VC	6/6	108
S. massa	Field	TP	AS	HC/VC	6/6	83
S. massa	Aquaria	GT	HL	HC/VC	12/12	107
S. massa	Aquaria	GT	LR	HC/VC	12/12	107
S. massa	Aquaria	CT	AS	HC/VC	6/6	45

TP: Teluk Pegametan, PP: Pulau Pari, GT: glass tank, CT: cement tank, HL (t/m/b): horizontal line (top/middle/bottom), LR: live rock, AS: artificial substrate, HC: horizontally cut, and VC: vertically cut.

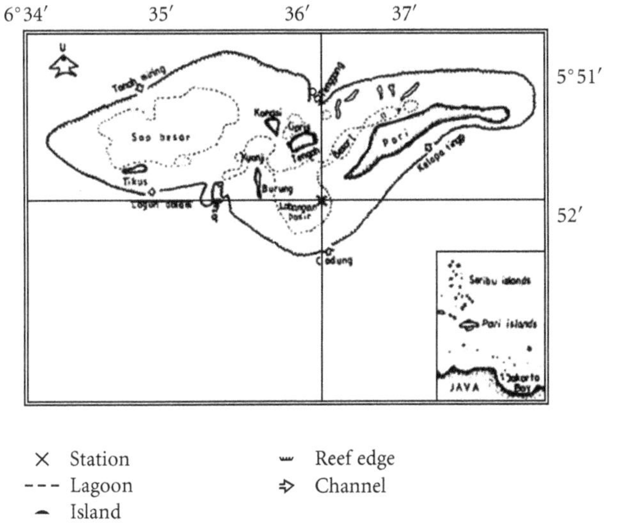

× Station   ⌣ Reef edge
--- Lagoon   ⇨ Channel
⌢ Island

FIGURE 2: Map of Pulau Pari, Thousand Islands, North of Jakarta [30] the X marks the study site.

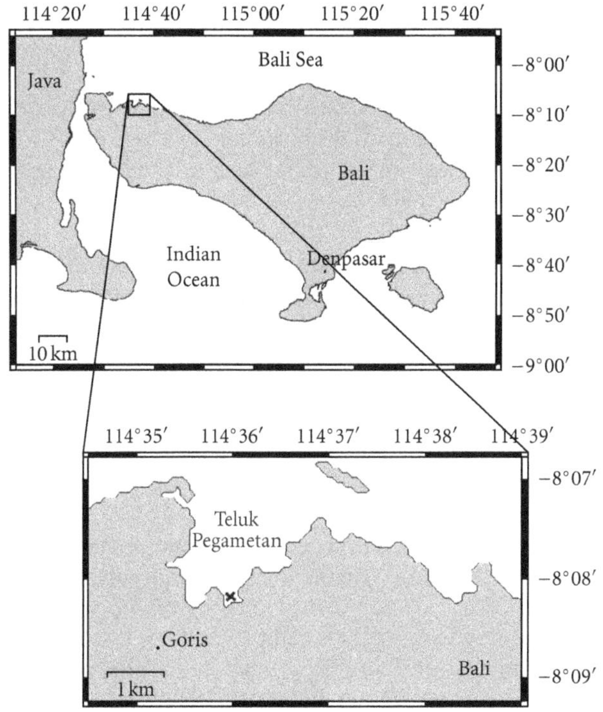

FIGURE 3: Map of the sampling and study site (X) at Teluk Pegametan, Northern Bali [31].

biological filter consisting of coral boulders. No additional nutrients were added to the aquarium seawater. A roof protected the aquaria from rainfall and solar radiation.

2.3. Substrata and Attachment Methods

Neopetrosia sp. The fragmentation of *Neopetrosia* sp. was always done in the same way by cutting horizontally to provide fragments of approximately equal size. Before the fragments were attached, the length and diameter of *Neopetrosia* sp. fragments were measured with a plastic calliper.

Stylissa massa. The fragmentation of *S. massa* was done in two different ways; half of the collected sponges were cut horizontally, the other half vertically. The fragments were approximately of equal size. Before the fragments were attached, the length, diameter, and cutting width of *S. massa* were measured with a plastic calliper.

Nine specimens of each species were left uncut as controls; three of each were placed in the field (Teluk Pegametan), glas and cement tank.

Three horizontal nylon ropes ($\varnothing$ = 0.7 cm) with the same length (120 cm) were tied on each end at two bamboo sticks (length = 130 cm, $\varnothing$ = 4 to 5 cm) that were fixed on the seabed. The three horizontal nylon ropes were tied to the sticks at different depths, the first rope started 30 cm above the sea bottom and the next two ropes followed every 30 cm. Twelve sponge fragments were tied to the rope with cable ties, four in each line (Figure 4, Table 1). To prevent shifting of the fragments every cable tie was tied to the nylon rope with a plastic ribbon.

Two horizontal ropes were placed parallel in one aquarium. Four sponge fragments of each of the two species were placed on each of the horizontal ropes in the same way as in the field (Table 1).

Four sponge fragments of each of the two species were attached to coral boulders by using toothpicks and elastics (Figure 4, Table 1). The toothpick was pushed through the fragment, placed on the boulder, and tied to it with elastics. In the aquaria eight sponge fragments of each of the two species on coral boulders were placed in two parallel lines.

In the field, four sponge fragments were tied on artificial substrates (round plates, made of pebbles and cement) in the same way as on coral boulders and placed on a metal grid (200 × 100 × 50 cm) that was set on the sea bottom (Figure 5, Table 1).

In the cement tank the four sponge fragments on artificial substrates were placed on the bottom (Table 1).

Three replicates were used for each method, both in the aquaria and at the field sites.

2.4. Abiotic Factors. Temperature, salinity, pH, and dissolved oxygen were measured with a WTW MultiLine P4 set (conductivity cell, pH combined electrode, D.O. probe) and turbidity with a turbidity meter (Lovibond). Ammonium, carbonate, and nitrite content were measured with a standard test from Merck (Merck KGaA, Darmstadt). Surface currents were measured in the field by recording the time required for an object to float a distance of 6 m.

2.5. Analyses of Data. Growth was expressed in an increase in volume; therefore, length, diameter and cutting width (cutting width only for *Stylissa massa*) of the fragments were measured with a plastic calliper every two weeks when visiting the sites at Teluk Pegametan and Goris, Northern Bali, and every two months at the site Pulau Pari.

For both species the volume was calculated with the following factors and formulas:

for *Neopetrosia sp.*:

$$V = \frac{1}{3} * \pi * l * r^2 \tag{1}$$

(l = length, r = radius of the fragment),

for *S. massa*:

$$V = l * d * w \tag{2}$$

(l = length, d = diameter, w = width of the fragment).

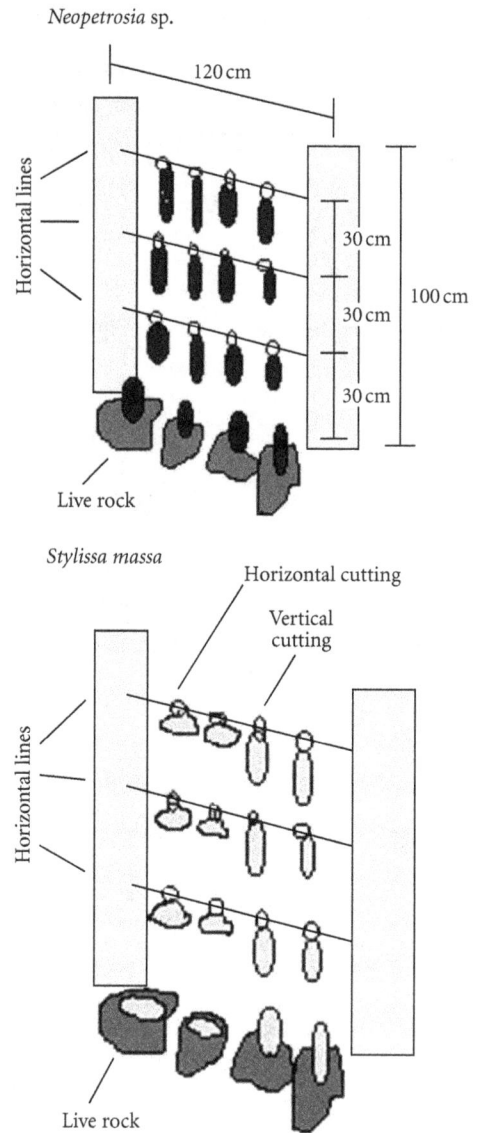

FIGURE 4: A scheme of *Neopetrosia sp.* (top) and *S. massa* (down) fragments on horizontal lines and live rock at the field sites.

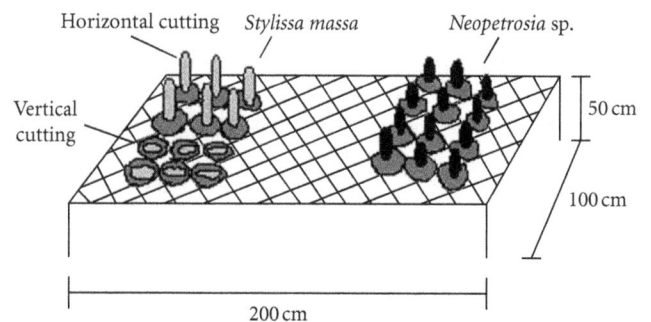

FIGURE 5: A scheme of *Neopetrosia sp.* (right) and *S. massa* (left) fragments on artificial substrate, placed on a metal grid at the field site Teluk Pegametan.

TABLE 2: Mean initial and end values of length and volume of fragments at the different sites for all methods.

Species	Place	Mean initial length [cm]	Mean end length [cm]	Mean initial volume [cm³]	Mean end volume [cm³]
Neopetrosia sp.	Field	5.9	10.8	10.8	48.3
Neopetrosia sp.	Aquaria	4.2	3.7	9.3	8.3
Stylissa massa	Field TP HC	7.1	8.2	39.7	95.4
Stylissa massa	Field TP VC	8.4	9.3	43.5	94
Stylissa massa	Field PP HC	5.9	5.6	40.9	37.1
Stylissa massa	Field PP VC	7.2	7.7	46.1	53.6
Stylissa massa	Aquaria HC	6.2	4.7	37.6	38.9
Stylissa massa	Aquaria VC	7.2	6.1	39.5	40.2

TP: Teluk Pegametan, PP: Pulau Pari, HC: horizontally cut, and VC: vertically cut.

In addition, every time visiting the study site, the following state of the fragments was considered: alive, partly dead, dead, damaged, fretted, covered with sediment, overgrown by, for example, fouling organisms, and attached to the substrates. The survival (S) and attachment (A) rate of S. *massa* and *Neopetrosia* sp. was calculated as the percentage of the number of fragments being alive and accordingly attached at the beginning of the study (N_0) divided by the fragments left in the end of the study (N_E):

$$S = \frac{N_E}{N_0} \times 100,$$

$$A = \frac{N_E}{N_0} \times 100. \tag{3}$$

To calculate the mean percentage growth (V) of fragments, the following formula was used, where N_0 is the initial volume and N_E is the volume of fragments in the end of the study period:

$$V = \frac{(N_E - N_0)}{N_0} \times 100. \tag{4}$$

Before starting statistical analysis, the homogeneity of variances of the different variables (species, places, methods, and cutting method) was tested with Levene's test. It was decisive for the statistical analysis method.

To identify significant differences in monthly length and volume growth between the different species, places, methods, and in case of S. *massa* different ways of cutting (horizontal cuttings and vertical cuttings) a t-test for independent samples with separate variance estimates was done with Statistica-Student Version (Version'99).

3. Results

3.1. Growth Rates. The experiment started in the field at Pulau Pari on 2 November 2002, at Teluk Pegametan, in the glass tanks at Goris in the facility of CV. Dinar on 3 December 2002, in the field at Teluk Pegametan with the artificial substrate on 28 December 2002 and on 03 February 2003 with the artificial substrate in the cement tank. The last measurement of sponge fragments at Pulau Pari took place

on 12 March 2003 and at Teluk Pegametan on 20 March 2003. In Table 1 the days of culture and number of fragments associated with the tested methods at the different study sites are listed.

In Table 2 mean initial and end values of length and volume are compared for the two species *Neopetrosia* sp. and *S. massa*, from aquaria and field sites and for *S. massa* between horizontal and vertical cuttings.

3.1.1. Mean Length and Volume Increase at the Field Sites Teluk Pegametan and Pulau Pari

Neopetrosia sp. *Neopetrosia* sp. fragments grew better on horizontal ropes. Mean length increase for all three transplantation methods was significantly different ($P = 0.02$), as was mean volume increase on horizontal rope and live rock or live rock and artificial substrate ($P = 0.002$).

Stylissa massa. At both field sites (Teluk Pegametan and Pulau Pari) *S. massa* fragments did not show any significant differences for mean length and volume increase for the different transplantation methods (horizontal rope, coral boulders, and artificial substrate) or for the different cutting methods (horizontal and vertical way of cutting) (Figures 6(b) and 6(c)), but fragments on horizontal ropes and coral boulders at Pulau Pari showed significant differences in both mean length and volume increment ($P = 0.001$). On horizontal ropes those fragments grew better. When comparing the two field sites, fragments at Teluk Pegametan showed higher mean length ($P = 0.02$) and volume ($P = 0.01$) increase than those at Pulau Pari.

S. massa fragments on horizontal ropes at Pulau Pari showed an initial decline in mean length and volume at the second measuring interval (after one week) before they finally started to grow.

For both mean volume and length increase *S. massa* and *Neopetrosia* sp. fragments on the individual depths of the horizontal ropes did not show any significant differences.

Comparing mean length and volume increase at the field sites, the two species *S. massa* and *Neopetrosia* sp. showed significant differences in mean length ($P = 0.001$) and volume ($P = 0.01$) increase, *Neopetrosia* sp. fragments grew faster.

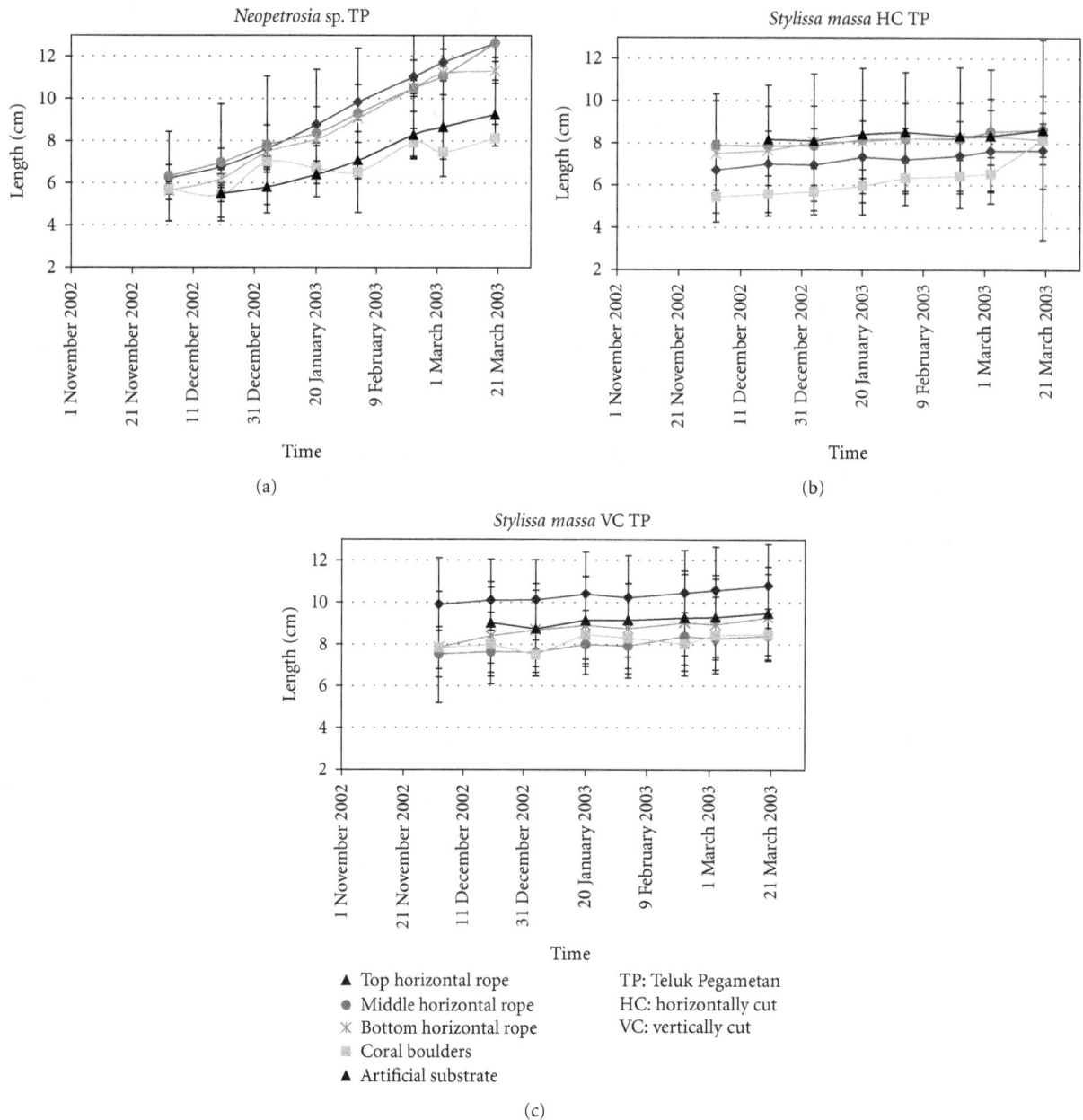

FIGURE 6: Fragments mean length at the field sites for different attachment methods and different places.

3.1.2. Mean Length and Volume Increase in the Aquaria. In the aquaria the fragments of both species showed a decline or steady state in mean length, except fragments of *S. massa* on artificial substrate in the big tank.

Neopetrosia sp. fragments were significantly different ($P = 0.01$) in mean length for all three transplantation methods, horizontal rope and coral boulders, horizontal rope and artificial substrate and coral boulders and artificial substrate and in mean volume increase only for fragments on horizontal ropes and coral boulders. The fragments of *Neopetrosia* sp. on horizontal ropes developed small tentacle-like branches on top of the fragments in horizontal directions.

The mean length and volume increase differed greatly between the study sites. Significant differences ($P = 0.001$) in mean length and volume increase were determined for *Neopetrosia* sp. fragments between field Teluk Pegametan and aquaria and *S. massa* fragments between Teluk Pegametan and Pulau Pari, whereas a significant difference was found in mean length increase between Teluk Pegametan and aquaria and in mean volume increase between Pulau Pari and aquaria (Figure 7).

In general, *Neopetrosia* sp. and *S. massa* fragments showed significant differences in mean length, but not in volume increase.

TABLE 3: Mean values for abiotic factors.

Abiotic factors	Teluk Pegametan	Pulau Pari	Glass tank	Cement tank
Salinity	33.52 ($\pm$0.09)	33.23 ($\pm$0.09)	39.17 ($\pm$2.5)	36.2 ($\pm$1.02)
Oxygen [%]	94.59 ($\pm$0.9)	102.25 ($\pm$1.5)	98.84 ($\pm$1.6)	102.62 ($\pm$4.24)
Oxygen [mg/L]	5.97 ($\pm$0.07)	6.47 ($\pm$0.13)	6.39 ($\pm$0.1)	6.7 ($\pm$0.29)
pH	8.15 ($\pm$0.02)	8.10 ($\pm$0.04)	8.23 ($\pm$0.07)	8.36 ($\pm$0.06)
Turbidity [NTU/FNU]	0.71 ($\pm$0.07)	0.50 ($\pm$0.3)	0.43 ($\pm$0.1)	X
Temperature [°C]	29 ($\pm$0.15)	30 ($\pm$0.7)	28 ($\pm$0.05)	27.6 ($\pm$0.30)
Ammonium [mg/L NH_4^+]	0.1	X	0	0.1
Nitrite [mg/L NO_2^-]	0	X	0	0
Hardness [°d]	8	X	10	10

Standard deviation in brackets; X: no measurements.

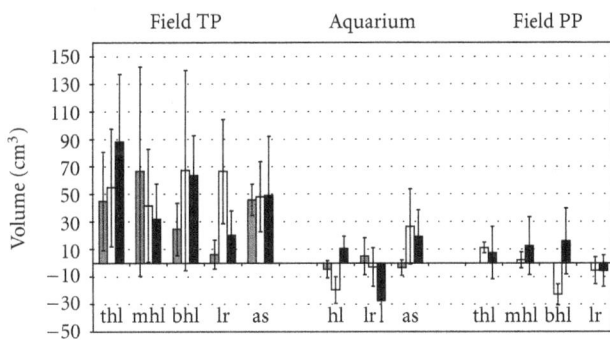

FIGURE 7: Mean final growth rate (percentage end volume of initial volume) for different methods. Error bars represent the standard deviation between fragments of specific methods (grey: Neopetrosia sp., white: horizontally cut Stylissa massa, black: vertically cut Stylissa massa, TP: Teluk Pegametan, PP: Pulau Pari, thl: top horizontal rope, mhl: middle horizontal rope, bhl: bottom horizontal rope, lr: coral boulders, as: artificial substrate, and hl: horizontal rope).

3.1.3. Mean Final Growth, Survival, and Attachment Rates. Similar results for mean length and volume increase were found in mean final growth rates (end volume minus initial volume) (Figure 7).

As shown in Figure 8, no significant differences for final survival and attachment were calculated between fragments of Neopetrosia sp. and S. massa. In the field Teluk Pegametan Neopetrosia sp. fragments on horizontal ropes and artificial substrate showed the highest final survival and attachment rates, but only one of 12 fragments on coral boulders survived (Figure 8(a)). In the case of S. massa the horizontal cuttings showed higher survival and attachment rates than the vertical cuttings for coral boulders and artificial substrate, while vertical cuttings survived and attached more easily on the horizontal ropes (Figures 8(b) and 8(c)).

For both species, the survival and attachment rates of fragments on the artificial substrate at the field site Teluk Pegametan and in the aquaria were highest at 100%, with the exception of 50% attachment of Neopetrosia sp. fragments in the big tank.

The mean final survival rates for horizontal ropes and coral boulders in the aquaria for both species were very low, zero percent for Neopetrosia sp. fragments and vertical cuttings of S. massa on coral boulders and horizontal cuttings of S. massa on horizontal ropes, 21% of Neopetrosia sp. and 12.5% of horizontally cut S. massa fragments on coral boulders, 8% vertically cut on horizontal ropes survived. Almost all fragments of S. massa (horizontal ropes: 50% for horizontal cuttings, 92% for vertical cuttings; coral boulders: 58% for horizontal cuttings, 83% for vertical cuttings) in the aquaria which did not survive until the final checking were already dead two weeks after the installation; in the case of Neopetrosia sp. fragments, 58% were dead one month after installation.

Although Neopetrosia sp. and S. massa fragments did not show significant differences in survival and attachment, it is obvious that almost all fragments of Neopetrosia sp. which survived attached well.

Already two weeks after Neopetrosia sp. fragments were placed on horizontal ropes, coral boulders, and artificial substrates in the field, all of them did attach. In contrast, S. massa fragments did not attach as easily.

One month after starting the experiments, 30 of 33 remaining Neopetrosia sp. fragments started to encrust the cable ties, attaching them to the horizontal ropes. After encrusting the cable ties, those fragments started to close the gap between them and the encrusted cable tie, while they were also starting to grow vertically towards the horizontal rope. The first two of 36 fragments reached the rope already two weeks after installation. At the end of the study, after 108 days, all 33 remaining fragments had closed the space between fragment and cable tie and 30 of 33 had attached to the rope, eight of them already grew along the rope and 20 above it.

One month after starting the experiments, eight of 12 Neopetrosia sp. fragments on artificial substrates at Teluk Pegametan started to encrust their substrate. After two months all of the 12 fragments had already encrusted parts of the artificial substrate.

3.2. Abiotic Factors. As seen in Table 3 salinity, oxygen, pH, turbidity, and temperature for the four different study sites did not vary significantly at the different measuring times. Even daily profiles did not show a great difference between the values. The aquaria showed a higher mean salinity value

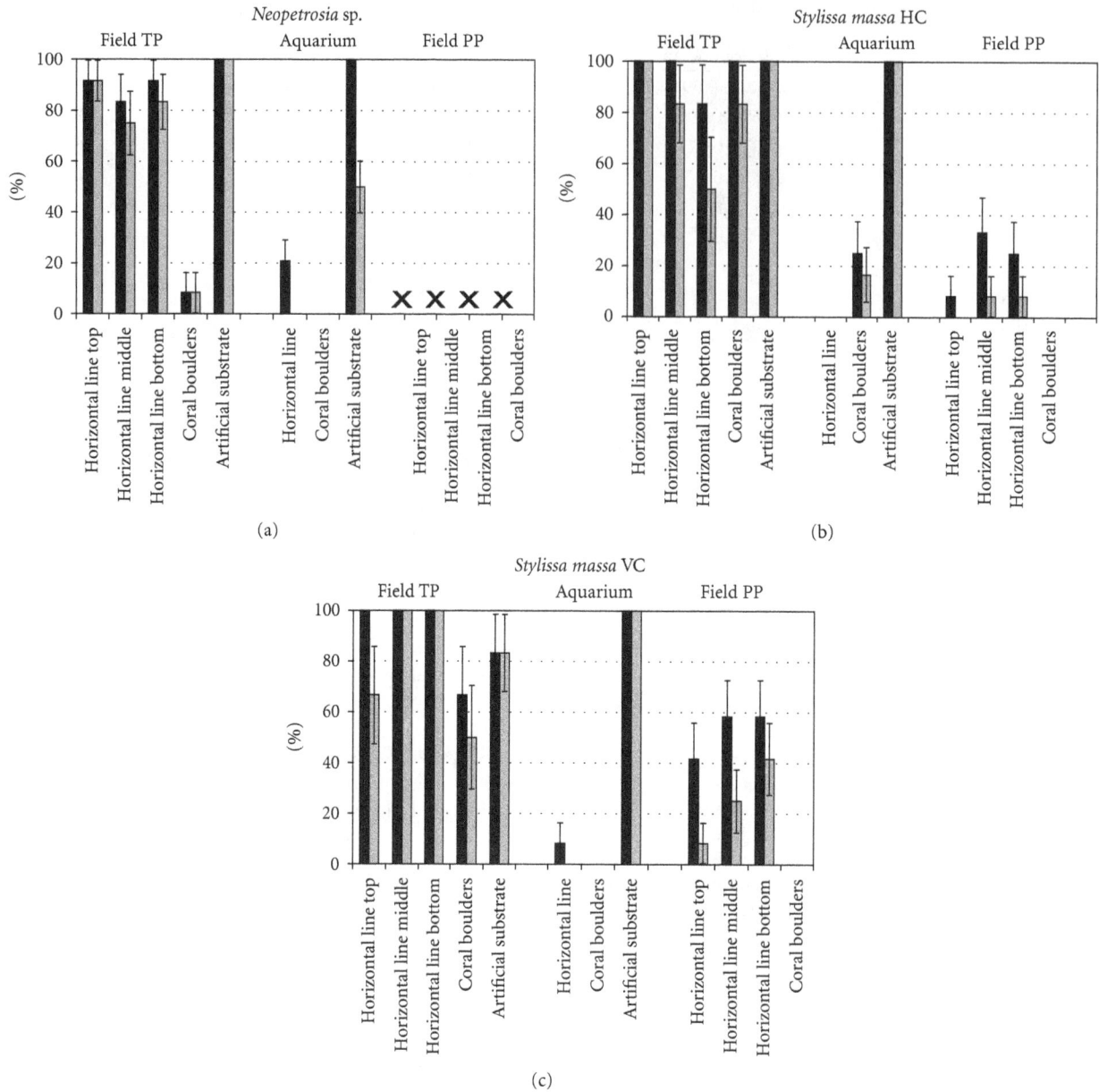

FIGURE 8: Mean final percentage survival and attachment in the end of the study for different methods. Error bars represent the standard deviation between fragments of specific methods. (X: no fragments were planted, TP: Teluk Pegametan, PP: Pulau Pari, HC: horizontally cut, VC: vertically cut, thl: top horizontal rope, mhl: middle horizontal rope, bhl: bottom horizontal rope, lr: coral boulders, as: artificial substrate, and hl: horizontal rope).

than the field sites, possibly due to evaporation. Ammonium, nitrite, and carbon hardness were very low (<0.1). The only difference between the two types of aquaria was the volume of water, glass tank 80 L and cement tank 23,520 L. The surface currents for both field locations were between 0.07 and 0.075 m/s (highest 0.12, lowest 0.048). Stormy conditions were observed at Pulau Pari in December 2002 and January 2003.

4. Discussion

4.1. Growth, Attachment, and Survival Rates at the Field Sites.
Growth experiments in this study tested farming methods on two sponge species, Neopetrosia sp. and S. massa. The growth behaviour of most sponge species is poorly understood; this study helps to contribute to a better understanding of sponge biology and also to provide information on suitable ways to reproduce sponges for reef rehabilitation, the aquarium trade, and biomedical research. In other studies, sponge farming experiments were conducted to produce bath sponges for cosmetic interest [5, 9, 22] or to receive bioactive metabolites [6, 8, 9, 18].

The horizontal rope method and artificial substrates in the field Teluk Pegametan were shown to be the best method for growth, attachment, and survival rates of Neopetrosia

TABLE 4: Mean percentage growth rates per month [%].

Method	Neopetrosia sp.	S. massa	Hippo spongia lachne [32]	Dysidea avara [6]	Latrunculia welling-tonensis [17]	Polymastia croceus [17]	Hippo spongia and Spongia spp. [33]	Mycale hentscheli [18]
HL-field	106 TP	34 TP	X	7	58	30	X	X
LR-field	12 TP	27 TP	X	X	X	X	X	X
AS-field	122 TP	31 TP	~13	X	X	X	~8	X
mesh arrays	X	X	X	28	23	11	X	409
HL-aquaria	−13 GT	−4	X	X	X	X	X	X
LR-aquaria	16 GT	−11	X	X	X	X	X	X
AS-aquaria	−7 CT	10	X	X	X	X	X	X

TP: Teluk Pegametan, GT: glass tank, CT: cement tank, HL (t/m/b): horizontal line (top/middle/bottom), LR: live rock, AS: artificial substrate, HC: horizontally cut, and VC: vertically cut.

sp. For *S. massa* all three methods (horizontal lines, coral boulders, and artificial substrate) have almost equal growth rates. The horizontal rope method was already successfully tested by de Voogd [8] on *Callyspongia biru*, an encrusting sponge, and for bath sponges by Corriero et al. [5] and Çelik et al. [22].

The success of farming sponges requires the knowledge of the optimal environmental conditions of the cultured sponge species [9]. The position of *Neopetrosia* sp. fragments on horizontal ropes and artificial substrates (50 cm above the sea bottom on a metal grid) is similar to its position in natural habitats, where they were observed to grow on other sponges or corals in order to have an exposed position to filter water. *Neopetrosia* sp. fragments on coral boulders had a low survival rate (8.33%). It is assumed that the fragments of this species probably did not like to be placed directly on the sea bottom. The higher sedimentation near the bottom leads to the block-up of ostia, causing the sponge to stop feeding and forcing all its energy toward cleaning [23]. For other sponge species like *Spongia officinalis* no significant differences in growth could be observed for different positions of transplants in the water column [22].

S. massa prefers to live on coral boulders or on the sandy bottom, in the transition zone between coral reef and sea grasses or mangroves. Its habitat preferences are therefore more widespread than those of *Neopetrosia* sp. In the Bay Teluk Pegametan, all *S. massa* fragments showed higher growth, attachment, and survival rates than in the Pulau Pari lagoon. Fragments on coral boulders were totally covered with sand due to wave action and crab movements. Newly built crab burrows can cover sponges and corals in nonelevated positions. Currents ripped off 38% of the fragments on horizontal ropes. Too much wave action and unsuitable farming methods for those areas were the reasons for the low survival and attachment rates. As described by Duckworth [9], in areas with strong water flow or partly stormy conditions the growth behavior of sponge species could be different, there is less feeding, and more energy is used to form a stronger skeletal structure.

Generally, *Neopetrosia* sp. showed higher growth rates on the different substrates than *S. massa*, but they had almost the same attachment and survival rates. It is unclear if there are any differences in growth rates of wild stocks between these two species. There were no differences in growth between uncut sponges (donor sponge) of both species at the field site Teluk Pegametan, but *Neopetrosia* sp. displayed signs of being fed on or taken away by current, while *S. massa* does not show any signs of damage. Like in the study by De Caralt et al. [6], it is difficult to compare the uncut sponges of the studied species due to their much more complex growth forms in comparison to the regularly cut fragments.

Fragments of *Neopetrosia* sp. grow faster after cutting than uncut sponges. Duckworth [24] tested regeneration and growth on wounded *Latrunculia wellingtonensis* sponges and demonstrated that damaged specimens showed better growth than nondamaged ones, which possibly invest their energy in reproduction. Wound size is greater for *S. massa* (massive growth form) than for *Neopetrosia* sp. (branching growth form) due to different morphological appearance. Wound recovery needs energy and reduces growth, and wound size influences survival [24]. Duckworth et al. [25] demonstrated wounds on some fragments of a massive sponge *Psammocinia hawere*, which did not fully recover after 80 days. This could be a reason why *S. massa* grew slower. However, it is difficult to compare both tested species for growth and attachment due to different skeletal structure, habitat preference, growth forms of the donor sponge, and wound size of the fragments.

Other studies have shown that growth of fragments on specific substrates can depend on sponge skeletal structure [7, 8, 18], soft sponges grow better in mesh bags and hard sponges on rope lines [17, 20]. In comparison with other studies (Table 4) growth rates of *S. massa* showed similar results using the suspended rope method, but *Neopetrosia* sp. showed even higher growth rates, 106% growth rate per month. Fragments in mesh bags showed higher growth (over 400% per month [18]) and survival rates, but the ones on the ropes showed higher attachment rates [19]. Fragments can be placed gently in mesh bags, which partly protect fragments against damage, especially for soft and fleshy sponges, but the bags also limit fragments growth and feeding possibility, especially if biofouling species like algae, bryozoans and ascidians were growing on the mesh [9, 17, 19]. In contrast the exclusion of sediment can lead to smaller growth rates; those sponges can grow 50% less than others [23]. In a different study, Duckworth and Battershill [17] showed that two different New Zealand sponge species (*L. wellingtonensis* and *P. croceus*) differed in survival and growth rates, greatest survival rates for meshes (meshes: 61 and 96%, ropes: 22 and 59%) and greatest growth for ropes (weight increment meshes for *P. croceus*: about 50%, ropes: over 100%) after nine months.

4.2. Growth, Attachment, and Survival Rates at the Aquaria Sites. Sponge fragments in the field showed higher growth, attachment, and survival rates than the ones in the aquaria. Inhibition of sponge growth can be caused not only by high sediment loads but also by low loads [2], which can be the problem in the aquaria. The aquaria fragments of both species even show a decrease in length and volume. Shrinking is evidence for insufficient food supply. Even if food like bacteria is present, also too gentle aeration, infrequent change of seawater, and the slowness of inflow can affect the availability of food [9, 26]. The higher survival and attachment rate, observed in the cement tank in comparison to the glass tank, may have been due to a higher water volume exchange.

Neopetrosia sp. fragments on horizontal ropes in the glass tank were shrinking, but they also have developed little tentacle-like branches. These branches might be a sign of starvation. However, adding food can help sponges to survive for several years in aquaria [26].

Another reason for the low survival rates in the glass tank is dying sponge fragments, which can be toxic [27] and affect the survival rate of healthy fragments.

Nickel et al. [28] suggest removing specimen with little parts of the attached substrate to have a better recovery in aquaria, because attachment will promote survival [19]. Further proper collection and transportation and more experiments in aquaria, for example, on light, current, and nourishment could help to discover how to maintain, feed, or reproduce species like *Neopetrosia* sp. and *S. massa* in aquaria.

Artificial substrate is easy to install and in the case of *Neopetrosia* sp. the encrusting growth form helps to cover the substrate in order to gain a marketable appearance in the aquaria.

4.3. Recommendations. The precision of calculated volumes is limited due to the nonuniformly morphological appearance. For a new series of studies it is recommended to use additional measuring methods to determine growth, for example weight [8], or other calculation methods depending on the growth form of the sponge species as described by Sipkema et al. [29]. Page et al. [18] measured the growth of sponge fragments of *Mycale hentscheli* via the surface area in correlation with wet weight.

As *Neopetrosia* sp. showed high survival, growth, and attachment rates, it would be interesting to test smaller cuttings. Duckworth [9] has shown that some sponge species (*Latrunculia wellingtonensis* and *Polymastia croceus*) can also recover totally from 90% removal of biomass. As other studies [5, 22] have also shown, growth rates of *Spongia officinalis* between different sized explants are similar. *Neopetrosia* sp. fragments show nearly no algae cover. In case few algae

attach, they are easily removable and do not affect their state of health.

As sponges are important for the reef ecosystem, this study could help to reproduce sponges to help a reef to recover, to make it more attractive again to other marine species after destruction, further offering new and hopefully sustainable exploitation for local fishermen.

To minimize harvesting impact, it would be advisable not to harvest the whole sponge stock, but to leave a small part behind to give the sponge the chance to recover and regenerate into a new sponge. Sampling sponges for export trade does not only place extreme pressure on the wild stock, also the attachment surface will be damaged, which could cause injury to other living creatures, mostly corals. The knowledge of how to reproduce sponges on different substrates will contribute to the reduction of wild harvest and its damage to the environment.

Acknowledgments

The authors are very grateful for the support given by the heads of the Indonesian company CV. Dinar, E. Wahyuono and A. Setiabudi, also T. Siswati and D. Timur Wahjuadi, and by their staff from Jakarta, Goris, and Denpasar. They wish to thank Mr. Soemarso and Dr. T. Hestirianoto from Institut Pertanian Bogor (IPB) for their support and N. de Voogd for her help to identify the species. Thanks go to Dr. D. Soedharma (IPB), the institutions Asosiasi Koral Kerang dan Ikan hias Indonesia (AKKII) and Lembaga Ilmu Pengetahuan Indonesia (LIPI) for logistic help.

References

[1] J. M. Gili and R. Coma, "Benthic suspension feeders: their paramount role in littoral marine food webs," *Trends in Ecology and Evolution*, vol. 13, no. 8, pp. 316–321, 1998.

[2] C. R. Wilkinson, "Role of sponges in coral reef structural processes," in *Perspectives on Coral Reefs*, D. J. Barnes, Ed., pp. 263–274, 1983.

[3] N. A. Campbell, *Biologie*, Spektrum Akademischer, Heidelberg, Germany, 2nd edition, 1998.

[4] S. A. Fossa and A. J. Nilsen, *The Modern Coral Reef Aquarium*, vol. 3, Birgit Schmettkamp Verlag, 1st edition, 2000.

[5] G. Corriero, C. Longo, M. Mercurio, C. Nonnis Marzano, G. Lembo, and M. T. Spedicato, "Rearing performance of *Spongia officinalis* on suspended ropes off the Southern Italian Coast (Central Mediterranean Sea)," *Aquaculture*, vol. 238, no. 1–4, pp. 195–205, 2004.

[6] S. De Caralt, J. Sánchez-Fontenla, M. J. Uriz, and R. H. Wijffels, "*In situ* aquaculture methods for *Dysidea avara* (demospongiae, porifera) in the Northwestern Mediterranean," *Marine Drugs*, vol. 8, no. 6, pp. 1731–1742, 2010.

[7] N. J. De Voogd, "An assessment of sponge mariculture potential in the Spermonde Archipelago, Indonesia," *Journal of the Marine Biological Association of the United Kingdom*, vol. 87, no. 6, pp. 1777–1784, 2007.

[8] N. J. de Voogd, "The mariculture potential of the Indonesian reef-dwelling sponge *Callyspongia* (*Euplacella*) *biru*: growth, survival and bioactive compounds," *Aquaculture*, vol. 262, no. 1, pp. 54–64, 2007.

[9] A. Duckworth, "Farming sponges to supply bioactive metabolites and bath sponges: a review," *Marine Biotechnology*, vol. 11, no. 6, pp. 669–679, 2009.

[10] R. Wehner and W. Gehring, *Zoologie*, Georg Thieme, Stuttgart, Germany, 23th edition, 1995.

[11] P. L. Colin and C. Arneson, *Tropical Pacific Invertebrates*, Coral Reef Press, Beverly Hills, Calif, USA, 2nd edition, 1995.

[12] S. M. MacMillan, "Starting a successful commercial sponge aquaculture farm," University of Hawaii Sea Grant College Program Communications Office, School of Ocean and Earth Science and Technology, CTSA publication 120, 2002.

[13] Wet Web Media, Sponges, Phylum Porifera, Part 3, 2003, http://www.wetwebmedia.com/spongesii.htm.

[14] P. Bethge, "Füllhorn der Meere," *Der Spiegel*, vol. 17, pp. 112–113, 2003.

[15] D. Tasdemir, R. Mallon, M. Greenstein et al., "Aldisine alkaloids from the Philippine sponge *Stylissa massa* are potent inhibitors of mitogen-activated protein kinase kinase-1 (MEK-1)," *Journal of Medicinal Chemistry*, vol. 45, no. 2, pp. 529–532, 2002.

[16] D. E. Williams, P. Austin, A. R. Diaz-Marrero et al., "Neopetrosiamides, peptides from the marine sponge *Neopetrosia* sp. that inhibit amoeboid invasion by human tumor cells," *Organic Letters*, vol. 7, no. 19, pp. 4173–4176, 2005.

[17] A. Duckworth and C. Battershill, "Sponge aquaculture for the production of biologically active metabolites: the influence of farming protocols and environment," *Aquaculture*, vol. 221, no. 1–4, pp. 311–329, 2003.

[18] M. J. Page, P. T. Northcote, V. L. Webb, S. Mackey, and S. J. Handley, "Aquaculture trials for the production of biologically active metabolites in the New Zealand sponge *Mycale hentscheli* (Demospongiae: Poecilosclerida)," *Aquaculture*, vol. 250, no. 1-2, pp. 256–269, 2005.

[19] A. R. Duckworth and C. N. Battershill, "Developing farming structures for production of biologically active sponge metabolites," *Aquaculture*, vol. 217, no. 1–4, pp. 139–156, 2003.

[20] P. Van Treeck, M. Eisinger, J. Müller, M. Paster, and H. Schuhmacher, "Mariculture trials with Mediterranean sponge species: the exploitation of an old natural resource with sustainable and novel methods," *Aquaculture*, vol. 218, no. 1–4, pp. 439–455, 2003.

[21] J. N. A. Hooper and R. W. M. Van Soest, *Systema Porifera. A Guide to the Classification of Sponges, Volume 1 Introductions and Demospongiae*, Kluwer Academic, Plenum Publishers, New York, NY, USA, 1st edition, 2002.

[22] I. Çelik, Ş. Cirik, U. Altnağaç et al., "Growth performance of bath sponge (*Spongia officinalis* Linnaeus, 1759) farmed on suspended ropes in the Dardanelles (Turkey)," *Aquaculture Research*, vol. 42, pp. 1807–1815, 2011.

[23] C. R. Wilkinson and J. Vacelet, "Transplantation of marine sponges to different conditions of light and current," *Journal of Experimental Marine Biology and Ecology*, vol. 37, no. 1, pp. 91–104, 1979.

[24] A. R. Duckworth, "Effect of wound size on the growth and regeneration of two temperate subtidal sponges," *Journal of Experimental Marine Biology and Ecology*, vol. 287, no. 2, pp. 139–153, 2003.

[25] A. R. Duckworth, C. N. Battershill, and P. R. Bergquist, "Influence of explant procedures and environmental factors on culture success of three sponges," *Aquaculture*, vol. 156, no. 3-4, pp. 251–267, 1997.

[26] W. C. Jones, "Process-formation by aquarium-kept sponges and its relevance to sponge ecology and taxonomy," in *Sponges in Time and Space*, R. W. M. Van Soest, T. M. G. Van Kempen, and J. C. Braekman, Eds., pp. 241–250, A.A. Balkema, Rotterdam, The Netherlands, 1994.

[27] J. C. Delbeek and J. Sprung, *The Reef Aquarium—A Comprehensive Guide to the Identification and Care of Tropical Marine Invertebrates*, Ricordea Publishing, Miami, Fla, USA, 1st edition, 1994.

[28] M. Nickel, S. Leininger, G. Proll, and F. Brümmer, "Comparative studies on two potential methods for the biotechnological production of sponge biomass," *Journal of Biotechnology*, vol. 92, no. 2, pp. 169–178, 2001.

[29] D. Sipkema, N. A. M. Yosef, M. Adamczewski et al., "Hypothesized kinetic models for describing the growth of globular and encrusting demosponges," *Marine Biotechnology*, vol. 8, no. 1, pp. 40–51, 2006.

[30] S. Juwana, "Crab culture technique at RDCO-LIPI, Jakarta, Indonesia, 2001," in *Proceedings of the Workshop on Mariculture in Indonesia*, W. W. Kastoro, S. Soemodihardjo, T. Ahmad, and S. Anggoro Putro Dwiono, Eds., pp. 49–60, Lombok, Indonesia, February 2002.

[31] Seaturtle.org Maptool, 2002, http://seaturtle.org/maptool/.

[32] L. R. Crawshay, "Studies in the market sponges: I. Growth from the planted cutting," *Journal of the Marine Biological Association of the United Kingdom*, vol. 23, pp. 553–574, 1939.

[33] H. F. Moore, "A practical method of sponge culture," *Bulletin of the Bureau of Fisheries*, vol. 28, pp. 545–585, 1910.

17

Glass Sponges off the Newfoundland (Northwest Atlantic): Description of a New Species of *Dictyaulus* (Porifera: Hexactinellida: Euplectellidae)

Francisco Javier Murillo,[1] Konstantin R. Tabachnick,[2] and Larisa L. Menshenina[3]

[1] *Instituto Español de Oceanografía, 36280 Vigo, Spain*
[2] *Department of Benthic Fauna, P. P. Shirshov Institute of Oceanology, Russian Academy of Sciences, Nakhimovsky Prospekt 36, Moscow 117997, Russia*
[3] *Physical Department, Moscow State University, Moscow 119991, Russia*

Correspondence should be addressed to Konstantin R. Tabachnick; tabachnick@mail.ru

Academic Editor: Jakov Dulčić

Three species of hexactinellid sponges: *Aphrocallistes beatrix beatrix* Gray, *Asconema foliata* (Fristedt), and *Dictyaulus romani* sp. n. were collected off the Flemish Cap in the Flemish Pass and from the Grand Banks off the Newfoundland (northwest Atlantic) during different surveys on board of Spanish RV *Vizconde de Eza* and RV *Miguel Oliver*.

1. Introduction

The hexactinellid fauna of the northwest Atlantic is remaining poorly investigated up to now. A unique notable population of *Vazella pourtalesi* Schmidt was recently described from the adjacent area (the shelf of Scotia) [1], a monospecific genus previously known by few specimens from two findings in the Caribbean Sea off Florida and off Azores [2].

The present work complements data on hexactinellid fauna of the northwest Atlantic (the Newfoundland area). It is based on materials collected on groundfish bottom trawl surveys and on NEREIDA cruises. We describe a new species of *Dictyaulus*, a genus known for a long time by a single representative from the Indian Ocean, and recently two species from the central Pacific [3] and the north Atlantic [4] were described.

2. Material and Methods

Material was collected by groundfish bottom trawl surveys by the RV *Vizconde de Eza* supported by the Instituto Español de Oceanografía and the European Union (methodology used is described in [5]) and from rock dredge samples on board the RV *Miguel Oliver* sustained by Spain's General Secretariat of the Sea (Secretaría General del Mar) in the framework of the NEREIDA (a multidisciplinary research project). Samples were fixed in 70° ethanol.

Spicules were dissociated using solution of $K_2Cr_2O_7$, fresh water, and H_2SO_4 then washed in fresh, water, and dried. Spicules were mounted in Canada balsam and analysed (including measurements) using light microscopy.

Type materials are stored in the National Museum of Natural Sciences in Madrid (Spain); other specimens are deposited in the Instituto Español de Oceanografía, Vigo (Spain); small fragments and slides for the light microscopy of all investigated specimens are treasured in the P.P. Shirshov Institute of Oceanology, Russian Academy of Sciences, Moscow (Russia).

Taxonomy

Hexactinellida Schmidt, 1870

Hexasterophora Schulze, 1886

Hexactinosida Schrammen, 1903

Aphrocallistidae Gray, 1867

Aphrocallistes aff. *beatrix beatrix* Gray, 1858

FIGURE 1: *Dictyaulus romani* sp. n. A: sieve plate of holotype. B: holotype, external shape. C: bottom basidictyonal plate of holotype. D: paratype, external shape. Scale bar 20 mm.

Material Examined. IEO: RV *Vizconde de Eza*, St. FC07 L19, 25.06.2007, 46°43,62′-42,02′ N 46°19,40′-18,58′ W, depth 404 m.

Description. Three small fragments of 20 × 15 mm, 15 × 10 mm, and 15 × 5 mm were studied.

Remarks. This specimen does not contain any loose spicules; hence its precise identification is impossible, while their morphology is typical for the genus. Nevertheless due to the location of this specimen and data on distribution of representatives of this genus [6], and the name *Aphrocallistes beatrix beatrix* is assigned with minimal hesitations Figures 1, 2, and Table 1.

Lyssacinosida

Euplectellidae Gray, 1867

Corbitellinae Gray, 1872

Dictyaulus Schulze, 1895

Dictyaulus romani sp. n.

Etymology. The species is named in the honor of Mrs. Esther Roman, whose activity in the Newfoundland investigations allowed the capturing of these specimens.

Material Examined. Holotype MNCN 1.04/8 RV *Vizconde de Eza*, St. FN3L06 L99, 18.08.2006, 46°08,25′-06,72′ N 46°50,74′-50,46′ W, depth 1394–1418 m. Paratype MNCN 1.04/9 RV *Vizconde de Eza*, St. FC07 L170, 18.07.2007, 46°09,21′-09,15′ N 45°46,11′-48,33′ W, depth 1332–1488 m.

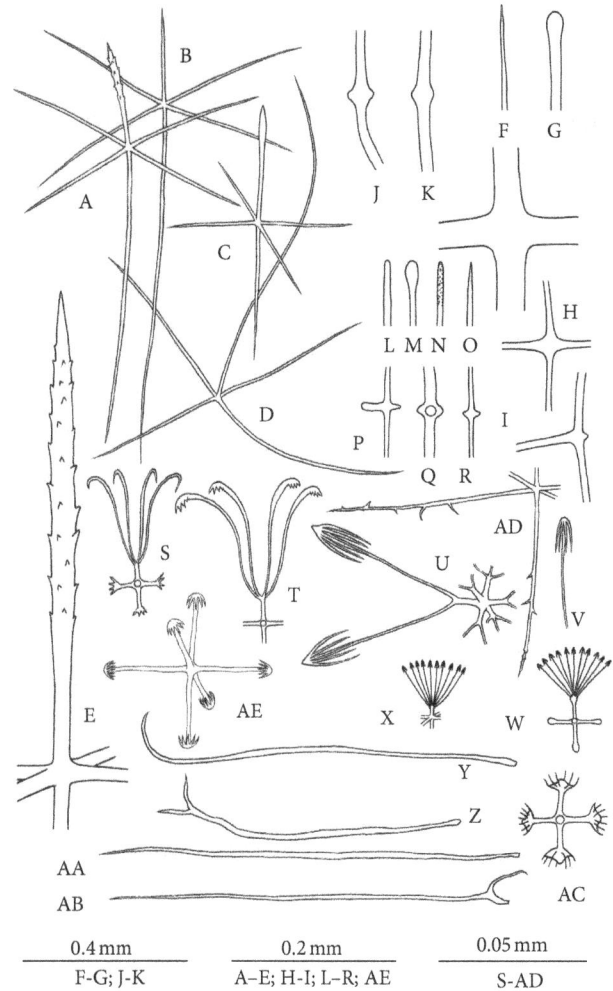

FIGURE 2: Spicules of *Dictyaulus romani* sp. n. A–AD: paratype. AE: holotype. A–C: dermal hexactins. D: atrial pentactin. E: hypodermal hexactin or prostalia oscularia. F-G: large choanosomal stauractin and its outer ends. H: small choanosomal stauractin. I: choanosomal tauactin. J-K: large choanosomal diactins. L–R: small choanosomal diactins. S: drepanocome. T: floricome. U-V: anchorate macrodiscohexaster and its secondary ray. W-X: spherical microdiscohexasters. Y–AC: fragments of oxyhexaster: secondary rays (Y–AB) and primary rosette (AC). AD: spiny oxyhexactin. AE: discohexactin.

Description

Body Morphology. The holotype is tubular 200 mm high, 47 mm in diameter, 18 mm in diameter at base (basiphytous), and 32 mm in diameter of upper part carrying the sieve plate. The paratype is 265 mm high, 50 mm in diameter, and at base it is broken. Numerous round or oval lateral oscula 0.5–3 mm in diameter penetrate thin 0.5–0.8 mm walls of the specimen. The sieve plate is colander in structure, and it is made of the beams with meshes usually square, sometimes rectangular, or triangular with sides 2–5 mm long. A fine, circular, about 5 mm high structure of small, loose spicules is situated above and around the main osculum of both specimens. No prostasia lateralia was observed in this specimen.

TABLE 1: Spicule dimensions (in mm) *Dictyaulus romani* sp. n.

| | Holotype | | | | | Paratype | | | |
	n	min.	max.	avg	std	*n*	min.	max.	avg	std
L dermal hexactin distal ray	24	0.037	0.252	0.151	0.044	35	0.104	0.211	0.148	0.024
L dermal hexactin tangential ray	25	0.104	0.185	0.143	0.018	28	0.111	0.241	0.159	0.035
L dermal hexactin proximal ray	18	0.141	0.481	0.310	0.104	18	0.093	0.525	0.284	0.147
L atrial pentactin tangential ray	6	0.278	0.414	0.335	0.050	7	0.200	0.500	0.370	0.119
L atrial pentactin ray directed inside body	4	0.352	0.444	0.389	0.040	4	0.252	0.444	0.352	0.107
D drepanocome	12	0.059	0.081	0.074	0.006	25	0.072	0.101	0.092	0.007
d drepanocome	12	0.015	0.019	0.017	0.002	25	0.014	0.020	0.017	0.002
D floricome	25	0.089	0.118	0.105	0.007	26	0.072	0.140	0.097	0.012
d floricome	25	0.015	0.030	0.020	0.004	26	0.014	0.018	0.017	0.001
D spiny hexactin	25	0.178	0.266	0.220	0.025	27	0.133	0.227	0.175	0.024
D pappocome	25	0.259	0.333	0.300	0.016	25	0.169	0.263	0.216	0.022
d pappocome	1	0.016	0.016	0.016		25	0.014	0.026	0.019	0.003
D large discohexaster	2	0.118	0.141	0.130	0.016	4	0.072	0.108	0.097	0.017
d large discohexaster	2	0.022	0.036	0.029	0.009	4	0.014	0.016	0.015	0.001
D small discohexaster	11	0.032	0.065	0.048	0.010	20	0.022	0.054	0.040	0.007
d small discohexaster	11	0.014	0.020	0.017	0.002	20	0.007	0.017	0.012	0.003
D discohexactin	3	0.185	0.215	0.195	0.017					

L: length; *D*: diameter; *d*: diameter of primary rosette.

Spicules. Choanosomal spicules are principalia (stauractins, some tauactins, and diactins), rare pinular hexactins, and additional spicules (comitalia: smaller stauractins, tauactins, diactins, and, maybe, pentactins). Principalia are large stauractins with rays 10/0.06–0.11 mm, and their outer ends are smooth, conically pointed, or rounded. The choanosomal spicules of smaller sizes have rays 0.48–2.8/0.004–0.015 mm, and their outer ends are conically pointed, or rounded, smooth or rough. The spicules of the sieve plate are similar to the choanosomal ones, both large and small, but most spicules of the sieve plate are curved diactins. The large diactins of the sieve plate have a widening in the middle, and they are about 3/0.02–0.08 mm. The small diactins of the sieve plate 0.9–3/0.004–0.009 mm have a widening in the middle or 2 or 4 tubercles. Some hexactins, found among the spicules of the sieve plate, have rays 0.08–0.3/0.020–0.025 mm. The large pinular hexactins have distal ray pinular 0.6–1.2/0.06 mm, and the diameter of the tangential rays is smaller—about 0.04 mm; only one broken spicule of this type was observed. Maybe this spicule is principalia of the sieve plate which forms the principal attachment of the sieve plate to the margin of the main osculum. Fusion between choanosomal spicules was observed in the holotype only in some large spicules at the basal part.

Dermalia are hexactins smaller than that described above; they have distal rays lanceolate, lanceolate-pinular, or stout, similar to other rays of these spicules. The distal ray of dermal hexactin is 0.037–0.252 mm long up to 0.03 mm in diameter at the thickest part; tangential rays are 0.104–0.241 mm long, their proximal ray is 0.093–0.525 mm long, the diameter of these rays is 0.004–0.007 mm, and their outer ends are conically pointed, smooth. Atrialia are pentactins with tangential rays 0.200–0.500 mm long, their ray directed

inside the body is 0.252–0.444 mm long, the diameter of these rays is 0.004–0.007 mm, and the outer ends are equal to those of dermal hexactins and to small choanosomal spicules. These pentactins are very rarely found, and it is possible that these pentactins are choanosomal spicules (common in many representatives of the subfamily).

Microscleres. Spicules of this type are numerous: drepanocomes, floricomes, pappocomes, large and small discohexasters, and spiny hexactins; discohexactins were unique (found in the holotype only). The drepanocomes with 4–7 secondary rays are 0.059–0.101 mm in diameter, and their primary rosette is 0.014–0.020 mm in diameter. The floricomes have 4–7 secondary rays, each secondary ray has 2–5 teeth, the diameter of the floricomes is 0.072–0.140 mm, and their primary rosette is 0.014–0.030 mm in diameter. The spiny hexactins are 0.133–0.266 mm in diameter. The pappocomes, spicules with secondary rays approximately 11–15 in number, are distributed in slightly widened tufts (as it may be suggested since these spicules were found always broken: secondary rays separated from the primary rosettes); the secondary rays are straight and oxyoidal, rarely onychoidal, curved oxyoidal and rarely spiny. The pappocomes are 0.169–0.333 mm in diameter, and their primary rosette is 0.014–0.026 mm in diameter. The large discohexasters with 2-3 secondary rays are rare, they are 0.072–0.141 mm in diameter, and their primary rosette is 0.014–0.036 mm in diameter. The large discohexasters have anchorate discs with umbel 0.015–0.034 mm long and 0.007–0.016 mm in diameter. The small discohexasters with numerous secondary rays are 0.022–0.065 mm in diameter, and their primary rosette is 0.007–0.020 mm in diameter. Their discs are also anchorate with umbel 0.004–0.007 mm long and 0.002–0.003 mm in diameter. The large discohexactins were found in the lower part of the holotype specimen, and they are 0.185–0.215 mm in diameter.

Remarks. Until now the genus *Dictyaulus* comprised 3 species: *D. elegans* Schulze, 1895 [7] (NW Indian Ocean); *D. starmeri* Tabachnick and Levi, 2004 [3] (SE Pacific); and *D. marecoi* Tabachnick and Collins, 2008 [4] (N Mid-Atlantic). Huge discoidal spicules in *Dictyaulus* are discohexasters in *D. elegans*, discasters in *D. starmeri*, and discohexactins in *D. romani*; in *D. marecoi* such spicules were not found as well as in the paratype of *D. romani*, described above. Specific feature of the new species is presence of pappocomes and their derivatives with onychoidal outer ends and secondary rays carrying some spines. As for other spicules the species from the Atlantic Ocean are very similar.

Rossellidae Schulze, 1885

Rossellinae Schulze, 1885

Asconema Kent, 1870

Asconema foliata (Fristedt, 1887)

Material Examined. IEO: RV *Vizconde de Eza*, St. PLA07 L55(1); L55(2); L55(3); L55(4), 10.06.2007, 42°51,59′-52,73′ N 49°53,61′-52,10′ W, depth 337–334 m. St. PLA07 L61(1); L61(2); L61(3), 11.06.2007, 43°05,86′-07,29′ N 49°29,92′-29,60′ W, depth 629–656 m. St. PLA07 L70(1); L70(2), 12.06.2007, 43°20,26′-21,57′ N 49°14,37′-13,52′ W, depth 640–656 m. St. PLA07 L72(1); L72(2), L72(3), 13.06.2007, 43°30,37′-30,40′ N 49°12,74′-14,88′ W, depth 390–418 m. St. PLA07 L93, 17.06.2007, 45°24,90′-23,83′ N 48°21,28′-22,77′ W, depth 1249–1266 m. St. PLA07 L94, 17.06.2007, 45°24,97′-26,11′ N 48°26,37′-25,10′ W, depth 1003–1000 m. St. FC07 L18, 25.06.2007, 46°51,49′-50,23′ N 46°27,82′-26,35′ W, depth 423–420 m. St. FC07 L20, 25.06.2007, 46°37,14′-35,87′ N 46°15,74′-14,01′ W, depth 407 m. St. FC07 L22(1); L22(2), 26.06.2007, 46°30,39′-29,33′ N 46°15,42′-13,42′ W, depth 494–498 m. St. FC07 L26, 26.06.2007, 46°26,92′-28,47′ N 45°45,24′-46,34′ W, depth 538–492 m. St. FC07 L44, 29.06.2007, 46°44,81′-46,23′ N 44°07,03′-05,52′ W, depth 480–471 m. St. FC07 L45, 29.06.2007, 46°55,01′-56,44′ N 43°54,86′-53,50′ W, depth 528–539 m. St. FC07 L48, 30.06.2007, 47°00,05′-01,76′ N 44°34,24′-34,57′ W, depth 138–148 m. St. FC07 L58, 01.07.2007, 47°14,28′-15,96′ N 43°45,39′-45,91′ W, depth 647–642 m. St. FC07 L59, 01.07.2007, 47°28,83′-30,51′ N 43°50,15′-50,46′ W, depth 638–645 m. St. FC07 L74, 03.07.2007, 48°13,60′-14,76′ N 44°43,85′-45,77′ W, depth 596–599 m. St. FC07 L75, 03.07.2007, 48°14,89′-15,66′ N 44°52,92′-55,22′ W, depth 567–570 m. St. FC07 L123, 09.07.2007, 48°17,51′-16,03′ N 45°45,05′-46,14′ W, depth 957–947 m. St. FN3L09 L5(1); L5(2), 26.07.2009, 48°03,90′-04,21′ N 47°17,77′-19,93′ W, depth 634–639 m. St. FN3L10 L44(1); L44(2); L44(3), 03.08.2010, 47°29,47′-28,02′ N 46°44,74′-45,35′ W, depth 1175 m. St. FN3L10 L64(1); L64(2); L64(3), 07.08.2010, 46°45,72′-44,23′ N 47°05,53′-05,22′ W, depth 1024–1085 m. RV *Miguel Oliver*, St. NEREIDA 0710 RD88-41 19.07.2010, 45°22,35′-22,06′ N 48°34,05′-33,57′ W, depth 676–700 m. St. NEREIDA 0710 DR93-60 24.07.2010, 44°32,46′-31,92′ N 48°56,45′-56,25′ W, depth 1167–1382 m.

Description

Body Morphology. These sponges are presented by broken fragments of thin walls 0.5–2 mm thick.

Spicules. Choanosomal spicules are diactins. Hypodermalia are smooth pentactins. Hypoatrialia are also pentactins with equal hypodermal ones; they were found only in some specimens in low amounts. Dermalia are pentactins with pinular rays distally directed, a short rudiment instead of the sixth ray has proximal direction, and the outer ray tips are rounded. Atrialia are large hexactins with pinular rays directed into the atrial cavity and pentactins smaller than dermal ones, with thinner rays and conically pointed outer ends, and their pinular rays are directed into the atrial cavity.

Microscleres. Oxyoidal microscleres are common, and they are oxyhexactins, oxyhexasters, and oxyhemihexasters. Discoidal microscleres are represented by spherical microdiscohexasters (found in all specimens) and low amounts of macrodiscasters (found in two specimens). In FC07 L48 a single macrodiscaster 0.089 mm in diameter with primary rosette 0.015 mm in diameter was found. In FC07 L26 we observed two macrodiscohexasters with very short primary rays, and they are 0.118–0.126 mm in diameter with primary rosettes 0.019–0.022 mm in diameter.

Remarks. A unique feature of the specimens collected off the Newfoundland is rarity of large discoidal microscleres (which were considered to be important in the species identification by [8]). Meanwhile, this is the largest series of representatives of this species, and this unique feature may be an important specific character as well.

Abbreviations

Avg:	Average
NEREIDA:	The project "NAFO" "potEntial vulnerable marine Ecosystems"-Impacts of Deep-seA fisheries
IEO:	Instituto Español de Oceanografía, Vigo (Spain)
MNCN:	National Museum of Natural Sciences in Madrid (Spain)
n:	Number of measurements
Std:	Standard deviation.

Acknowledgments

The authors would like to acknowledge the scientific staff involved in NEREIDA surveys and NAFO groundfish bottom trawl surveys and the heads of these surveys, especially Mrs. Esther Roman, for facilitating the data collection, and the research vessels crews (RV *Vizconde de Eza* and RV *Miguel Oliver*) for assistance at the sea. They appreciate Dr. D. Janussen for her critical notes and the partial financial support of RFBR (research Grant 13-04-01332a).

References

[1] S. D. Fuller, *Diversity of marine sponges in the Northwest Atlantic [Ph.D. thesis]*, Dalhousie University, Halifax, Canada, 2011.

[2] K. R. Tabachnick, "Rossellidae Schulze, 1885," in *Systema Porifera: A Guide to the Classification of Sponges*, J. N. A. Hooper and R. W. M. Van Soest, Eds., pp. 1441–1504, Kluwer Academic/Plenum Publishers, New York, NY, USA, 2002.

[3] K. R. Tabachnick and C. Lévi, "Lyssacinosida du Pacifique sud-ouest (Porifera: Hexactinellida). Bruce A Marshall & Bertran Richer de Forges," in *Tropical Deep-Sea Benthos*, vol. 23, pp. 11–70, 2004, Mémoires du Museum national d'Histoire naturelle.

[4] K. R. Tabachnick and A. G. Collins, "Glass sponges (Porifera, Hexactinellida) of the northern Mid-Atlantic Ridge," *Marine Biology Research*, vol. 4, no. 1-2, pp. 25–47, 2008.

[5] F. J. Murillo, P. D. Muñoz, J. Cristobo et al., "Deep-sea sponge grounds of the Flemish Cap, Flemish Pass and the Grand Banks of Newfoundland (Northwest Atlantic Ocean): distribution and species composition," *Marine Biology Research*, vol. 8, no. 9, pp. 842–854, 2012.

[6] H. M. Reiswig, "Family Aphrocallistidae Gray, 1867," in *Systema Porifera: A Guide to the Classification of Sponges*, J. N. A. Hooper and R. W. M. Van Soest, Eds., pp. 1182–1286, Kluwer Academic/Plenum Publishers, New York, NY, USA, 2002.

[7] F. E. Schulze, *Hexactinelliden des indischen Oceanes. II. Die Hexasterophora*, Abhandlungen der Preussischen Akademie der Wissenschaften, Berlin, Germany, 1895.

[8] K. R. Tabachnick and L. L. Menshenina, "Revision of the genus Asconema (Porifera: Hexactinellida: Rossellidae)," *Journal of the Marine Biological Association of the United Kingdom*, vol. 87, no. 6, pp. 1403–1429, 2007.

18

Spatial Distributions of Picoplankton and Viruses in the Changjiang Estuary and Its Adjacent Sea Area during Summer

Yun Li[1,2] and Daoji Li[2]

[1] East China Sea Fisheries Research Institute, Chinese Academy of Fishery Sciences, Shanghai 200090, China
[2] State Key Laboratory of Estuarine and Coastal Research, East China Normal University, Shanghai 200062, China

Correspondence should be addressed to Yun Li, salixly@yahoo.com.cn

Academic Editor: Oscar Schofield

Simultaneous determination of picoplankton (i.e., *Synechococcus* spp., *Prochlorococcus* spp., picoeukaryotes, and heterotrophic bacteria) and viruses in the Changjiang (Yangtze) River estuary and its adjacent sea area was made using flow cytometry during a cruise in June 2006. The results show that *Prochlorococcus* in all samples was below detectable level. The abundances of *Synechococcus*, picoeukaryotes, heterotrophic bacteria, and viruses ranged from 0.00 to 1.22×10^8 cell L^{-1}, 0.01×10^6 to 1.42×10^7 cells L^{-1}, 8.40×10^7 to 4.29×10^9 cells L^{-1}, and 1.20×10^7 to 1.06×10^{10} particles L^{-1}, respectively. The determined picoplankton groups and viruses distinctly increased with the distance off the estuary and where the maximum abundance that occurred in these groups was different somewhat due to the individual sensitivity to environmental changes. Viral abundance showed a positive correlation with salinity and negative correlations with turbidity and inorganic nutrient concentrations. Positive linear relations were found between *Synechococcus*, heterotrophic bacteria, and viruses.

1. Introduction

Autotrophic *Synechococcus* spp., *Prochlorococcus* spp., and picoeukaryotes, together with heterotrophic bacteria, are the principal components of marine picoplankton communities (0.2–2 μm in diameter) [1–4]. These picoplankton groups in the Changjiang estuary have been the subject of research in the past [5–14]. However, few data have been published on picoplankton community structure after the Three Gorges Dam construction. Previous studies were mainly focused on only one picoplankton group rather than simultaneously observing all picoplankton groups, and most of the data in the literature were obtained by epifluorescence microscope, which was not as efficient, sensitive, and precise as flow cytometry (FCM) adopted in this study [12, 13]. Marie et al. [15] reported the method of FCM measurement on marine viruses, providing a glorious prospect on studying natural viral community. Compared with other oceanic regions, the field investigation on viruses in the Changjiang estuary was much less, and, up to now, only a few reports were found [12, 14, 16, 17].

With the construction of the huge Three Gorges Dam (height 180 m, width 2 km) in the middle reaches of Changjiang River, both the amount and the timing of the terrigenous input into the East China Sea will be different [18]. It has been suggested that after the water storage of the Three Gorges reservoir, light availability required for photosynthesis enhanced in the river mouth area due to less discharge of water and sediments from the river [7, 8]. Studies indicate that picoplankton and viruses are more sensitive to environmental changes than larger organisms because of their small size [19, 20]. Therefore, the quantifications of picoplankton and viruses are of great importance as it may reflect the changes in estuarine water quality associated with the construction of the dam.

In this study, a simultaneous determination of picoplankton groups and viruses was made in June 2006. The main objective is to observe the spatial distributions of picoplankton and viruses in the Changjiang estuary and its adjacent sea area and to discern the relationships between them and the environmental factors.

TABLE 1: List of the measured environmental parameters in the Changjiang estuary and its adjacent sea area in June 2006.

Parameter	Section I	Section II	Section III	Total station
Number of sample	39	45	15	96
Salinity (psu)				
Range	0.00–33.92	14.86–34.42	29.79–34.13	0.00–34.42
Mean ± SD	20.49 ± 14.09	30.13 ± 4.41	32.41 ± 1.58	26.62 ± 10.69
CV	69%	15%	5%	40%
Turbidity (FTU)				
Range	0.07–122.13	0.45–122.13	0.00–32.43	0.00–122.13
Mean ± SD	57.84 ± 57.09	22.77 ± 27.97	3.68 ± 8.29	31.78 ± 44.77
CV	99%	123%	225%	141%
Nitrate ($\mu g\,L^{-1}$)				
Range	11.89–1645.36	19.47–342.04	11.41–128.93	11.41–1645.36
Mean ± SD	493.43 ± 503.65	162.73 ± 84.10	45.65 ± 38.18	275.06 ± 372.35
CV	103%	52%	84%	135%
Nitrite ($\mu g\,L^{-1}$)				
Range	0.11–34.44	0.95–12.69	0.11–11.55	0.11–34.44
Mean ± SD	5.70 ± 6.86	4.90 ± 2.64	3.31 ± 3.21	5.11 ± 5.48
CV	120%	54%	97%	107%
Ammonia ($\mu g\,L^{-1}$)				
Range	3.62–63.93	26.31–130.22	17.97–123.74	3.62–130.22
Mean ± SD	39.23 ± 17.28	55.09 ± 17.83	56.20 ± 25.69	50.01 ± 20.37
CV	44%	32%	46%	41%
Phosphate ($\mu g\,L^{-1}$)				
Range	2.53–43.57	2.21–30.96	0.98–12.45	0.98–43.57
Mean ± SD	17.90 ± 11.91	13.28 ± 7.06	5.34 ± 3.14	13.63 ± 10.20
CV	67%	53%	59%	75%
Silicate ($\mu g\,L^{-1}$)				
Range	55.22–3027.20	157.57–1144.37	33.43–243.76	33.43–3027.20
Mean ± SD	936.83 ± 898.12	405.36 ± 195.53	151.69 ± 69.83	576.29 ± 659.66
CV	96%	48%	46%	114%

2. Materials and Methods

2.1. Field Sampling. The cruise was conducted from June 12 to 22, 2006, on board R/V *Dong Fang Hong 2*. Thirty-two stations, between 29°00′ N and 32°19′ N and 120°46′ E and 124°01′ E, were located along three different but representative sections identified as I, II, and III (Figure 1). Section I extended from the freshwater river to the saltwater sea covering the whole salinity gradient. Section II ran across the edge of the bar area and covered the whole river mouth. Section III represented the area off the river mouth. Water sample collections were guided by a Sea-Bird CTD (conductivity-temperature-depth) probe and carried out with Niskin bottles at three different depths (one meter below the surface, a few meters above the sea bed, and midway between, according to the depth at each station). Salinity and turbidity were recorded in the water column using the CTD assembly.

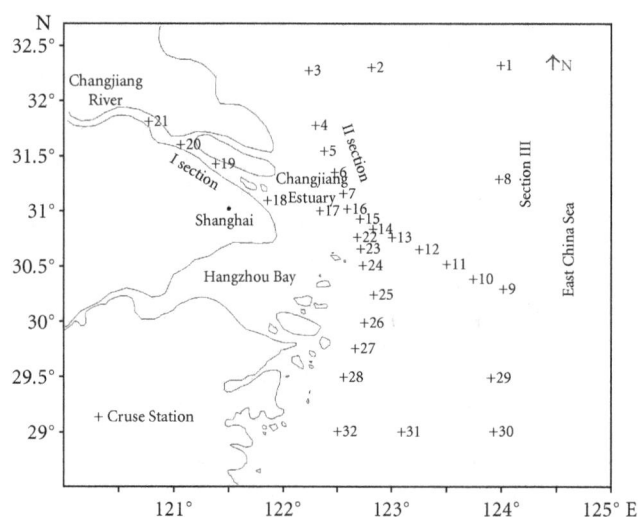

FIGURE 1: Location of sampling stations in June 2006.

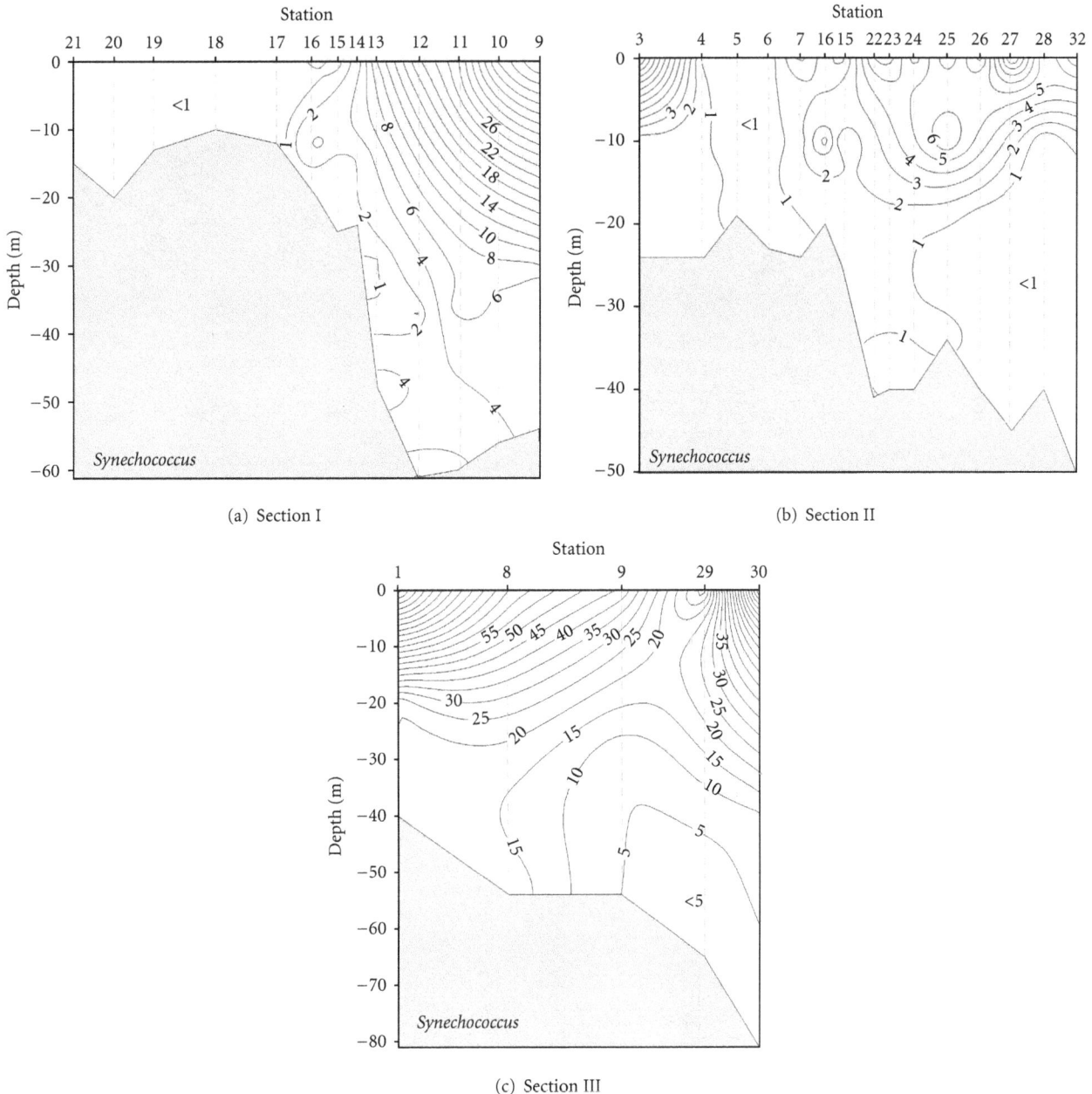

(a) Section I

(b) Section II

(c) Section III

FIGURE 2: Spatial distribution of *Synechococcus* ($\times 10^6$ cells L^{-1}) in the Changjiang estuary and its adjacent sea area in June 2006.

2.2. Inorganic Nutrients Analysis. Water samples were filtered immediately after collection through precleaned 0.45 μm pore-size cellulose filters. Subsequently, 1-2 drips of saturated HgCl were added to the filtrate, which was preserved at room temperature in brown bottles for later analysis. Inorganic nutrients (nitrate, nitrite, ammonia, phosphate, and silicate), were determined using spectrophotometry with a SKALAR San^plus Segmented Flow Analyzer (Breda, The Netherlands).

2.3. Picoplankton and Viral Analysis. Samples were fixed for 15 min with paraformaldehyde (final concentration: 1% v/v) in the dark and then kept frozen in liquid nitrogen until they were analyzed in the laboratory. All of the

samples were analyzed with a FACScan instrument (Becton Dickinson, San Jose, CA) within a month of collection. The picophytoplankton was separated into groups according to their specific autofluorescence properties and side scatter differences and their abundance was recorded [12, 21–23]. For the enumeration of heterotrophic bacteria, samples were stained with 1/10000 (v/v) SYBR Green I (Molecular Probes, Inc.) and incubated in the dark for 15 min before FCM analysis [21]. For virus enumeration, natural samples were diluted using TE buffer (10 mmol/L Tris, 1 mmol/L EDTA, pH 8) to 100–1000 virus s^{-1} using flow cytometry [24]. The diluted virus samples were stained with SYBR Green I (final concentration: 0.5×10^{-4}, v/v) for 10 min in the dark at

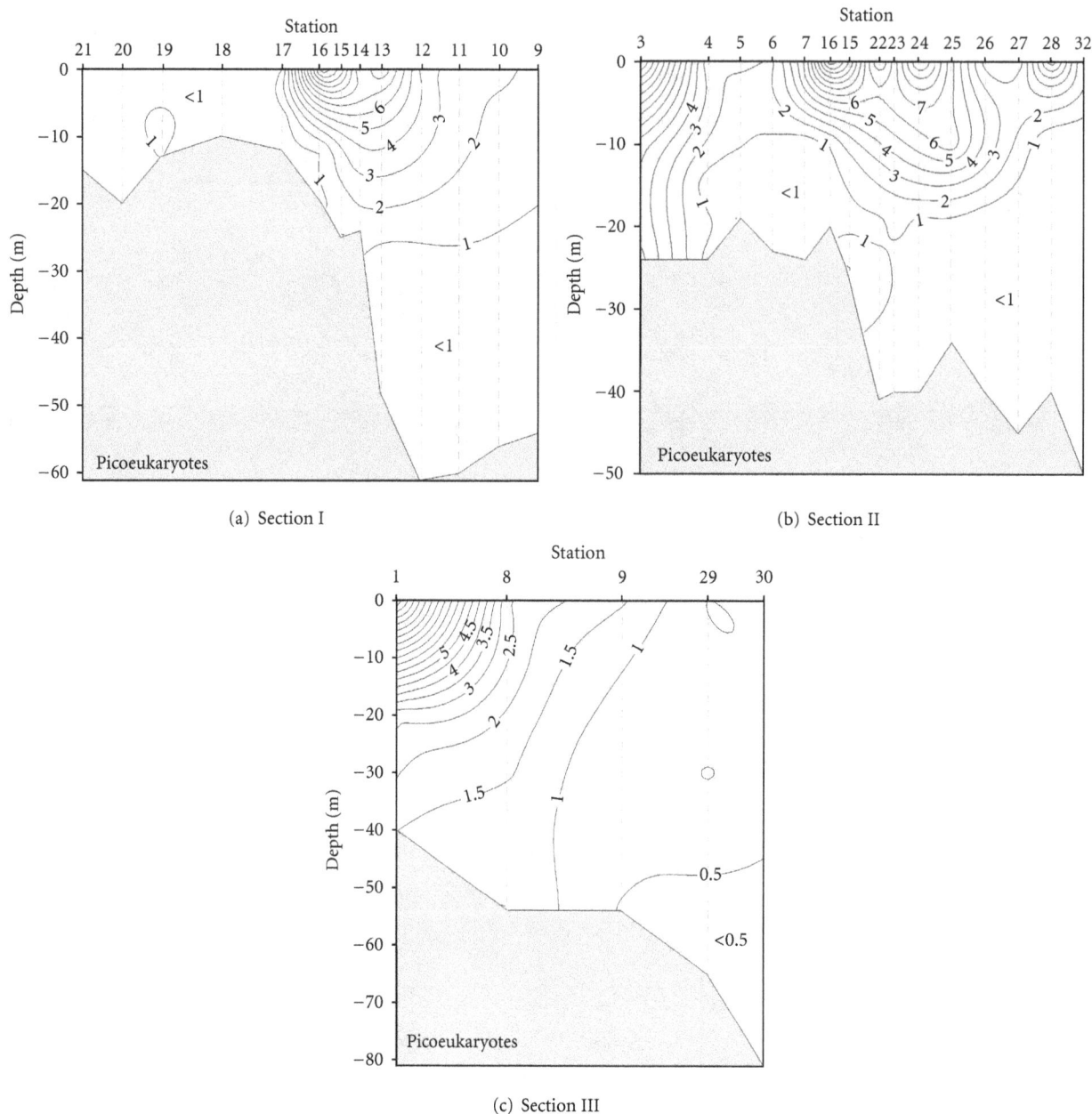

(a) Section I

(b) Section II

(c) Section III

FIGURE 3: Spatial distribution of picoeukaryotes ($\times 10^6$ cells L^{-1}) in the Changjiang estuary and its adjacent sea area in June 2006.

80°C before FCM analysis [15, 24]. Bacteria and viruses were characterized according to their distinctive side scatter and green fluorescence signals, which are related to their size and nucleic acid content, respectively [15, 21].

Yellow-green fluorescent beads, 1.002–2.139 μm in diameter (Polysciences, Inc.) were added as internal references to calibrate cell fluorescence emissions and light scatter signals and to allow comparisons of fluorescence and cell sizes among samples. All stock solutions were prefiltered through a filter (0.2 μm pore size) before use to avoid contamination. The raw data were processed using CELLQuest software (Becton Dickinson, San Jose, CA). Samples were measured in triplicate to give an estimated precision greater than 8.0% (relative standard deviation).

2.4. Data Analysis. The data in tables are expressed as mean ± SD. Coefficient of variation (CV) was calculated by CV = [SD/Mean] × 100%. Regression analysis using the software SPSS 13.0 (SPSS Inc., Chicago, IL) was applied to test if there was any significant relationship between the detected biological groups and the environmental parameters. Pearson's correlation analysis was used to assess major linear relations between the detected biological groups.

3. Results

3.1. Environmental Parameters. High values of CV indicated strong changes in salinity, turbidity, and nutrients in the area studied (Table 1). Salinities ranged from 0 to

(a) Section I

(b) Section II

(c) Section III

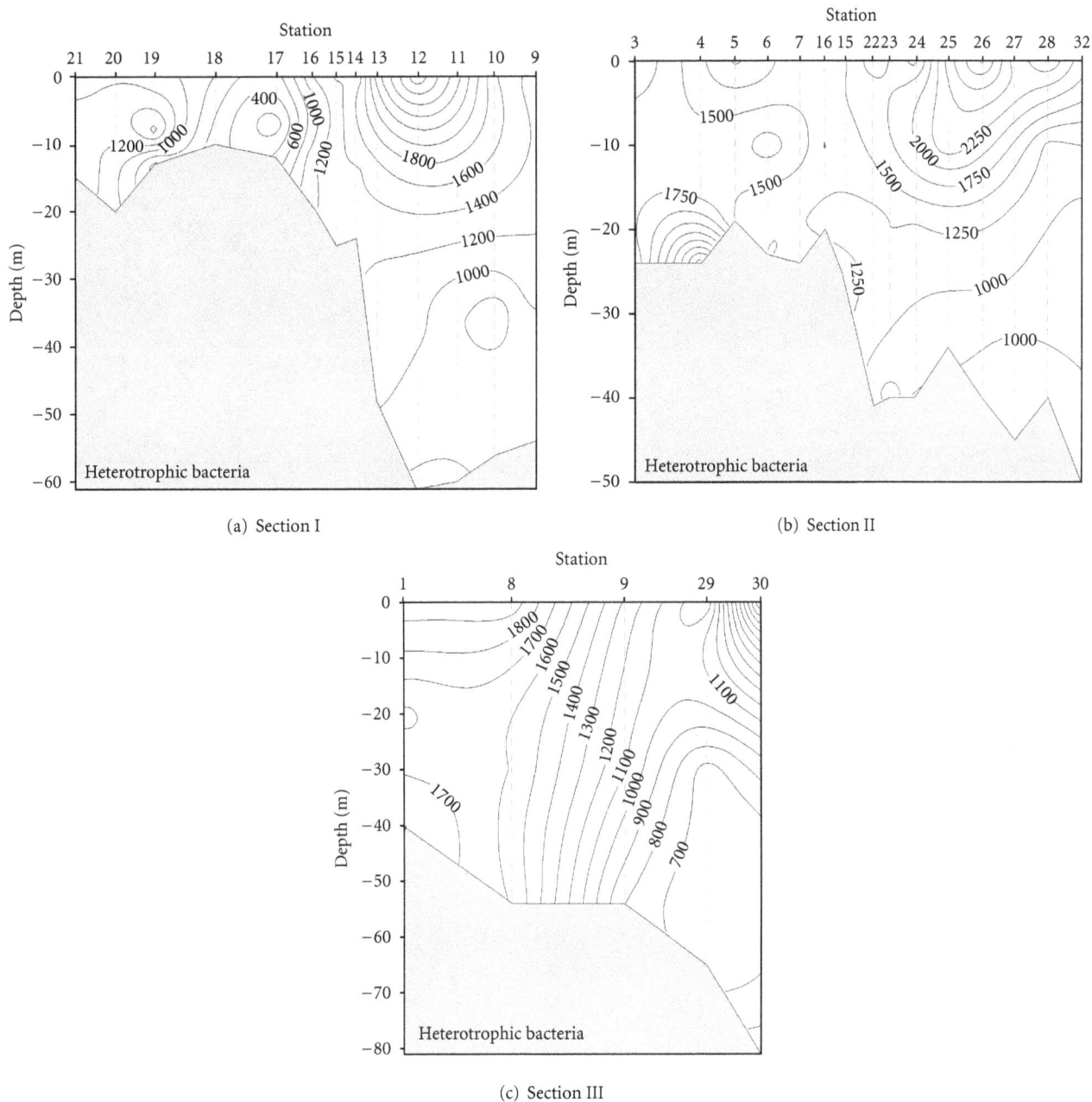

FIGURE 4: Spatial distribution of heterotrophic bacteria ($\times 10^6$ cells L^{-1}) in the Changjiang estuary and its adjacent sea area in June 2006.

34.42 psu (practical salinity unit) and obviously increased seaward. Turbidity and concentrations of inorganic nutrients decreased rapidly with the distance off the river mouth. The changes in these environmental parameters showed that freshwater from the Changjiang River dramatically affected the water quality in the estuary area, particularly at the stations near the river mouth.

3.2. Spatial Distributions of Picoplankton and Viruses. In the area studied, heterotrophic bacteria and viruses were ubiquitous. *Synechococcus* and picoeukaryotes could be detected except for a very few samples from stations in the river mouth

(Stations 16–18), while *Prochlorococcus* in all samples was below detectable level.

Average abundance of *Synechococcus* was 7.82×10^6 cells L^{-1}, with a range of 0–122.22×10^6 cells L^{-1} (Table 2). Its abundance exhibited distinct spatial differences (Figure 2). The abundance values were $<1 \times 10^6$ cells L^{-1} in the water around stations 17–21 and stations 5 and 6. At these stations near the river mouth, the abundance differences between the surface and bottom layers were relatively small because their shallow depth allows water to be mixed. At other stations outside the river mouth, the abundance was high at the surface layers and decreased with water depth.

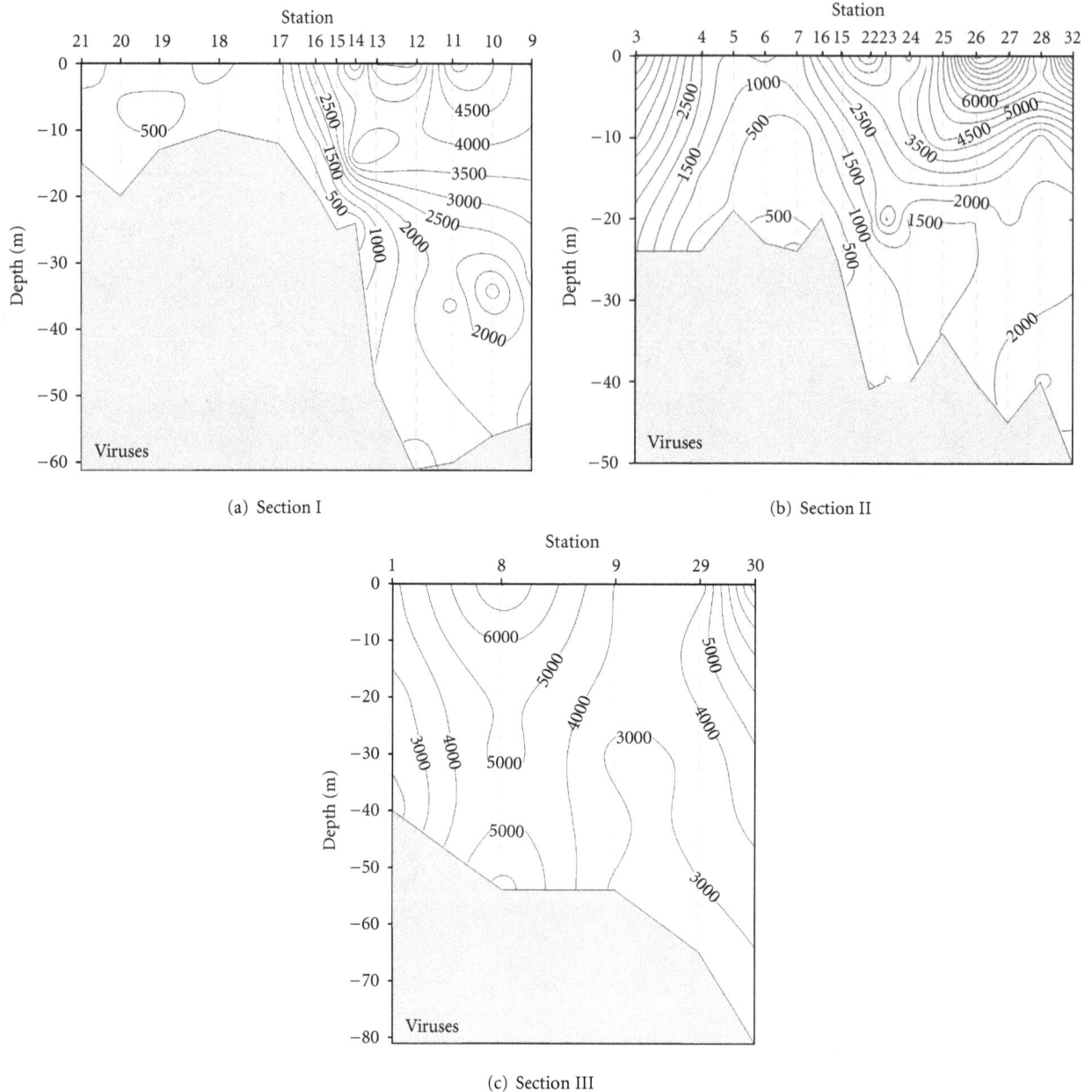

(a) Section I

(b) Section II

(c) Section III

FIGURE 5: Spatial distribution of viruses ($\times 10^6$ particles L^{-1}) in the Changjiang estuary and its adjacent sea area in June 2006.

Picoeukaryotes showed a lower average abundance (2.13×10^6 cells L^{-1}) than *Synechococcus* (Table 2). Its higher abundances ($>6 \times 10^6$ cells L^{-1}) were mostly recorded from the surface layers at stations 13–16 and 23–25 and the lower abundances ($<1 \times 10^6$ cells L^{-1}) in the water around stations 17–21 and 29–31 in the near bottom water around stations 9–13 and 22–28 (Figure 3). As a whole, the distribution of picoeukaryotes was uneven in different regions and water layers.

Heterotrophic bacteria were quite abundant, with two to three orders of magnitude higher in average abundance (1387×10^6 cells L^{-1}) than those of small autotrophs and showed least variation in abundant (Table 2). Viruses were nearly 2 times more abundance on average (2663×10^6 particles L^{-1}) than heterotrophic bacteria (Table 2). Spatial distribution of heterotrophic bacteria was similar to that of viruses (Figures 4 and 5). Commonly, their vertical distributions were almost even for the stations 17–21 near the river mouth, and for the other stations the abundances decreased with water depth.

4. Discussion

Spatial distributions of detected picoplankton groups and viruses featured distinct spatial variability, with different factors exerting control on their growths in different regions. In

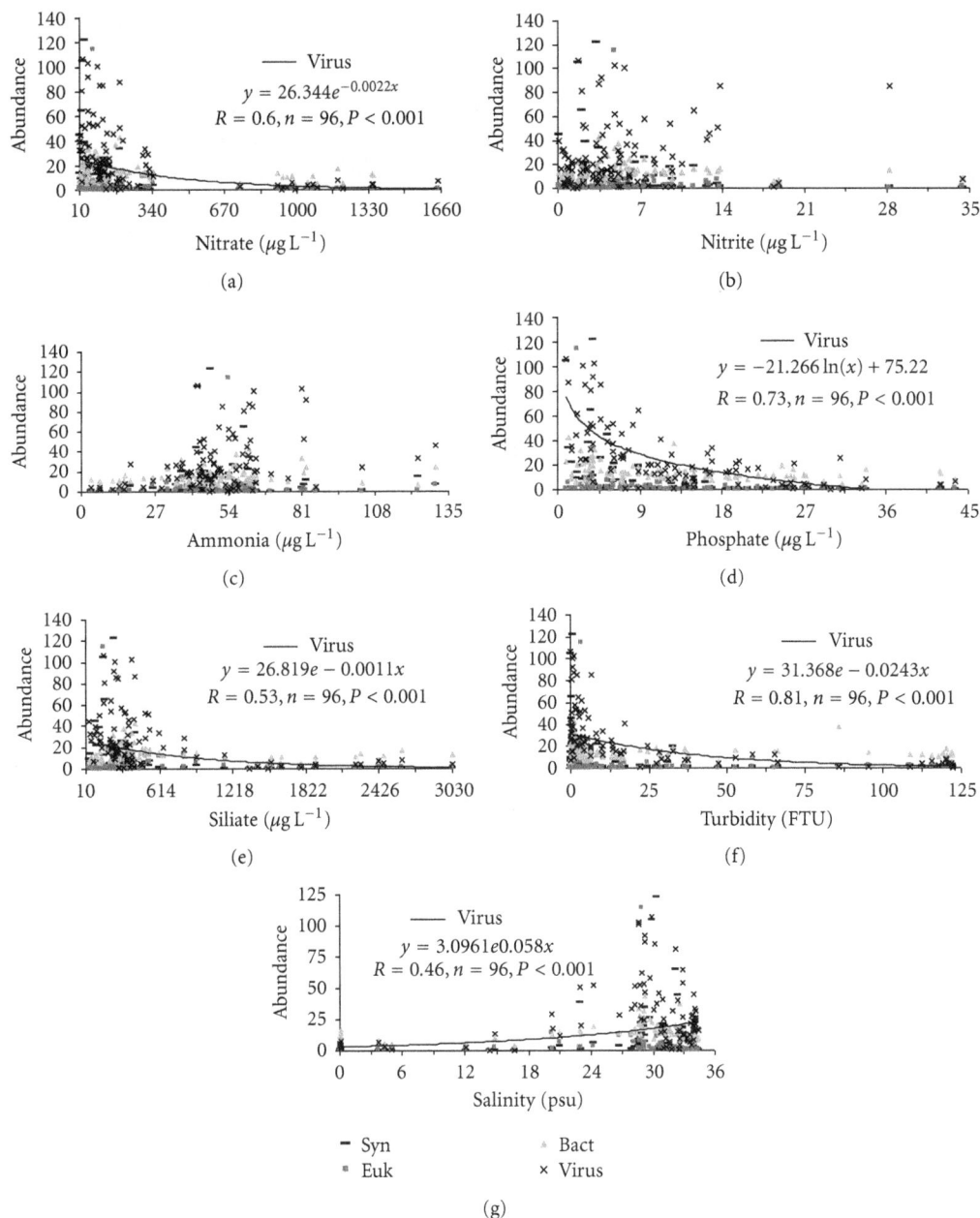

FIGURE 6: Relationships between the determined biological groups and the environmental parameters (Syn: *Synechococcus*; Euk: picoeukaryotes; Bact: heterotrophic bacteria).

the turbid river mouth, picoplankton and viral abundances were low, although the concentrations of nutrients were high (Figures 6(a)–6(e)). Unfavorable light conditions because of high turbidity exerted strong limitations to picoplankton [14, 25]. Further offshore, the abundances increased gradually, with suspended matter settling out. The maximum abundances occurred at about 150 km (this distance differs somewhat in different groups due to the difference among the groups in the sensitivity to environmental changes) from the river mouth, where turbidity was less than 5 FTU (Formazan Turbidity Units, Figure 6(f)), salinity varied from 28 to 32 psu (Figure 6(g)), and nutrients largely depleted

(Figures 6(a)–6(e)). Ning et al. [25] pointed out that the front of microzooplankton just appears in these regions because of the availability of abundant preys, which are not tested in this study, however.

Positive linear relations were found between *Synechococcus*, heterotrophic bacteria, and viruses (Figure 7), and these relations have been reported in many previous studies. Heterotrophic bacteria are dependent on substrates produced by the small primary producers like picophytoplankton (e.g., *Synechococcus*) due to a tight link between picophytoplankton and dissolved organic matter (DOM) production [13, 26, 27]. Viral lysis of heterotrophic bacteria

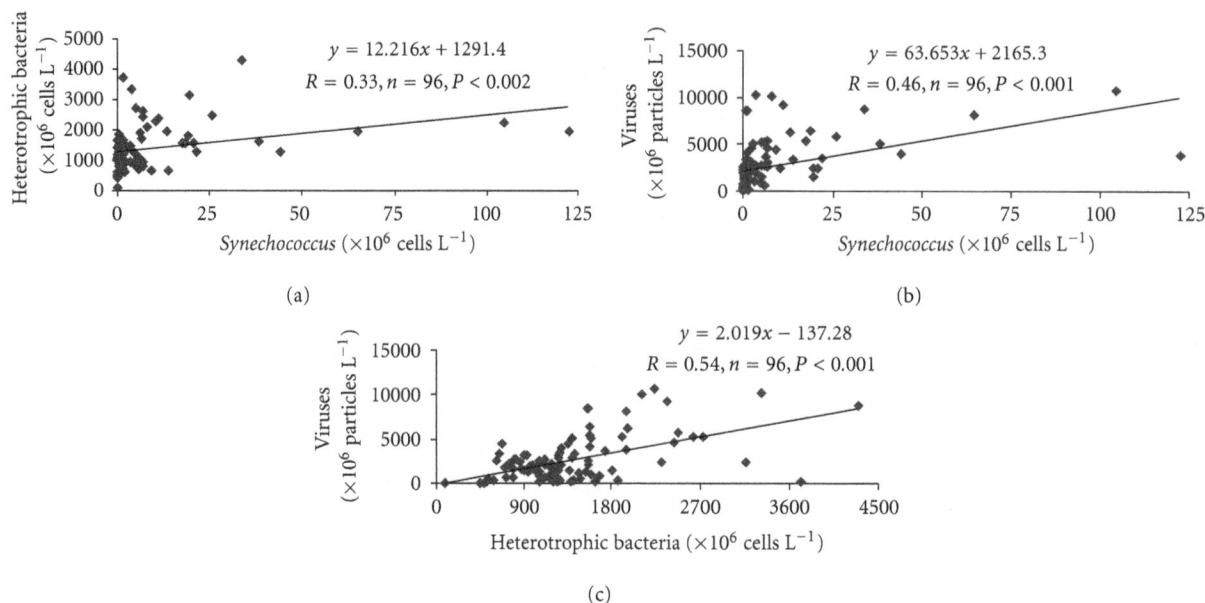

$$y = 12.216x + 1291.4$$
$$R = 0.33, n = 96, P < 0.002$$

(a)

$$y = 63.653x + 2165.3$$
$$R = 0.46, n = 96, P < 0.001$$

(b)

$$y = 2.019x - 137.28$$
$$R = 0.54, n = 96, P < 0.001$$

(c)

FIGURE 7: Linear relations between *Synechococcus*, heterotrophic bacteria, and viruses.

TABLE 2: List of the measured picoplankton ($\times 10^6$ cells L^{-1}) and viruses ($\times 10^6$ particles L^{-1}) in the Changjiang estuary and its adjacent sea area in June 2006.

Item	Section I	Section II	Section III	Total station
Number of sample	39	45	15	96
Prochlorococcus	Undetected	Undetected	Undetected	Undetected
Synechococcus				
Range	0.00–44.24	0.11–13.23	1.50–122.22	0.00–122.22
Mean ± SD	4.98 ± 10.03	2.35 ± 3.02	30.09 ± 38.02	7.82 ± 18.64
CV	201%	129%	126%	238%
Picoeukaryote				
Range	0.11–13.97	0.01–14.21	0.38–10.84	0.01–14.21
Mean ± SD	2.02 ± 3.11	2.81 ± 3.84	1.78 ± 2.61	2.13 ± 3.11
CV	154%	137%	147%	146%
Heterotrophic bacteria				
Range	84–3159	543–3706	626–2229	84–4287
Mean ± SD	1158 ± 567	1533 ± 637	1309 ± 541	1387 ± 694
CV	49%	42%	41%	50%
Virus				
Range	12–5762	85–10266	1479–10642	12–10642
Mean ± SD	1604 ± 1615	2523 ± 2498	4247 ± 2471	2663 ± 2574
CV	101%	99%	58%	97%

potentially benefits *Synechococcus* by supplying nutrients while simultaneously removing their potential competitors (heterotrophic bacteria) for inorganic and organic nutrients, which is consistent with the view that nutrient cycling by viral lysis of heterotrophic bacteria may control phytoplankton growth and ecosystem scale carbon production [28].

Meanwhile, viral abundance showed a positive correlation with salinity and negative correlations with turbidity and inorganic nutrient (nitrate, phosphate, and silicate) concentrations (Figure 6), indicating that viruses are sensitive to environmental changes [29, 30] and their abundance and dynamics may be influenced by these parameters to a large extent [19]. Over the past 20 years, considerable progress has been made in revealing the ecological role of viruses in aquatic ecosystems, but further studies are needed for understanding of the environmental controls on viral

TABLE 3: Abundances of picoplankton groups and viruses in the Changjiang estuary and its adjacent sea area from different studies.

Location	Year	Month	Abundances ($\times 10^6$ cells L^{-1})					Detection technique	Source
			Syn[a]	Pro[a]	Euk[a]	Bact[a]	Virus		
31–32°N, 121–124°E	1986	Jul	0.1–200					Epifluorescence microscope	[5]
25–33°N, 120–130°E	1998	Jul	<15	<120	<14	<540		Flow cytometry	[6]
27–32°N, 122–130°E	1998	Jul	0.6–19.7					Epifluorescence microscope	[9]
25–33°N, 120–128°E	2000	Oct-Nov	7.6–31.1			690 ± 1780		Epifluorescence microscope	[10, 11]
	2000	Oct-Nov	21 ± 45	6 ± 15	14 ± 15				
28–32°N, 122–129°E	2003	Sep	97 ± 335	53 ± 98	5 ± 10	733 ± 583		Flow cytometry	[13]
29–32°N, 122–124°E	2004	Sep-Oct	23.5 (1.83–270)	11.3 (0–210)	2.62 (0.14–9.34)	1000 (411–2530)	19000 (3440–45600)	Flow cytometry	[12]
29–32°N, 120–124°E	2006	Jun	7.82 (0–122.22)	0	2.13 (0.01–14.21)	1387 (84–4287)	2663 (12–10642)	Flow cytometry	This study

a Syn: *Synechococcus*; Pro: *Prochlorococcus*; Euk: picoeukaryotes; Bact: heterotrophic bacteria.

abundance, impacts of viral infection upon host community structure, and roles of viruses in biogeochemical cycles [20].

Prochlorococcus was below detectable level, which may be because of high turbidity and nutrients in this area studied. *Prochlorococcus* has been found to be more abundant in oligotrophic waters than in eutrophic waters [31]. Pan [12] indicated that *Prochlorococcus* only presents in waters of salinity >32.6 psu and suspended sediment concentration <72 g m^{-3} during a summer cruise in the Changjiang estuary and its adjacent sea area. The key limiting factor for the coastward distribution of *Prochlorococcus* in the East China Sea was considered to be the movements of the warm water currents, the Kuroshio and the Taiwan Warm Water Currents [6, 7].

The abundances of different picoplankton groups obtained in this study were different to a certain extent in the other studies conducted in the Changjiang estuary and its adjacent sea area (Table 3). Particularly, the abundance of viruses remarkably decreased compared to that reported in the literature [12]. The virus-to-bacterium ratio (VBR) was only 1.92, which is very low given the Changjiang estuary is a eutrophic coastal ecosystem, and a high VBR of 8.7 has been reported in previous study [32]. Jiao et al. [8] suggested that, although causes for the changes in the microbial community structure in the Changjiang estuary could be multiple, the sudden decrease of river runoff and an ensuing intrusion of East China Sea ocean currents were postulated to be among the major ones. The information obtained from this study provides an important and valuable base for evaluating the effects of environmental changes in this area, associated with the construction of the Three Gorges Dam, and long-term monitoring of picoplankton and viruses is necessary.

Acknowledgments

This work was jointly supported by the Ministry of Science and Technology of China (2002CB412405, 2004CB720505), the Ministry of Education of China (IRT0427), and the Project no. 2008M13 supported by the Special Research Fund for the National Non-Profit Institutes (East China Sea Fisheries Research Institute).

References

[1] F. Azam and R. E. Hodson, "Size distribution and activity of marine microheterotrophs," *Limnology and Oceanography*, vol. 22, pp. 492–501, 1977.

[2] J. Sieburth, V. Smetacek, and J. Lenz, "Pelagic ecosystem structure: heterotrophic compartments of the plankton and their relations to plankton size fractions," *Limnology and Oceanography*, vol. 23, pp. 1256–1263, 1978.

[3] J. B. Waterbury, S. W. Watson, R. R. L. Guillard, and L. E. Brand, "Widespread occurrence of a unicellular, marine, planktonic, cyanobacterium," *Nature*, vol. 277, no. 5694, pp. 293–294, 1979.

[4] S. W. Chisholm, R. J. Olson, E. R. Zettler, R. Goericke, J. B. Waterbury, and N. A. Welschmeyer, "A novel free-living prochlorophyte abundant in the oceanic euphotic zone," *Nature*, vol. 334, no. 6180, pp. 340–343, 1988.

[5] D. Vaulot and N. Xiuren, "Abundance and cellular characteristics of marine *Synechococcus* spp. in the dilution zone of the Changjiang (Yangtze River, China)," *Continental Shelf Research*, vol. 8, no. 10, pp. 1171–1186, 1988.

[6] N. Jiao, Y. Yang, H. Koshikawa, and M. Watanabe, "Influence of hydrographic conditions on picoplankton distribution in the East China Sea," *Aquatic Microbial Ecology*, vol. 30, no. 1, pp. 37–48, 2002.

[7] N. Jiao, Y. Yang, N. Hong et al., "Dynamics of autotrophic picoplankton and heterotrophic bacteria in the East China Sea," *Continental Shelf Research*, vol. 25, no. 10, pp. 1265–1279, 2005.

[8] N. Jiao, Y. Zhang, Y. Zeng et al., "Ecological anomalies in the East China Sea: impacts of the Three Gorges Dam?" *Water Research*, vol. 41, no. 6, pp. 1287–1293, 2007.

[9] T. Xiao, H. D. Yue, W. C. Zhang, and R. Wang, "Distribution of *Synechococcus* spp. and its role in the microbial food loop in the East China Sea," *Oceanologia ET Limnologia Sinica*, vol. 34, pp. 33–43, 2003.

[10] S. Sun, T. Xiao, and H. D. Yue, "Distribution character of *Synechococcus* spp. in the East China Sea and the Yellow Sea in autumn and spring," *Oceanologia ET Limnologia Sinica*, vol. 34, pp. 161–168, 2003.

[11] S. J. Zhao, T. Xiao, and H. D. Yue, "The distribution of heterotrophic bacteria and some related factors in the East China and Yellow Seas during fall," *Oceanologia ET Limnologia Sinica*, vol. 34, pp. 295–305, 2003.

[12] L. A. Pan, *Preliminary study of microbial community structure in the East China Sea shelf area and the frontal region of the northern part of the South China Sea*, Dissertation of Master Candidate, East China Normal University, 2005.

[13] L. A. Pan, L. H. Zhang, J. Zhang, J. M. Gasol, and M. Chao, "On-board flow cytometric observation of picoplankton community structure in the East China Sea during the fall of different years," *FEMS Microbiology Ecology*, vol. 52, no. 2, pp. 243–253, 2005.

[14] Y. Li, D. Li, T. Fang, L. Zhang, and Y. Wang, "Tidal effects on diel variations of picoplankton and viruses in the Changjiang estuary," *Chinese Journal of Oceanology and Limnology*, vol. 28, no. 3, pp. 435–442, 2010.

[15] D. Marie, C. P. D. Brussaard, R. Thyrhaug, G. Bratbak, and D. Vaulot, "Enumeration of marine viruses in culture and natural samples by flow cytometry," *Applied and Environmental Microbiology*, vol. 65, no. 1, pp. 45–52, 1999.

[16] Y. H. Yang, *Distribution of one group of virus in the eastern East China Sea: results from flow cytometry measurement*, Dissertation of Doctoral Candidate, Institute of Oceanology Chinese Academy of Sciences, 2000.

[17] N. Jiao, Y. Zhao, T. Luo, and X. Wang, "Natural and anthropogenic forcing on the dynamics of virioplankton in the Yangtze river estuary," *Journal of the Marine Biological Association of the United Kingdom*, vol. 86, no. 3, pp. 543–550, 2006.

[18] S. Gao and Y. P. Wang, "Changes in material fluxes from the Changjiang River and their implications on the adjoining continental shelf ecosystem," *Continental Shelf Research*, vol. 28, no. 12, pp. 1490–1500, 2008.

[19] H. Slováčková, *Study of the ecological role of viruses and bacteria in aquatic ecosystems*, Dissertation thesis, Masaryk University, Brno, Czech, 2008.

[20] S. Jacquet, T. Miki, R. Noble, P. Peduzzi, and S. Wilhelm, "Viruses in aquatic ecosystems: important advancements of the last 20 years and prospects for the future in the field of microbial oceanography and limnology," *Advances in Oceanography and Limnology*, vol. 1, no. 1, pp. 97–141, 2010.

[21] D. Marie, F. Partensky, S. Jacquet, and D. Vaulot, "Enumeration and cell cycle analysis of natural populations of marine picoplankton by flow cytometry using the nucleic acid stain SYBR Green I," *Applied and Environmental Microbiology*, vol. 63, no. 1, pp. 186–193, 1997.

[22] K. K. Cavender-Bares, E. L. Mann, S. W. Chisholm, M. E. Ondrusek, and R. R. Bidigare, "Differential response of equatorial Pacific phytoplankton to iron fertilization," *Limnology and Oceanography*, vol. 44, no. 2, pp. 237–246, 1999.

[23] G. A. Tarran, M. V. Zubkov, M. A. Sleigh, P. H. Burkill, and M. Yallop, "Microbial community structure and standing stocks in the NE Atlantic in June and July of 1996," *Deep-Sea Research Part II*, vol. 48, no. 4-5, pp. 963–985, 2001.

[24] C. P. D. Brussaard, D. Marie, and G. Bratbak, "Flow cytometric detection of viruses," *Journal of Virological Methods*, vol. 85, no. 1-2, pp. 175–182, 2000.

[25] X. R. Ning, J. X. Shi, Y. M. Cai, and C. G. Liu, "Biological productivity front in the Changjiang Estuary and the Hangzhou Bay and its ecological effects," *Acta Oceanologica Sinica*, vol. 26, pp. 96–106, 2004.

[26] H. Liu, K. Imai, K. Suzuki, Y. Nojiri, N. Tsurushima, and T. Saino, "Seasonal variability of picophytoplankton and bacteria in the western subarctic Pacific Ocean at station KNOT," *Deep-Sea Research Part II*, vol. 49, no. 24-25, pp. 5409–5420, 2002.

[27] H. Liu, M. Dagg, L. Campbell, and J. Urban-Righ, "Picophytoplankton and bacterioplankton in the Mississippi River plume and its adjacent waters," *Estuaries*, vol. 27, no. 1, pp. 147–156, 2004.

[28] M. G. Weinbauer, O. Bonilla-Findji, A. M. Chan et al., "*Synechococcus* growth in the ocean may depend on the lysis of heterotrophic bacteria," *Journal of Plankton Research*, vol. 33, no. 10, pp. 1465–1476, 2011.

[29] S. W. Wilhelm and C. A. Suttle, "Viruses and nutrient cycles in the sea," *BioScience*, vol. 49, no. 10, pp. 781–788, 1999.

[30] K. E. Wommack and R. R. Colwell, "Virioplankton: viruses in aquatic ecosystems," *Microbiology and Molecular Biology Reviews*, vol. 64, no. 1, pp. 69–114, 2000.

[31] F. Partensky, W. R. Hess, and D. Vaulot, "Prochlorococcus, a marine photosynthetic prokaryote of global significance," *Microbiology and Molecular Biology Reviews*, vol. 63, no. 1, pp. 106–127, 1999.

[32] N. Jiao, Y. Zhao, T. Luo, and X. Wang, "Natural and anthropogenic forcing on the dynamics of virioplankton in the Yangtze river estuary," *Journal of the Marine Biological Association of the United Kingdom*, vol. 86, no. 3, pp. 543–550, 2006.

19

Kelp Forests versus Urchin Barrens: Alternate Stable States and Their Effect on Sea Otter Prey Quality in the Aleutian Islands

Nathan L. Stewart[1] and Brenda Konar[2]

[1] School of Fisheries and Ocean Sciences, University of Alaska Fairbanks, 905 N. Koyukuk Drive, 245 O'Neill Building, Fairbanks, AK 99775, USA
[2] Global Undersea Research Unit, University of Alaska Fairbanks, 905 N. Koyukuk Drive, 217 O'Neill Building, Fairbanks, AK 99775, USA

Correspondence should be addressed to Nathan L. Stewart, nathan.stewart@tufts.edu

Academic Editor: Andrew McMinn

Macroalgal and urchin barren communities are alternately stable and persist in the Aleutians due to sea otter presence and absence. In the early 1990s a rapid otter population decline released urchins from predation and caused a shift to the urchin-dominated state. Despite increases in urchin abundance, otter numbers continued to decline. Although debated, prey quality changes have been implicated in current otter population status. This study examined otter prey abundance, size, biomass, and potential energy density in remnant kelp forest and urchin-dominated communities to determine if alternate stable states affect prey quality. Findings suggest that although urchin barrens provide more abundant urchin prey, individual urchins are smaller and provide lower biomass and potential energy density compared to kelp forests. Shifts to urchin barrens do affect prey quality but changes are likely compensated by increased prey densities and are insufficient in explaining current otter population status in the Aleutians.

1. Introduction

Natural communities can exist at multiple stable points in time or space [1]. Stable points are characterized by a specific structural and functional species assemblage recognizably different from other assemblages that can occur under the same set of environmental conditions. Such states are non-transitory, persist over ecologically relevant timescales, and are therefore considered domains of stable equilibrium [2, 3]. Although multiple stable states can exist simultaneously, communities typically alternate from one stable state to another, a shift often conveyed by a large perturbation applied directly to the state variables (e.g., population densities; [4]). Significant changes in the abundance of key species are widely cited as evidence of phase shifts ([5, 6] but see [7]) and have been documented both experimentally [8] and empirically [9, 10] in coastal marine ecosystems. In general, predator removal causes prey community shifts enabling one or few algal or invertebrate competitive dominants to proliferate.

In ecological studies in the Aleutian Islands, the presence and absence of dense sea otter populations can instigate state shifts between two alternately stable nearshore communities, one dominated by kelp and the other by sea urchins [11–13]. With sea otters present, sea urchins are reduced to sparse populations enabling kelps to flourish. With sea otters absent, dense sea urchin populations overgraze and exclude foliose macroalgae. In the early 1990s, a rapid sea otter population decline (ca. 25% per year) caused a shift in alternate stable states in the region, resulting in much of the nearshore rocky ecosystem to be dominated by urchin barrens and largely devoid of macroalgae [14]. Although urchin biomass increased during the decline [15], the sea otter population continued to decline (from ca. 77,435 in 1990 to 17,036 in 1997) and has remained at low densities (ca. 4.3 otters per km² in 1965 to 0.5 otters per km² in 2000) in the two decades since the decline [14, 16, 17]. The cause of the initial decline remains debated (starting with [15, 18, 19]), and has manifested in a debate involving two fundamentally different processes, bottom-up and top-down forcing [20]. In general, bottom-up forcing hypotheses posit that the sea otter decline is due to changes in the availability or quality of prey (i.e., the nutritional limitation hypothesis). In contrast, top-down hypotheses posit that the decline is predator-mediated (i.e., the killer whale predation hypothesis). Sea otter diets at the population level are diverse [21] and it is argued that nutritional stress arising from changes in prey quality rather than prey quantity has not been sufficiently tested [22]. Nutritional limitation is one explanation for the decline of Steller sea lions [23, 24] and other marine predators [25] in the North Pacific and Bering Sea based on a shift from energy-rich prey to abundant energy-poor prey (the junk food hypothesis; [23, 26]). The degree to which shifts between kelp forests and urchin barrens affect prey quality and whether or not such changes could have initiated sea otter population declines and continue to limit sea otter recovery in the central and western Aleutians remains to be evaluated.

Kelp forest systems provide critical resources to nearshore marine communities in the central and western Aleutian Islands and throughout the temperate coastal zones [27, 28]. Principal resources include physical structure (habitat) and food (both directly and indirectly). Kelp forests dampen wave propagation and can mitigate the associated processes of coastal erosion, sedimentation, benthic productivity, and recruitment [29]. In addition, kelp canopies can influence interspecific algal competition by attenuating sunlight [30] and creating habitat for low-light-adapted species [31]. The structural complexity of macroalgal systems provides substratum for numerous sessile animals and algae [32, 33] and habitat for mobile organisms specialized to live and feed directly on the kelp or kelp-associated assemblages [34, 35]. Although kelps are highly productive, nutrients are primarily made available through macroalgal detritus [36, 37], while relatively little kelp production (≤10%) is consumed directly by herbivores [38]. Thus, kelp systems affect the abundance and biomass of associated species and mitigate ecological and oceanographic processes important to nutrient transfer to higher trophic levels.

Broad scale kelp deforestation can result from disease, herbivory, or physiological stress [32, 39, 40]. At lower latitudes, periodic kelp forest deforestation results from oceanographic anomalies in temperature, salinity, or nutrients that either kill kelps directly or trigger diseases that become lethal to algae [30, 38]. Coastal warming can also lead to increases in herbivory at lower latitudes [41–43]. In contrast, at higher latitudes sea urchin herbivory has been the most common agent of kelp deforestation and, despite morphological and chemical defenses in kelps, often leads to the formation of barren grounds [40, 44–47]. Intensive sea urchin grazing has both immediate, direct effects on the algal assemblage and numerous complex indirect effects on the greater community [11, 45, 48]. Although constituent species may remain the same, kelp forest and urchin barren systems support notably different assemblages in terms of species abundance, biomass, size distribution, and individual health [15, 49]. In general, relatively few epibenthic invertebrates succeed in urchin barrens and sea urchins themselves, the competitive dominant, are likely food ([50], but see [51, 52]) and size limited [53, 54]. Sea urchins size limitation in urchin barrens has

been attributed to both the natural organization of urchin feeding aggregations (e.g., larger urchins lead feeding fronts in kelp beds and smaller urchins occupy adjacent barren zones; [54]) and poor nutritional resources in barrens [52]. The lack of structural habitat complexity associated with urchin barrens can lead to increased predation and further affect prey abundance, biomass, and size. Experimental studies with tethered crabs and observational studies of fishes in kelp beds of varying complexity have shown that predation rates are a function of both kelp presence and architectural complexity. In general, larger and more abundant crabs and fishes are associated with more complex algal structure [55, 56]. Consequently habitats lacking kelp harbor smaller prey and are relatively unproductive compared to those with kelp [13].

Shifts to urchin barren stable states often entail an ecosystem service and function loss (for review see [6]). This is seen in the nearshore where decreases in the proportion of kelp to barrens have led to coastal consumer decreases [37], reduced interaction strengths between predatory sea stars and their invertebrate prey [57], and altered fish abundances and diets [58]. In the Aleutian Islands, kelp removal by sea urchins had negative effects on bald eagle, glaucous-winged gull, benthic-feeding sea duck, harbor seal and fish abundance [11, 59, 60], declines attributed to poor nearshore energy returns, and kelp forest habitat loss. Predator declines initiated by phase shifts have been linked to diminished prey resources in many nearshore marine systems [9, 61, 62]. To date, research focusing on alternate stable states has predominantly used predator abundance, diet analyses, or behavior to describe cascading effects associated with shifts to "less-desirable" states [6]. Very few studies have focused directly on individual prey attribute changes associated with phase shifts. Notable exceptions include documented declines in gamete production [63], prey palatability [64], and altered growth rates [65]. Although several studies have used sea urchin gonad indices to test for food limitation in urchins [50, 52], no studies to date have described changes in the biomass, size, and potential energy density of prey associated with shifts between kelp forests (productive systems) and urchin barrens (less-productive systems). Sea urchins have been shown to exhibit slower growth rates in barrens compared to kelp beds [66]. The persistence of remnant kelp forests in the in the current urchin barren dominated stable state is likely maintained by physical processes such as algal whipping [50]; however the long-term stability of these communities is largely unknown. The cooccurrence of both remnant kelp forests and urchin barrens in the central and western Aleutians provides an opportunity to evaluate prey quality in each community and to evaluate the hypothesis that prey quality changes initiated the sea otter decline and continue to limit their recovery.

This study quantified sea otter prey quality in remnant kelp forest and urchin barren communities across a longitudinal gradient in the central and western Aleutians to determine if prey quality is affected by phase shifts and if these changes could feasibly limit sea otter recovery. Three hypotheses were developed: (1) remnant kelp forests will provide greater individual prey biomass than urchin barrens, (2) kelp forests will provide greater prey energy density per unit area than urchin barrens, and (3) sea otters foraging in kelp forests require less predicted feeding effort to meet daily energy requirements than sea otters foraging in urchin barrens. To test these hypotheses, and the feasibility of nutritional limitation, sea otter prey abundance, biomass, size, and energy density were evaluated and then related to a foraging sea otter's daily energetic costs and to prey values from elsewhere in the sea otters range where populations are increasing or stable.

2. Methods

This study was carried out at eight central and western Aleutian Islands in Alaska (Figure 1). The study spanned a 460 nm longitudinal gradient, from Atka Island (52° 20′ 6N, 174° 7′ 1W) to Alaid Island (52° 45′ 2N, 186° 5′ 8E), and was sampled in June of 2009 and July of 2010. Sites ($n = 8$) were selected based on a definitive kelp-barren interface (≥ 30 m long) containing dense understory kelp (≥ 5 stipes·m^{-2}). Sites were sampled at mean sea otter foraging depths (10–15 m, [67]) and consisted of continuous bedrock or large stable boulder substratum. Cryptic habitats such as deep crevices or loosely piled boulders capable of harboring small sea otter prey species were rare or absent.

To determine if alternate stable state communities provide similar sea otter prey abundance, size, and biomass, forty randomly placed 0.25 m^2 quadrats were sampled within urchin barrens ($n = 20$) and adjacent kelp forests ($n = 20$) at each island. All kelp stipes, including *Laminaria saccharina*, *Agarum cribrosum*, *Thalassiophyllum clathrus*, *Cymathere triplicate*, and *Laminaria yezoensis*, occurring within quadrats were counted in kelp forests before sampling to ensure minimum kelp density requirements were met. Sea otter prey species, which included sea urchins (*Strongylocentrotus polyacanthus*), mussels (*Mytilus trossulus*), rock jingles (*Pododesmus machrochisma*), discordant mussels (*Musculus discors*), and hairy tritons (*Fusitriton oregonensis*), occurring within quadrats were counted. Scat analysis and focal observations suggest that these species are the dominant sea otter prey in the region [13, 45]. Fishes are a component of sea otter diets in the region but are generally less preferred prey [45, 60] and, in the case of smooth lumpsuckers, are episodic in their contribution sea otter diet [68] and were therefore not included in this analysis. Prey size was determined by measuring the maximum test or shell linear diameter of all prey encountered within each quadrat. In addition to counts and size measurements within quadrats, subsamples of sea urchins ($n = 10$ per community per island) and other prey species ($n = 5$ per community per island due to lower abundances) were collected from each quadrat (ADFG Permit No. CF-08-016 and CF-09-028). Only the largest individuals of each species occurring within quadrats were collected to simulate size selective foraging behavior exhibited by sea otters [69, 70]. Biomass per individual prey species was determined using test or shell-free wet weight from collected prey and is expressed in terms of g wet mass (WM) per individual. Biomass per unit area (g WM·0.25 m^{-2}) was calculated using species-specific size to biomass conversion factors [71, 72] and calibrated using size to biomass values from specimens collected in this study.

To determine if alternate stable state communities provide similar sea otter prey energy density per individual and per unit area, the caloric content of sea otter prey species was determined using bomb calorimetry. In preparation for ash weighing and caloric content analysis, a random subsample of test and shell-free wet samples from collected prey ($n = 3$ per species per community per island) were oven dried at 110° for 24 h and finely pulverized into powder. Ashing was carried out in a muffle furnace at 500° for 4 h. Weight loss from ashing was regarded as organic content and used to express the caloric content in terms of ash-free dry weight (AFDW). Homogeneous dry samples were formed into pellets and calorimetric determinations were made with both a Parr model 6200 Isoperibol bomb calorimeter with an 1108 oxygen bomb and 6510 water handling system. Energy from dry matter (cal·g^{-1} DM) was then multiplied by the proportion of dry matter in the wet mass to express potential energy density in terms of wet mass (kcal g^{-1} WM) per individual prey. Potential prey energy density per unit area (kcal·g^{-1} WM·0.25 m^{-2}) was calculated using species-specific biomass to energy density conversion factors [71, 72] and calibrated using values from specimens collected in this study. Although wet mass is influenced by ash and water dilution, it is a better representation of the actual prey biomass consumed by sea otters [72].

To determine if sea otters foraging in kelp forests require less predicted feeding effort to meet the daily energy requirements than sea otters foraging in urchin barrens, sea otter prey variables were compared to the activity budget and metabolic rate of a typical 34 kg male sea otter from the central and western Aleutians [73]. Prey abundance, size, and energy density were related to sea otter daily energy requirements to calculate predicted feeding effort required to meet daily caloric needs foraging in each community. Predicted feeding effort was calculated in terms of both percent time needed in a 24-hour period to meet daily caloric needs and in terms of the number of individual urchins needed to meet daily caloric needs. Sea otter prey assimilation efficiency was standardized at 82% efficiency [74] and feeding rates were standardized at 1.9 urchins·min^{-1} for kelp forests and 3 urchins·min^{-1} for urchin barrens, respectively (USGS, unpublished data).

To determine if prey quality values in the central and western Aleutians are limiting current sea otter recovery, potential prey energy density values measured in this study were compared to species-specific energy density determinations from locations where sea otter populations are currently increasing (Kachemak Bay Alaska; Stewart and Konar, unpubl. data, [75], Glacier Bay Alaska; [72, 76], and San Nicholas Island California; [72, 77]) or stable (Monterey Bay California; [21, 72]). Direct comparisons between sea otter prey species were made at the species level with the exception of *Strongylocentrotus polyacanthus* (Aleutians) and *S. droebachiensis* (elsewhere in Alaska and California) due to the similarities in the size and potential energy density of these two species [78].

2.1. Data Analysis. Differences in prey abundance, biomass, size, and energy density between communities were examined using ANOVA ($\alpha = 0.05$) with communities as

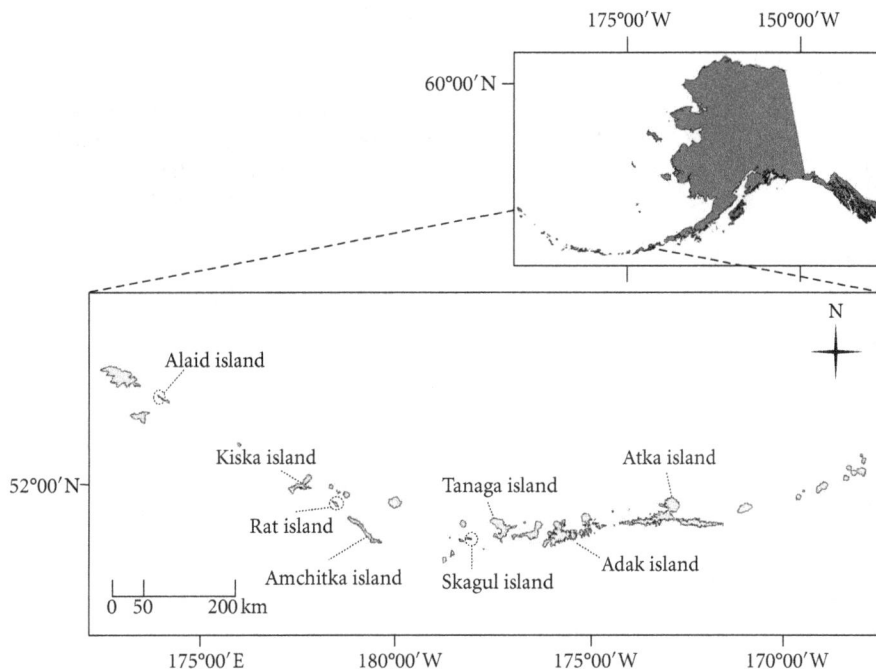

FIGURE 1: Map of the central and western Aleutian Islands, Alaska (inset), indicating the eight islands sampled in this study.

TABLE 1: Kelp density and sea otter prey abundance (individuals per 0.25 m^2) in kelp forest and urchin barren communities in the central and western Aleutian Islands, Alaska. N is the mean of prey species counted within 0.25 m^2 quadrats in each community (n = 8 islands with 20 quadrats per community per island). Prey species include *Strongylocentrotus polyacanthus (Strongylo.)*, *Pododesmus machrochisma (Pododes.)*, *Mytilus trossulus (Mytilus)*, *Musculus discors (Muscul.)*, and *Fusitriton oregonensis (Fusitrit.)*. Significant differences from Tukey post hoc comparisons $P \leq 0.05$ level are indicated by ($*$). NP: Not present.

	Kelp density (inds.$\cdot$0.25 m^{-2})	N	Abundance (inds.$\cdot$0.25 m^{-2})				
			Strongylo.	*Pododes.*	*Mytilus*	*Muscul.*	*Fusitrit.*
Kelp	5.89 ± 1.14	1.6 ± 1.1	3.8 ± 1.9*	2.3 ± 1.2	0.9 ± 0.7	0.8 ± 1.1	0.2 ± 0.1
Barren	NP	6.5 ± 2.3	28.6 ± 5.8	2.5 ± 1.5	1.1 ± 0.6	NP	0.3 ± 0.1

treatments (e.g., remnant kelp forests and urchin barrens) and means from quadrats within communities as replicates. When significant effects were found in ANOVA, post hoc comparisons were made using Tukey's Honestly Significant Difference (HSD) test. Mulitvariate analysis was used to illustrate differences in urchin barren and kelp prey communities attributable to prey availability, quality, and size (PRIMER-E v.6, [79, 80]). Prior to analyses, data were square root transformed to reduce the dominant contributions of abundant species and a similarity matrix of all samples was produced using a Bray-Curtis index. The similarity between urchin barren and kelp communities was assessed in terms of prey variables using multidimensional scaling ordination. Similarity percentages analysis (SIMPER) was used to determine which taxa contributed most to the observed dissimilarity among urchin barren and kelp communities represented by the euclidean distances between sites.

3. Results

Kelp forest and urchin barren prey communities were clearly delineated by differences in kelp and macroinvertebrate abundance (Table 1). Urchin barrens provided significantly more abundant prey than remnant kelp forests (ANOVA, n = 8, F = 132.1, $P < 0.001$). Tukey post hoc comparisons indicated that *Strongylocentrotus polyacanthus* abundance was significantly different between kelp forests and urchin barrens. Dense *S. polyacanthus* populations comprised the bulk of available prey (45.5 to 87.9%) in both communities but were seven times more abundant in barrens (28.6 ± 5.8 inds.$\cdot$0.25 m^{-2}) than in remnant kelp forests (3.8 ± 1.9 inds.$\cdot$0.25 m^{-2}). When present, *Musculus discors* was more abundant in kelp forests (0.8 ± 1.1 inds.$\cdot$0.25 m^{-2}) than urchin barrens (none present); however this species only occurred at three of the eight islands. The rock jingle, *Pododesmus machrochisma*, was consistently abundant at low densities in both communities. The remaining prey species, *Mytilus trossulus* and *Fusitriton oregonensis*, showed patchy distributions or were equally present in both kelp forests and urchin barrens. No significant differences were detected in nonurchin prey size kelp forest and urchin barren communities (ANOVA, n = 8, F = 2.02, P = 0.33).

Kelp forests supported significantly larger urchins (54.1 ± 21.4 mm) than barren habitats (47.1 ± 17.3 m; ANOVA,

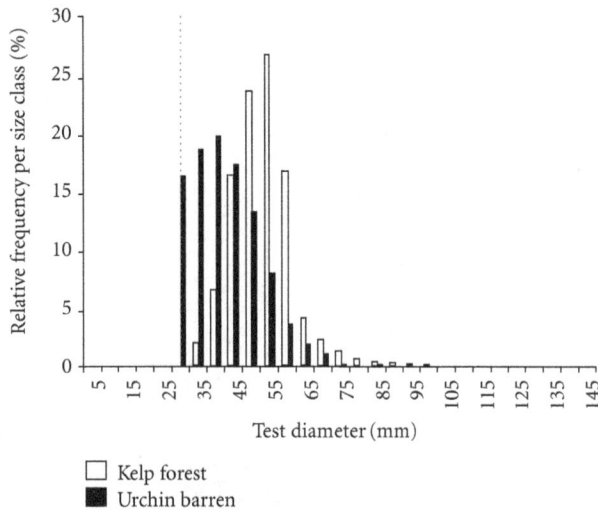

FIGURE 2: Relative frequency distributions of sea urchins (*Strongylocentrotus polyacanthus*) collected in remnant kelp forest and urchin barren communities in the central and western Aleutian Islands. Relative frequency percentages are determined using counts from sea urchins in kelp forests ($n = 308$) and urchin barrens ($n = 4569$). The dotted line indicates the minimum size threshold for sea otter predation on urchins (30 mm).

$n = 8$, $F = 5.34$, $P = 0.02$; Figure 2). Kelp forests provided significantly higher biomass per individual urchin than barrens (ANOVA, $n = 8$, $F = 39.1$, $P = 0.016$; Figure 3). Urchin barrens, however, provided significantly higher biomass per unit area compared to remnant kelp forests (ANOVA, $n = 8$, $F = 97.9$, $P < 0.001$). Although constituent reproductive tissue biomass was significantly greater in individual urchins in kelp forests (0.05 ± 0.02 kcal·g^{-1} per urchin) than urchin barrens (0.02 ± 0.01 kcal·g^{-1} per urchin, ANOVA, $n = 8$, $F = 22.2$, $P = 0.04$), differences in available reproductive tissue per unit area were not significant (ANOVA, $n = 8$, $F = 0.84$, $P = 0.31$). Individual urchins in kelp forests provided significantly higher potential energy density (0.21 ± 0.02 kcal·g^{-1} per urchin) compared to urchin barrens (0.14 ± 0.08 kcal·g^{-1} per urchin, ANOVA, $n = 8$, $F = 26.6$, $P = 0.03$; Figure 4). In contrast, urchin barrens provided significantly greater potential energy per unit area than kelp forests (ANOVA, $n = 8$, $F = 107.2$, $P < 0.001$). Potential energy density values of individual prey species (kcal·g^{-1} WM·ind.$^{-1}$) from both communities in this study were comparable to values from other studies conducted elsewhere in the sea otters range (Table 2).

A typical sea otter in the central and western Aleutians could easily meet daily energy requirements foraging in either kelp forest or urchin barren communities. A 34 kg male sea otter has a daily energy requirement of approximately 4600 kcal·day^{-1} [73]. Due to differences in the abundance, size, and energy density of urchins from remnant kelp forests and urchin barrens, a typical Aleutians sea otter would need to consume 484 urchins in a kelp forest (18% time, feeding rate of 1.9 urchins·min^{-1}) versus 1085 urchins (25% time, feeding rate of 3 urchins·min^{-1}) in an urchin barren to

meet daily caloric needs [72, 73]. The differences in percent time required to meet daily caloric needs foraging in either community are well below those seen in populations where food resources are limiting, such as central California where male sea otters spent 25–40% time feeding [73, 77]. In addition, given comparable feeding rates foraging in either community, the number of urchins required to meet daily caloric needs is well within the actual number observed in empirical studies of foraging sea otters (e.g., [68]).

Sea otter prey abundance, size, biomass, and energy density contribute to the separation in urchin barren and kelp forest prey communities in multidimensional scaling analyses (MDS; Figure 5). The separation between prey communities was driven by significantly higher total prey biomass and potential energy per unit area associated with dense sea urchin populations in urchin barrens (Figure 6, SIMPER, 87%). Individual urchin energy density did not contribute significantly to the separation in sites (SIMPER, 9%).

4. Discussion

In its current stable state, the nearshore community in the central and western Aleutian Islands is dominated by abundant but low-quality prey. Expansive urchin barrens support dense sea urchin populations that are generally contain smaller individuals and provide less biomass and energy density per individual than kelp forest urchins. Interspersed in the system is a patchwork mosaic of remnant kelp forests that support relatively few but large, calorically rich sea urchins. Though statistically significant, the difference in individual sea urchin potential energy density in kelp forests and urchin barrens is likely to be biologically inconsequential to foraging sea otters. Potential urchin energy density values measured in this study indicate that an average kelp forest urchin is equal to approximately one and a half barren urchins in terms of edible wet biomass and energy content. Given sea urchin feeding rates and assimilation efficiency [73], sea otter daily energy requirements are easily met foraging in either community. Although individual prey quality changes likely occurred during the shift from kelp forests to urchin barrens during the 1990s sea otter decline, these changes are unlikely to have caused the sea otter decline nor are they limiting current sea otter recovery. Both kelp forest and urchin barren urchins sampled in this study are comparable to potential energy density values of individual urchins elsewhere in the sea otter's range [72]. In addition, all other sea otter prey evaluated in this study, with the exception of *Musculus discors*, did not vary in abundance, size, biomass, or energy density between kelp and barren communities. Consequently, though changes in prey quality associated with phase shifts represent an ecosystem service loss to predators, this loss is likely compensated by increases in prey abundance and total available biomass.

Sea urchins competitively dominate nearshore communities in the central and western Aleutian Islands. Their dominance in the absence of top-down control is typical of urchin barren phase states elsewhere [9, 43, 81] and is comparable to other competitive dominants in marine systems where

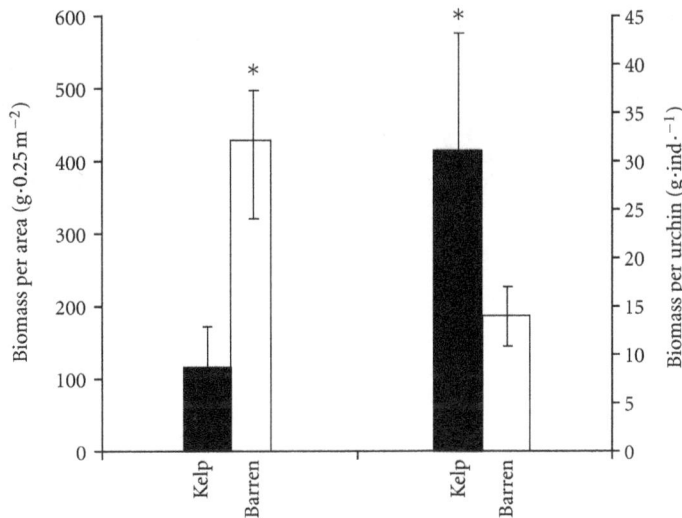

FIGURE 3: Sea urchin biomass per unit area and biomass per individual urchin from remnant kelp forest and urchin barren communities throughout the central and western Aleutian Islands. Significant differences at $P \leq 0.05$ level are indicated by (*).

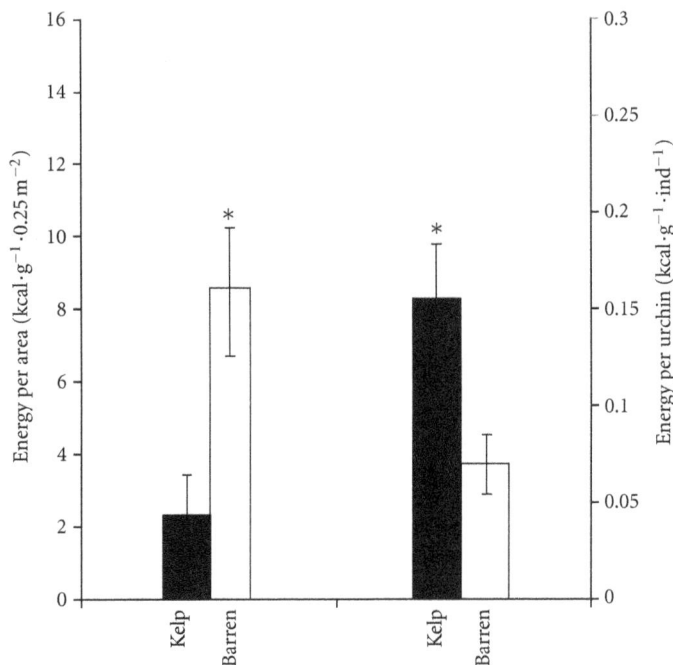

FIGURE 4: Sea urchin potential energy density per unit area and potential energy density per individual urchin from remnant kelp forest and urchin barren communities throughout the central and western Aleutian Islands. Significant differences at $P \leq 0.05$ level are indicated by (*).

predators have been experimentally removed, including barnacles [82] and mussels [83]. Urchin densities on barrens as sampled in this study were approximately seven times greater than those in remnant kelp forests, a pattern common throughout the Aleutian Islands during phase shifts to urchin barrens [13, 15]. Kelp forest-associated urchins were often found in the kelp blades and less commonly observed on the substrate, were larger, and provided significantly more biomass per urchin than urchins in barrens. Urchins associated with barrens, in addition to being smaller, were notably diminished in wet tissue mass and generally contained very

little to no reproductive tissue compared to kelp forest urchins. Mass differences between individual urchins were attributed both to size differences and to differences in the ratio of reproductive to nonreproductive tissue, a variable known to decrease with increasing urchin density [63] and increase with increasing macroalgal food sources [50]. Urchin reproductive tissue is significantly more energy dense than other tissues [84] and in this study translated into significantly higher energy density per individual urchin. Consequently, although urchins in kelp forests are significantly less abundant than those in urchin barrens, they provided more

TABLE 2: Individual prey quality values from remnant kelp forest and urchin barrens sampled in this study and from locations elsewhere in the sea otters range. Values indicate the potential energy density per individual species (kcal·g^{-1} wet mass) ± 1 S.D. Prey species include *Strongylocentrotus polyacanthus*, *Mytilus trossulus*, and *Fusitriton oregonensis*. *S. polyacanthus* is most common in the Aleutians but is compared to *S. droebachiensis* from other regions in this table given their similarity in mitochondrial DNA analysis [78]. Current population trends were referenced in Estes et al. 2005 [17] (C. and W. Aleut., AK: central and western Aleutians, AK), Gill et al. 2009 [75] (KBay, AK: Kachemak Bay, AK), Bodkin et al. 2003 [76] (GBNP, AK; Glacier Bay, AK), Estes et al. 2003 [21] (MBNMS, CA: Monterey Bay, CA), and Tinker et al. 2008 [77] (SNI, CA: San Nicholas Is., CA). Sources of regional prey values are (A) this study; (B) Stewart and Konar, unpublished data; and (C) Oftedal et al. 2007. Sea otter population status from each location is either (D) declining, (I) increasing, or (S) stable. NP: Not present.

Location	Source	Sea otter pop. status	Prey Species		
			Strongyloc.	*Mytilus*	*Fustitrit.*
C., W. Aleut., AK					
kelp forest	A	D	0.21 ± 0.02	0.41 ± 0.06	1.36 ± 0.06
urchin barren	A	D	0.14 ± 0.08	0.47 ± 0.04	1.41 ± 0.03
KBay, AK	B	I	0.26 ± 0.06	0.36 ± 0.02	1.09 ± 0.08
GBNP, AK	C	I	0.24 ± 0.04	0.33 ± 0.06	1.11 ± 0.23
MBNMS, CA	C	S	0.39 ± 0.04	0.55 ± 0.04	NP

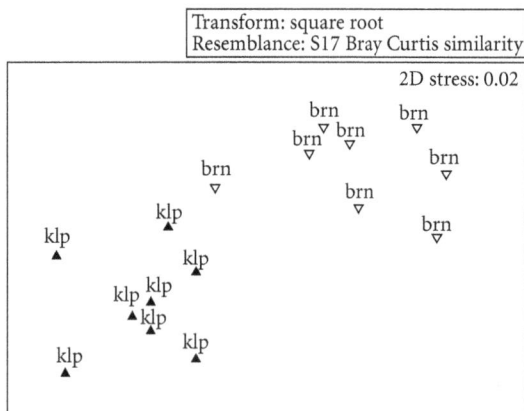

FIGURE 5: MDS ordination of the contribution of sea otter prey size, abundance, biomass, and energy density to the gradient in separation of remnant kelp forest and urchin barren communities in the central and western Aleutian Islands, Alaska. Kelp forests (klp) and urchin barrens (brn) are indicated.

biomass and potential energy density per individual. Greater sea urchin densities in urchin barrens provide greater total available prey biomass and total potential energy density due to the total mass of tissue available not the mass or quality of tissue per individual. Thus, the relationship between prey availability and quality in these two phase states is more complex than suggested by species abundances alone.

The absence of kelp did not have an effect on the distribution or density of four of the six sea otter prey species sampled during this study. Two prey species, including urchins themselves, varied in abundance with kelp presence and absence. In contrast to the inverse relationship urchins exhibited with kelp, *Musculus discors* was more abundant in kelp due to its preferred association with kelp [85]. *Musculus discors* had a nonuniform distribution and occurred in a dense but patchy distribution as seen in recruitment studies elsewhere in the North Pacific [86]. The remaining sea otter prey species sampled in this study did not show any variation in abundance, biomass, energy density, or size as a function of community type. *Pododesmus machrochisma* provided relatively high biomass and energy density per unit area but did not vary significantly in abundance between communities. This species is conspicuous, often occurring in dense aggregations on the edges of boulders and on ledges, and is easily removed from the substrate. Both *M. discors* and *P. machrochisma* are utilized by sea otters in the central and western Aleutians (Estes and Tinker unpubl. data) and likely supplement sea urchin energy density when preferred food items such as large sea urchins are scarce [87]. *Mytilus trossulus* and *Fusitriton oregonensis* exhibited patchy distributions and did not vary significantly between communities. With the exception of the patchy distribution of the kelp-associated *M. discors*, the availability and quality of nonurchin sea otter prey sampled in this study did not vary significantly with kelp presence or absence. The covariation between *M. discors* abundance and kelp abundance, and *S. Polyacanthus* and kelp abundance, suggests that the degree to which phase shifts effect prey quality depends on the interaction strength between a particular prey species, kelp, kelp subsidies, and the distance to kelp forest-urchin barren interfaces. As a result, sea urchins, a preferred prey of sea otters and directly linked with kelp forest-urchin barren dynamics, are a strong indicator of phase shifts in the Aleutians [88]. Findings from this study further support the claim that increases in sea urchin biomass during the sea otter decline are evidence against the nutritional limitation hypothesis [15]. It also addresses the concerns by Kuker and Barrett-Lennard [22] that additional abundance data for non-urchin sea otter prey species may refute Estes et al. [15].

Given sea urchin importance to sea otter diets in the central and western Aleutians [11, 13, 89], the potential impacts of changes in sea urchin abundance, biomass, size, and energy density between phase states detected in this study deserve closer evaluation. Potential calorie availability in the central and western Aleutians varies by the prey unit exploited (e.g., individual urchin versus aggregations of urchins) and by the type of community being targeted. Currently

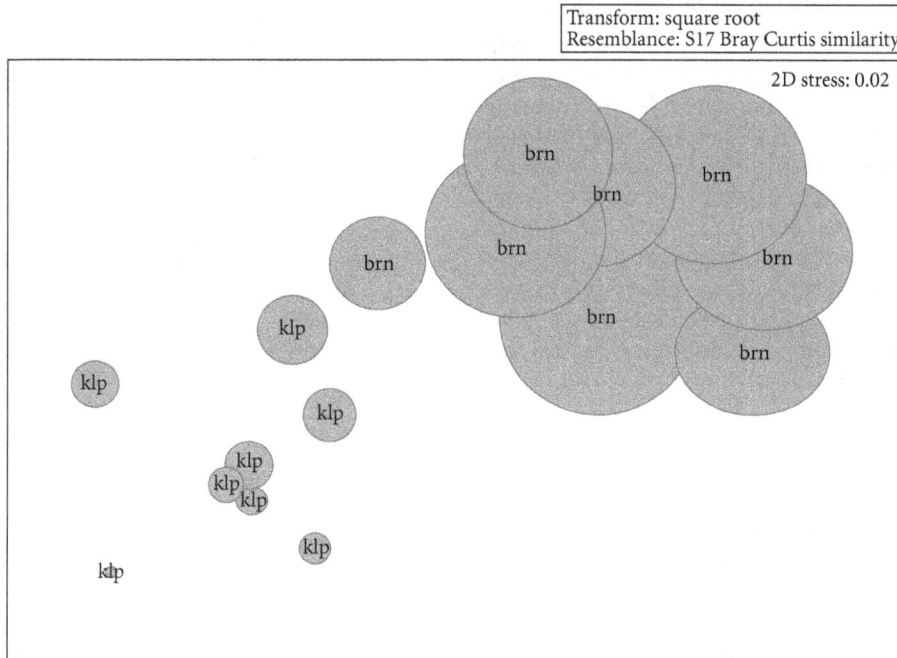

FIGURE 6: MDS bubble plot of the contribution of sea otter prey biomass to the separation in kelp forest and urchin barren communities in the central and western Aleutian Islands, Alaska. Bubble size scales with $g \cdot 0.25 \, m^{-2}$ from remnant kelp forest (klp) and urchin barren (brn) sites.

the spatially dominant community in the region, urchin barrens, supplies more nutrition per unit area but less nutrition per individual than kelp forests. Depending on a predator's foraging strategy, foraging for sea urchins in remnant kelp forest patches versus expansive urchin barrens could provide significantly different potential energy returns [90]. A predator that preferentially consumes larger and more calorically rich individual prey at the cost of increased search time in kelp understory would benefit from the selective use of kelp forest patches. Cormorants exhibit this foraging strategy in their selection of dense kelp-forested areas as opposed to recently kelp-harvested areas in Norway, despite significant increases in foraging times associated with locating fish in kelp [91]. This strategy is only feasible until the point at which the nutritional advantages of targeting prey in complex environments are outweighed by the cost of increased search time [90]. In contrast, a predator that exhibits general foraging behavior would likely exploit urchin barrens habitually and opportunistically forage in kelp forests. This strategy is exhibited by fish-eating killer whales that generally hunt in open water but occasionally specialize their foraging behavior and work cooperatively to take salmonid prey seeking refuge in dense kelp beds [92]. Sea otters are size selective foragers that generally select the largest and most calorically rich prey first before switching to smaller or less preferred prey species [69, 93]. Theoretically, a community dominated by small, low-quality prey could alter predator movement [94], lead to abandonment for areas with greater potential energy density [95, 96], or result in starvation and population decline; however, given the subtle differences in individual and areal potential energy density, these scenarios

are unlikely with sea otters. Sea otters have large energy requirements due to an elevated metabolic rate [89, 97] and as a result ingest 20 to 25% of their body mass in prey per day [74, 89] and spend 23 to 50% of the day foraging [98–100]. Although sea otter feeding rates on sea urchins differ in kelp forests (ca. $1.9 \, urchins \cdot min^{-1}$) and urchin barrens (ca. $3 \, urchins \cdot min^{-1}$) due to increased search times associated with foraging in kelp (Estes and Tinker, unpublished data), daily energy requirements are easily met foraging in either remnant kelp forest or urchin barren communities. Given sea otter activity budgets, metabolic rates, and distances travelled during foraging [73, 101, 102], the differences in individual sea urchin size and available biomass among kelp forest and urchin barrens are likely negligible to foraging sea otters. Numerous examples of expanding sea otter populations have reported otters continuing to forage in areas of depleted prey rather than moving to adjacent sites with larger individual prey and higher overall prey abundances [103, 104]. Furthermore, though depleted individual sea urchin quality values detected in urchin barrens in this study were lower than urchin values from locations where sea otter populations are currently increasing or stable, a difference of such small magnitude (ca. $0.07 \, kcal \cdot g^{-1}$) is not likely to affect sea otter resource selection. Consequently, despite the differences in individual prey abundance, biomass, size, and energy density between kelp forests and urchin barrens in the central and western Aleutians today, nutritional limitation is not likely to affect potential sea otter recovery to the region.

In conclusion, the phase shift between kelp forest and urchin barrens not only has an effect on kelp and urchin abundance and biomass but has also further effects on both individual and total potential energy density provided by urchins. In support of the first hypothesis, remnant kelp forests provide greater individual prey biomass than urchin barrens. Kelp forests do not, however, provide greater energy density per unit area than urchin barrens as was predicted in the second hypothesis. Prey quality differences were, however, not significant enough to explain the rate of sea otter population declines reported during the 1990s (ca 25% per year; [15]) nor are they sufficient to explain the persistent limitation of sea otter recovery in the two decades the decline. Given what is known about sea otter foraging behavior, gross daily metabolic needs, and prey availability, it is not feasible that sea otters were or are currently nutritionally limited in the central and western Aleutians. Although the capacity of sea otters to exploit sea urchin hyperabundance and recolonize their historical range is indisputable from both practical [104, 105] and conceptual standpoints [106], the ecosystem wide effects of alternating between energy-poor and energy-rich equilibrium points likely have effects on resource selection and ultimately the carrying capacity of other consumers in the central and western Aleutians. This study indicates that the overall potential energy density provided by kelp forests is diminished when urchin barrens are temporally and spatially dominant, as has been speculated for urchin barrens elsewhere [52]. Phase shifts in kelp forest-urchin barren systems have effects on the potential prey energy density available to higher trophic levels and, in addition to statistical differences in the abundance of key species, could provide a further means to differentiate between equilibrium states.

References

[1] R. C. Lewontin, "The meaning of stability," *Brookhaven Symposia in Biology*, vol. 22, pp. 13–24, 1969.

[2] C. S. Holling, "Resilience and stability of ecological systems," *Annual Review of Ecological Systems*, vol. 4, pp. 1–23, 1973.

[3] R. M. May, "Thresholds and breakpoints in ecosystems with a multiplicity of stable states," *Nature*, vol. 269, no. 5628, pp. 471–477, 1977.

[4] J. P. Sutherland, "Multiple stable points in natural communities," *American Naturalist*, vol. 108, no. 964, pp. 859–873, 1974.

[5] B. E. Beisner, D. T. Haydon, and K. Cuddington, "Alternative stable states in ecology," *Frontiers in Ecology and the Environment*, vol. 1, pp. 376–382, 2003.

[6] C. Folke, S. Carpenter, B. Walker et al., "Regime shifts, resilience, and biodiversity in ecosystem management," *Annual Review of Ecology, Evolution, and Systematics*, vol. 35, pp. 557–581, 2004.

[7] M. D. Bertness, G. C. Trussell, P. J. Ewanchuk, and B. R. Silliman, "Do alternate stable community states exist in the Gulf of Maine rocky intertidal zone?" *Ecology*, vol. 83, no. 12, pp. 3434–3448, 2002.

[8] R. T. Paine, "Food web complexity and species diversity," *The American Naturalist*, vol. 100, pp. 65–75, 1966.

[9] N. Knowlton, "Thresholds and multiple stable states in coral reef community dynamics," *American Zoologist*, vol. 32, no. 6, pp. 674–682, 1992.

[10] P. J. Mumby, A. Hastings, and H. J. Edwards, "Thresholds and the resilience of Caribbean coral reefs," *Nature*, vol. 450, no. 7166, pp. 98–101, 2007.

[11] J. A. Estes and J. F. Palmisano, "Sea otters: their role in structuring nearshore communities," *Science*, vol. 185, no. 4156, pp. 1058–1060, 1974.

[12] P. K. Dayton, "Experimental studies of algal canopy interactions in a sea otter–dominated community at Amchitka Island, Alaska," *Fisheries Bulletin U.S.*, vol. 73, pp. 230–237, 1975.

[13] C. A. Simenstad, J. A. Estes, and K. W. Kenyon, "Aleuts, sea otters, and alternate stable-state communities," *Science*, vol. 200, no. 4340, pp. 403–411, 1978.

[14] J. A. Estes, E. M. Danner, D. F. Doak et al., "Complex trophic interactions in kelp forest ecosystems," *Bulletin of Marine Science*, vol. 74, no. 3, pp. 621–638, 2004.

[15] J. A. Estes, M. T. Tinker, T. M. Williams, and D. F. Doak, "Killer whale predation on sea otters linking oceanic and nearshore ecosystems," *Science*, vol. 282, no. 5388, pp. 473–476, 1998.

[16] A. M. Doroff, J. A. Estes, M. T. Tinker, D. M. Burn, and T. J. Evans, "Sea otter population declines in the Aleutian archipelago," *Journal of Mammalogy*, vol. 84, no. 1, pp. 55–64, 2003.

[17] J. A. Estes, M. T. Tinker, A. M. Doroff, and D. M. Burn, "Continuing sea otter population declines in the aleutian archipelago," *Marine Mammal Science*, vol. 21, no. 1, pp. 169–172, 2005.

[18] A. M. Springer, J. A. Estes, G. B. Van Vliet et al., "Sequential megafaunal collapse in the North Pacific Ocean: an ongoing legacy of industrial whaling?" *Proceedings of the National Academy of Sciences of the United States of America*, vol. 100, no. 21, pp. 12223–12228, 2003.

[19] D. P. DeMaster, A. W. Trites, P. Clapham et al., "The sequential megafaunal collapse hypothesis: testing with existing data," *Progress in Oceanography*, vol. 68, no. 2–4, pp. 329–342, 2006.

[20] National Research Council, *The Decline of the Steller Sea Lion in Alaskan Waters: Untangling Food Webs and Fishing Nets*, National Academy Press, Washington, DC, USA, 2003.

[21] J. A. Estes, M. L. Riedman, M. M. Staedler, M. T. Tinker, and B. E. Lyon, "Individual variation in prey selection by sea otters: patterns, causes and implications," *Journal of Animal Ecology*, vol. 72, no. 1, pp. 144–155, 2003.

[22] K. Kuker and L. Barrett-Lennard, "A re-evaluation of the role of killer whales *Orcinus orca* in a population decline of sea otters *Enhydra lutris* in the Aleutian Islands and a review of alternative hypotheses," *Mammal Review*, vol. 40, no. 2, pp. 103–124, 2010.

[23] D. L. Alverson, "A review of commercial fisheries and the Steller sea lion (*Eumetopias jubatus*): the conflict arena," *Reviews in Aquatic Sciences*, vol. 6, pp. 203–256, 1992.

[24] A. W. Trites and C. P. Donnelly, "The decline of Steller sea lions Eumetopias jubatus in Alaska: a review of the nutritional stress hypothesis," *Mammal Review*, vol. 33, no. 1, pp. 3–28, 2003.

[25] H. Österblom, O. Olsson, T. Blenckner, and R. W. Furness, "Junk-food in marine ecosystems," *Oikos*, vol. 117, no. 7, pp. 967–977, 2008.

[26] D. A. S. Rosen and A. W. Trites, "Pollock and the decline of Steller sea lions: testing the junk-food hypothesis," *Canadian Journal of Zoology*, vol. 78, no. 7, pp. 1243–1250, 2000.

[27] K. H. Mann, "Seaweeds: their productivity and strategy for growth," *Science*, vol. 182, no. 4116, pp. 975–981, 1973.

[28] A. Cowles, J. E. Hewitt, and R. B. Taylor, "Density, biomass and productivity of small mobile invertebrates in a wide range of coastal habitats," *Marine Ecology Progress Series*, vol. 384, pp. 175–185, 2009.

[29] D. O. Duggins, C. A. Simenstad, and J. A. Estes, "Ecology of understory kelp environments. II. Effects of kelps on recruitment of benthic invertebrates," *Journal of Experimental Marine Biology and Ecology*, vol. 143, no. 1-2, pp. 27–45, 1990.

[30] P. K. Dayton, "Ecology of kelp communities," *Annual Review of Ecology and Systematics*, vol. 16, pp. 215–245, 1985.

[31] B. Santelices and F. P. Ojeda, "Effects of canopy removal on the understory algal community structure of coastal forests of Macrocystis pyrifera from southern South America," *Marine Ecology Progress Series*, vol. 14, pp. 165–173, 1984.

[32] D. O. Duggins, "Kelp beds and sea otters: an experimental approach," *Ecology*, vol. 61, pp. 447–453, 1980.

[33] K. H. Dunton and D. M. Schell, "Dependence of consumers on macroalgal (*Laminaria solidungula*) carbon in an arctic kelp community: δ13C evidence," *Marine Biology*, vol. 93, no. 4, pp. 615–625, 1987.

[34] B. B. Bernstein and N. Jung, "Selective pressures and co-evolution in a kelp canopy community in Southern California," *Ecological Monographs*, vol. 49, pp. 335–355, 1980.

[35] P. A. X. Bologna and R. S. Steneck, "Kelp beds as habitat for American lobster *Homarus americanus*," *Marine Ecology Progress Series*, vol. 100, no. 1-2, pp. 127–134, 1993.

[36] G. M. Branch and C. L. Griffiths, "The Benguela ecosystem, Part V. The coastal zone," *Oceanography and Marine Biology Annual Review*, vol. 26, pp. 395–486, 1988.

[37] D. O. Duggins, C. A. Simenstad, and J. A. Estes, "Magnification of secondary production by kelp detritus in coastal marine ecosystems," *Science*, vol. 245, no. 4914, pp. 101–232, 1989.

[38] K. H. Mann, *Ecology of Coastal Waters, with Implications for Management*, vol. 2, Blackwell Science, Oxford, UK, 2000.

[39] D. L. Leighton, L. G. Jones, and W. North, "Ecological relationships between giant kelp and sea urchins in southern California," in *Proceedings of the 5th International Seaweed Symposium*, E. G. Young and J. L. McLachlan, Eds., pp. 141–153, Pergamon Press, Oxford, UK, 1966.

[40] J. M. Lawrence, "On the relationships between marine plants and sea urchins," *Oceanography and Marine Biology Annual Review*, vol. 13, pp. 213–286, 1975.

[41] M. W. Hart and R. E. Scheibling, "Heat waves, baby booms, and the destruction of kelp beds by sea urchins," *Marine Biology*, vol. 99, no. 2, pp. 167–176, 1988.

[42] S. D. Ling, C. R. Johnson, S. Frusher, and C. K. King, "Reproductive potential of a marine ecosystem engineer at the edge of a newly expanded range," *Global Change Biology*, vol. 14, no. 4, pp. 907–915, 2008.

[43] S. D. Ling, C. R. Johnson, S. D. Frusher, and K. R. Ridgway, "Overfishing reduces resilience of kelp beds to climate-driven catastrophic phase shift," *Proceedings of the National Academy of Sciences of the United States of America*, vol. 106, no. 52, pp. 22341–22345, 2009.

[44] P. A. Breen and K. H. Mann, "Destructive grazing of kelp by sea urchins in Eastern Canada," *Journal of the Fisheries Research Board of Canada*, vol. 33, pp. 1278–1283, 1976.

[45] J. A. Estes, N. S. Smith, and J. F. Palmisano, "Sea otter predation and community organization in the western Aleutian Islands, Alaska," *Ecology*, vol. 59, pp. 822–833, 1978.

[46] B. B. Bernstein, B. E. Williams, and K. H. Mann, "The role of behavioral responses to predators in modifying urchins' (*Strongylocentrotus droebachiensis*) destructive grazing and seasonal foraging patterns," *Marine Biology*, vol. 63, no. 1, pp. 39–49, 1981.

[47] S. D. Ling, "Range expansion of a habitat-modifying species leads to loss of taxonomic diversity: a new and impoverished reef state," *Oecologia*, vol. 156, no. 4, pp. 883–894, 2008.

[48] J. A. Kitching and F. A. Ebling, "The ecology of Lough Ine. XI. The control of algae by *Paracentrotus lividus* (Echinoidea)," *Journal of Animal Ecology*, vol. 30, pp. 373–383, 1961.

[49] P. D. Steinberg, J. Estes, and F. C. Winter, "Evolutionary consequences of food chain length in kelp forest communities," *Proceedings of the National Academy of Sciences of the United States of America*, vol. 92, no. 18, pp. 8145–8148, 1995.

[50] B. Konar and J. A. Estes, "The stability of boundary regions between kelp beds and deforested areas," *Ecology*, vol. 84, no. 1, pp. 174–185, 2003.

[51] A. R. Russo, "Dispersion and food differences between two populations of the sea urchin *Strongylocentrotus franciscanus*," *Journal of Biogeography*, vol. 6, pp. 407–414, 1979.

[52] C. Harrold and D. C. Reed, "Food availability, sea urchin grazing, and kelp forest community structure," *Ecology*, vol. 66, no. 4, pp. 1160–1169, 1985.

[53] R. E. Scheibling, A. W. Hennigar, and T. Balch, "Destructive grazing, epiphytism, and disease: the dynamics of sea urchin—kelp interactions in Nova Scotia," *Canadian Journal of Fisheries and Aquatic Sciences*, vol. 56, no. 12, pp. 2300–2314, 1999.

[54] P. Gagnon, J. H. Himmelman, and L. E. Johnson, "Temporal variation in community interfaces: kelp-bed boundary dynamics adjacent to persistent urchin barrens," *Marine Biology*, vol. 144, no. 6, pp. 1191–1203, 2004.

[55] K. A. Hovel and R. N. Lipcius, "Habitat fragmentation in a seagrass landscape: patch size and complexity control blue crab survival," *Ecology*, vol. 82, no. 7, pp. 1814–1829, 2001.

[56] J. Hamilton and B. Konar, "Implications of substrate complexity and kelp variability for south-central Alaskan nearshore fish communities," *Fishery Bulletin*, vol. 105, no. 2, pp. 189–196, 2007.

[57] K. Vicknair, *Sea otters and asteroids in the western Aleutian Islands*, M.S. thesis, University of California, Santa Cruz, Calif, USA, 1996.

[58] S. D. Gaines and J. Roughgarden, "Fish in offshore kelp forests affect recruitment to intertidal barnacle populations," *Science*, vol. 235, no. 4787, pp. 479–481, 1987.

[59] D. B. Irons, R. G. Anthony, and J. A. Estes, "Foraging strategies of glaucous-winged gulls in a rocky intertidal community," *Ecology*, vol. 67, no. 6, pp. 1460–1474, 1986.

[60] S. E. Reisewitz, J. A. Estes, and C. A. Simenstad, "Indirect food web interactions: sea otters and kelp forest fishes in the Aleutian archipelago," *Oecologia*, vol. 146, no. 4, pp. 623–631, 2006.

[61] P. S. Petraitis and S. R. Dudgeon, "Experimental evidence for the origin of alternative communities on rocky intertidal shores," *Oikos*, vol. 84, no. 2, pp. 239–245, 1999.

[62] B. A. Menge and G. M. Branch, "Rocky intertidal communities," in *Marine Community Ecology*, M. D. Bertness, S. D. Gaines, and M. Hay, Eds., pp. 221–251, Sinauer Associates, Sunderland, Mass, USA, 2001.

[63] D. R. Levitan, "Influence of body size and population density on fertilization success and reproductive output in a free-spawning invertebrate," *Biological Bulletin*, vol. 181, no. 2, pp. 261–268, 1991.

[64] A. Barkai and C. McQuaid, "Predator-prey role reversal in a marine benthic ecosystem," *Science*, vol. 242, no. 4875, pp. 62–64, 1988.

[65] J. Van De Koppel, P. M. J. Herman, P. Thoolen, and C. H. R. Heip, "Do alternate stable states occur in natural ecosystems? Evidence from a tidal flat," *Ecology*, vol. 82, no. 12, pp. 3449–3461, 2001.

[66] S. D. Ling and C. R. Johnson, "Population dynamics of an ecologically important range-extender: kelp beds versus sea urchin barrens," *Marine Ecology Progress Series*, vol. 374, pp. 113–125, 2009.

[67] J. L. Bodkin, G. G. Esslinger, and D. H. Monson, "Foraging depths of sea otters and implications to coastal marine communities," *Marine Mammal Science*, vol. 20, no. 2, pp. 305–321, 2004.

[68] J. Watt, D. B. Siniff, and J. A. Estes, "Inter-decadal patterns of population and dietary change in sea otters at Amchitka Island, Alaska," *Oecologia*, vol. 124, no. 2, pp. 289–298, 2000.

[69] R. S. Ostfeld, "Foraging strategies and prey switching in the California sea otter," *Oecologia*, vol. 53, no. 2, pp. 170–178, 1982.

[70] J. A. Estes and D. O. Duggins, "Sea otters and kelp forests in Alaska: generality and variation in a community ecological paradigm," *Ecological Monographs*, vol. 65, no. 1, pp. 75–100, 1995.

[71] T. A. Dean, J. L. Bodkin, A. K. Fukuyama et al., "Food limitation and the recovery of sea otters following the "Exxon Valdez" oil spill," *Marine Ecology Progress Series*, vol. 241, pp. 255–270, 2002.

[72] O. T. Oftedal, K. Ralls, M. T. Tinker, and A. Green, "Nutritional constraints on the southern sea otter in the Monterey Bay National Marine Sanctuary and a comparison to sea otter populations at San Nicholas Island, California and Glacier Bay, Alaska," Joint Final Report to the Monterey Bay National Marine Sanctuary, Monterey Bay, Calif, USA, 2007.

[73] L. C. Yeates, T. M. Williams, and T. L. Fink, "Diving and foraging energetics of the smallest marine mammal, the sea otter (*Enhydra lutris*)," *Journal of Experimental Biology*, vol. 210, no. 11, pp. 1960–1970, 2007.

[74] D. P. Costa and G. L. Kooyman, "Contribution of specific dynamic action to heat balance and thermoregulation in the sea otter, *Enhydra lutris*," *Physiological and Biochemical Zoology*, vol. 57, pp. 199–203, 1984.

[75] V. A. Gill, A. M. Doroff, and D. Burn, "Aerial surveys of sea otters (*Enhydra lutris*) in Kachemak Bay Alaska, 2008," Marine Mammals Management Technical Report: MMM, U.S. Fish and Wildlife Service, 2009.

[76] J. L. Bodkin, K. A. Kloecker, G. G. Esslinger, D. H. Monson, H. A. Coletti, and J. Doherty, "Sea Otter studies in Glacier Bay National Park and Preserve," Annual Report 2002, USGS Alaska Biological Science Center, Anchorage, Alaska, USA, 2003.

[77] M. T. Tinker, G. Bentall, and J. A. Estes, "Food limitation leads to behavioral diversification and dietary specialization in sea otters," *Proceedings of the National Academy of Sciences of the United States of America*, vol. 105, no. 2, pp. 560–565, 2008.

[78] C. H. Biermann, B. D. Kessing, and S. R. Palumbi, "Phylogeny and development of marine model species: strongylocentrotid sea urchins," *Evolution and Development*, vol. 5, no. 4, pp. 360–371, 2003.

[79] K. R. Clarke and R. W. Warwick, *Change in Communities: An Approach to Statistical Analysis and Interpretation*, PRIMER-E, Plymouth, UK, 2nd edition, 2001.

[80] K. R. Clarke and R. N. Gorley, *PRIMER v6: User Manual/Tutorial*, PRIMER-E, Plymouth, UK, 2006.

[81] N. L. Andrew and A. J. Underwood, "Density-dependent foraging in the sea urchin *Centrostephanus rodgersii* on shallow subtidal reefs in New South Wales, Australia," *Marine Ecology Progress Series*, vol. 99, no. 1-2, pp. 89–98, 1993.

[82] P. K. Dayton, "Competition, disturbance, and community organization: the provision and subsequent utilization of space in a rocky intertidal community," *Ecological Monographs*, vol. 41, pp. 351–389, 1971.

[83] R. T. Paine, "Intertidal community structure—Experimental studies on the relationship between a dominant competitor and its principal predator," *Oecologia*, vol. 15, no. 2, pp. 93–120, 1974.

[84] N. L. Andrew, "The interaction between diet and density in influencing reproductive output in the echinoid *Evechinus chloroticus* (Val.)," *Journal of Experimental Marine Biology and Ecology*, vol. 97, no. 1, pp. 63–79, 1986.

[85] E. Waage-Nielsen, H. Christie, and E. Rinde, "Short-term dispersal of kelp fauna to cleared (kelp-harvested) areas," *Hydrobiologia*, vol. 503, pp. 77–91, 2003.

[86] C. Bégin, L. E. Johnson, and J. H. Himmelman, "Macroalgal canopies: distribution and diversity of associated invertebrates and effects on the recruitment and growth of mussels," *Marine Ecology Progress Series*, vol. 271, pp. 121–132, 2004.

[87] J. A. Estes, R. J. Jameson, and A. M. Johnson, "Food selection and some foraging tactics of sea otters," in *Proceedings of the Worldwide Furbearer Conference*, J. A. Chapman and D. Pursley, Eds., vol. 1, pp. 606–641, University of Maryland Press, Baltimore, Md, USA, 1981.

[88] J. A. Estes, M. T. Tinker, and J. L. Bodkin, "Using ecological function to develop recovery criteria for depleted species: sea otters and kelp forests in the Aleutian archipelago," *Conservation Biology*, vol. 24, no. 3, pp. 852–860, 2010.

[89] K. W. Kenyon, *The Sea Otter in the Eastern Pacific Ocean*, Dover Publications, New York, NY, USA, 1969.

[90] D. W. Stephens and J. R. Krebs, *Foraging Theory*, Princeton University Press, Princeton, NJ, USA, 1986.

[91] S. H. Lorentsen, K. Sjøtun, and D. Grémillet, "Multi-trophic consequences of kelp harvest," *Biological Conservation*, vol. 143, no. 9, pp. 2054–2062, 2010.

[92] J. K. B. Ford and G. M. Ellis, "Selective foraging by fish-eating killer whales *Orcinus orca* in British Columbia," *Marine Ecology Progress Series*, vol. 316, pp. 185–199, 2006.

[93] D. L. Garshelis, *Ecology of sea otters in Prince William Sound, Alaska*, Ph.D. thesis, University of Minnesota, Minneapolis, Minn, USA, 1983.

[94] E. Cruz-Rivera and M. E. Hay, "Can quantity replace quality? Food choice, compensatory feeding, and fitness of marine mesograzers," *Ecology*, vol. 81, no. 1, pp. 201–219, 2000.

[95] E. L. Charnov, "Optimal foraging, the marginal value theorem," *Theoretical Population Biology*, vol. 9, no. 2, pp. 129–136, 1976.

[96] P. A. Abrams, "Foraging time optimization and interactions in food webs," *American Naturalist*, vol. 124, no. 1, pp. 80–96, 1984.

[97] J. A. Iversen, "Basal energy metabolism of mustelids," *Journal of Comparative Physiology*, vol. 81, no. 4, pp. 341–344, 1972.

[98] J. A. Estes, K. E. Underwood, and M. J. Karmann, "Activity-time budgets of sea otters in California," *Journal of Wildlife Management*, vol. 50, no. 4, pp. 626–636, 1986.

[99] K. Ralls and D. B. Siniff, "Time budgets and activity patterns in California sea otters," *Journal of Wildlife Management*, vol. 54, no. 2, pp. 251–259, 1990.

[100] M. T. Tinker, *Sources of variation in the foraging behavior and demography of the sea otter, Enhydra lutris*, Ph.D. thesis, University of California, Santa Cruz, Calif, USA, 2004.

[101] D. L. Garshelis and J. A. Garshelis, "Movements and management of sea otters in Alaska," *Journal of Wildlife Management*, vol. 48, no. 3, pp. 665–678, 1984.

[102] J. A. Estes, "Growth and equilibrium in sea otter populations," *Journal of Animal Ecology*, vol. 59, no. 2, pp. 385–401, 1990.

[103] R. G. Kvitek, C. E. Bowlby, and M. Staedler, "Diet and foraging behavior of sea otters in southeast Alaska," *Marine Mammal Science*, vol. 9, no. 2, pp. 168–181, 1993.

[104] K. L. Laidre and R. J. Jameson, "Foraging patterns and prey selection in an increasing and expanding sea otter population," *Journal of Mammalogy*, vol. 87, no. 4, pp. 799–807, 2006.

[105] A. M. Doroff and A. R. DeGange, "Sea otter, *Enhydra lutris*, prey composition and foraging success in the northern Kodiak Archipelago," *U S National Marine Fisheries Service Fishery Bulletin*, vol. 92, pp. 704–710, 1994.

[106] R. T. Paine, "Controlled manipulations in the marine intertidal zone and their contributions to ecological theory," in *The Changing Scenes in the Natural Sciences, 1776–1976*, Clyde E. Goulden, Ed., pp. 245–270, Academy of Natural Sciences, Philadelphia, Pa, USA, 1977.

20

Incidence and Spatial Distribution of Caribbean Yellow Band Disease in La Parguera, Puerto Rico

Francisco J. Soto-Santiago[1, 2] and Ernesto Weil[1]

[1] *Department of Marine Sciences, University of Puerto Rico, Mayagüez, P.O. Box 9000, San Juan P.R. 00681, Puerto Rico*
[2] *Department of Environmental Sciences, University of Puerto Rico, Río Piedras Campus, P.O. Box 70377, San Juan P.R. 00936, Puerto Rico*

Correspondence should be addressed to Francisco J. Soto-Santiago, franciscoj_soto@yahoo.com

Academic Editor: Horst Felbeck

The incidence and spatial distribution patterns of Caribbean Yellow Band Disease (CYBD) on the important frame-builder coral *Montastraea faveolata* were assessed by counting, tagging, and mapping all diseased and healthy colonies for one year in each of three 100 m^2 quadrats on two inner, mid-shelf, and shelf-edge reefs off La Parguera, Puerto Rico. Healthy colonies were checked every month from January to December of 2009 to monitor disease spread within each quadrant. Incidence increased significantly from winter ($0.7 \pm 0.8\%$ SE) to summer ($1.5 \pm 1.1\%$ SE, $n = 23$, Sign Test; $Z = 2.40$; $P = 0.01$). Mid-shelf reefs had the highest host abundance and showed significantly higher CYBD incidence ($2.1 \pm 1.4\%$ SE, $n = 14$) compared to the other zones ($H = 9.74$; $df = 2$; $P = 0.04$). The increased incidence in the summer suggests that warmer months favor development of CYBD on *M. faveolata*. Results showed aggregated patterns of CYBD when all colonies (i.e., healthy + diseased) at the spatial scales sampled were analyzed on each reef. This suggests facilitation of disease spread between aggregated colonies within populations. Similar stressful conditions then might trigger the disease in susceptible, aggregated colonies harboring the potential pathogens.

1. Introduction

Coral diseases seem to be one of the main causes of the decline of Caribbean coral reefs. Over the last three decades, fast emergence and high virulence of coral reef diseases have produced substantial declines in live coral tissue and colonies [1–9]. The recent increase in diseases in the marine environment has been associated with different stressors such as elevated sea surface temperatures [3, 7, 9, 10] and nutrient enrichment [11, 12]. Certain coral diseases (particularly tissue-loss diseases) seem to be more prevalent where there is human disturbance [13–15], but not all (e.g., *Acropora* tumors, [13]). Caribbean Yellow Band Disease (CYBD) is one of the most damaging coral diseases affecting important reef builders in the wider Caribbean [5, 16]. In Puerto Rico, CYBD is one of the most prevalent, persistent, and detrimental diseases affecting important reef-building species [8, 9, 17, 18]. CYBD has shown a high prevalence in near-pristine areas far from anthropogenic disturbance

[5, 18]. However, very little is known about the incidence and the spatial distribution patterns of this disease.

Spatial distribution patterns describe how organisms are arranged in a particular habitat or community. Spatial patterns can be regular (i.e., uniform distribution), random (i.e., random distribution) or aggregated (i.e., clumped distribution). Patterns of disease infections can be studied by documenting spatial distribution patterns which could help in characterizing the etiology of the disease and identifying potential mechanisms of infection [19]. Clustered or aggregated disease patterns in a population along a reef could mean that the disease is infectious and water borne; the closer the hosts, the faster the rate of infection (incidence) and the higher the number of infected colonies (prevalence). A noninfectious disease should in fact display a Poisson (purely random) distribution within the host population (but of course the host population itself could still display a clumped or regular distribution). Spatial distribution patterns of aspergillosis affecting *Gorgonia ventalina* were highly aggregated and prevalence was high in the Florida Keys [19]. It has been stated that predictive models should be used to study spatial distribution patterns of coral diseases [20, 21]. Also, coral diseases should be investigated for environmental relationships separately until they are known to have a common cause [21]. Foley et al. [22] showed that *M. annularis* colonies infected with CYBD were less clustered (i.e., more regular) than were healthy colonies between 10–30 m in a shallow back-reef in Akumal Bay, Mexico. Clustered patterns of coral disease occurrence could also result if there are transmitted by a vector and the vector displays a clumped distribution (e.g., *Porites* trematodiasis, [14, 21]).

Prevalence is the proportion of infected colonies in a population. Thus, higher prevalence might suggest higher susceptibility of host colonies, higher pathogenicity of the disease agent, higher densities of susceptible colonies, or a combination of these [19]. However, not all diseases are necessarily caused by pathogens, they can be in response to many abiotic/biotic triggers [23] and, therefore, pathogen virulence is not universally relevant when discussing coral diseases. Prevalence of CYBD in the *Montastraea annularis* species complex increased dramatically to 52% in reefs off the western end of Mona Island [17] and from 1% to 55% between 1999 and 2007 in La Parguera, on the southwest coast of Puerto Rico [9]. After the 2005 bleaching event, CYBD prevalence increased 40% in *M. faveolata* with high coral mortality attributed to this disease. CYBD, white plague outbreak and bleaching have caused 60% of damage on some reefs [9, 18]. Even though CYBD seems to be caused by a group of *Vibrio* bacteria [24, 25], little is known about the mechanisms of spread of the disease and its pathogenicity of the disease agent. Several experimental attempts in the field and laboratory have failed to demonstrate infectiousness [26], suggesting that the diseased colonies in natural populations should not to be aggregated. None of the previous studies provides information on CYBD incidence on *M. faveolata* populations.

The goals of this study were to assess the spatial and temporal variability in CYBD incidence on *M. faveolata*, to characterize the spatial distribution patterns of CYBD infected colonies, and to explore the relationship between spatial pattern and incidence of CYBD within and across reefs off La Parguera, Puerto Rico. Results from this study help clarify the etiology and potential mechanisms of infection of CYBD, the rate of infection over time and how the distribution patterns of the population (susceptible colonies) may affect the spread and potential impact of the disease.

2. Materials and Methods

Coral reefs in La Parguera Natural Reserve on the southwest coast are considered the best developed reefs of Puerto Rico [18]. A broad insular shelf, moderate water energy, favorable environmental conditions (high temperatures and low rainfall), and low human impact over long periods of time have allowed extensive development of diverse coral reef communities [18]. To assess the CYBD incidence and spatial variability in CYBD diseased and healthy colonies, six reefs

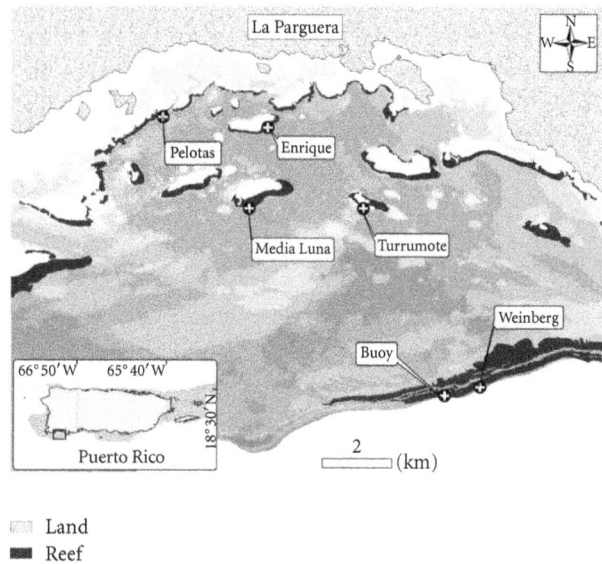

FIGURE 1: Map of the study sites in La Parguera, Puerto Rico.

were selected on an inshore-offshore gradient: two fringing reefs on the inner-shelf (Pelotas and Enrique), two fringing reefs on the mid-shelf (Media Luna and Turrumote), and two deep bank reefs on the shelf-edge (Weinberg and Buoy) (Table 1, Figure 1).

2.1. Temporal and Spatial Variability of CYBD Incidence in M. Faveolata. Incidence (i.e., proportion of newly infected colonies per month) of CYBD affecting *Montastraea faveolata* in La Parguera was assessed through winter and summer of 2009. Total incidence per site and season was also calculated. Incidence of CYBD was assessed using three 100 m^2 permanently marked areas (300 m^2) positioned between 3–9 m depths and separated by at least 20 m on each of the fringing reefs, and at a depth of 20 m on the offshore bank reefs. Each quadrant corner was marked with tagged rebar and, before each survey; the perimeter was marked with measuring tapes. Parallel tape lines were extended every two meters to form five $2 \times 10 \text{ m}$ bands (20 m^2 each), and all diseased and healthy colonies of *M. faveolata* were checked, counted, and mapped so the location and status of every single colony within the 100 m^2 was known after each survey. Healthy colonies were checked and photographed every month from January to December 2009 to monitor new CYBD infections. Temperature data for the year was obtained from Hobo Pro v.1 and v.2 temperature loggers that have been deployed on these reefs since 2005. The loggers record temperature every 2 hours. Disease incidence data (i.e., proportion of newly infected colonies per month) did not meet the assumptions of parametric statistical tests and could not be normalized with arcsine transformations. Differences in incidence between seasons (winter and summer) were therefore, evaluated using Sign Tests. Differences in incidence (median ± SE) between reef zones and sites were evaluated using nonparametric Kruskal-Wallis tests (Levene's test: $P > 0.05$). Significant differences

TABLE 1: Characteristics of the study sites in La Parguera, Puerto Rico (modified from [27]).

| Zone | Reef | Location | | Distance | Depth |
		N	W	from shore (km)	range (m)
Inner-shelf	Enrique	17°56.658	67°02.213	1.5	1–14
Inner-shelf	Pelotas	17°57.442	67°04.176	1	1–12
Mid-shelf	M. Luna	17°56.093	67°02.931	2	4–20
Mid-shelf	Turrumote	17°56.097	67°01.130	2	2–20
Shelf-edge	Weinberg	17°53.429	66°59.320	6	18–23
Shelf-edge	Buoy	17°53.110	66°59.510	6	18–23

(a)   (b)

FIGURE 2: Photographic time series of a colony at Enrique showing signs of YBD. (a) January 2009; (b) September 2009.

were followed by multiple comparison tests. Statistica 7 software was used to run the analyses.

2.2. Spatial Distribution of CYBD in M. Faveolata.

Single colonies were defined and selected as continuous patches of tissue of *M. faveolata*, including isolated patches of tissue showing partial mortality. This was carefully inspected to avoid the inclusion of ramets of a different genet pooled together as a single colony. Spatial distribution patterns were assessed for the whole population and CYBD-infected *M. faveolata* within each of the 100 m² quadrants on each of the six reefs using the nearest neighbor method [28, 29]. The method is based on comparing the distribution of distances from one individual to its nearest neighbor in space [28, 29]. This distribution can be defined as

$$R = \frac{r_A}{r_E}, \tag{1}$$

where R is the index of aggregation, r_A is the mean distance to the nearest neighbor and can be defined as

$$r_A = \frac{\sum r_i}{n}, \tag{2}$$

where r_i is the distance to nearest neighbor for individual I and n is the number of individuals in the study area; r_E is the expected distance to nearest neighbor and can be defined as

$$r_E = \frac{1}{2\sqrt{\rho}}, \tag{3}$$

where ρ is the density of organisms (number of organisms/size of the study area).

If the spatial pattern is random, $R = 1$, when clumping occurs, R approaches zero and in a regular pattern R increases >1. A simple test of significance for deviation from randomness was used, as the standard error of the expected distance is known from plane geometry [29]. This test of significance can be defined as

$$z = \frac{(r_A - r_E)}{S_r}, \tag{4}$$

where z is the standard normal deviate and S_r is the standard error of the expected distance to the nearest neighbor and can be defined as

$$S_r = \frac{0.26136}{\sqrt{n\rho}}. \tag{5}$$

3. Results

Significant differences in CYBD incidence in *M. faveolata* were found among reef sites ($H = 12.13$; $df = 4$; $P = 0.02$). Pelotas (an inner-shelf reef) showed no sign of CYBD over the study period. Turrumote and Media Luna (mid-shelf reefs) showed the highest total incidence levels (median ± SE%) in the study, each with 2.1 ± 1.5% ($n = 14$), followed by Enrique (inner-shelf reef) with 1.5 ± 1.2% ($n = 5$). Buoy and Weinberg (shelf-edge reefs) showed the lowest incidence levels in the study with each site showing 0.6 ± 1.4% ($n = 4$). Time series photographs of tagged colonies at the different sites showed healthy colonies in the winter developing the CYBD signs during the summer (Figure 2). Disease signs were still visible at December 2009, but no tissue mortality was observed.

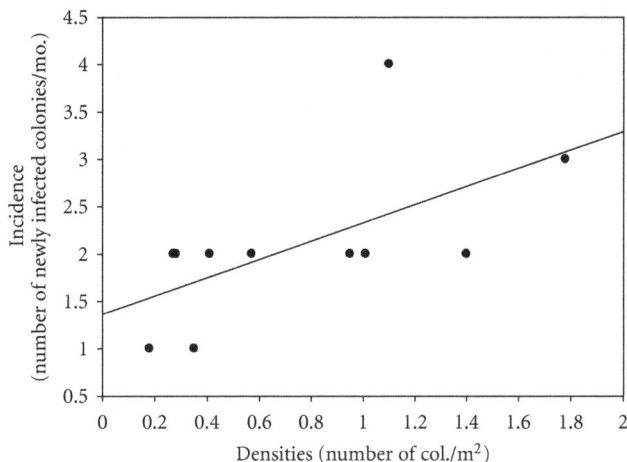

FIGURE 3: Spearman correlation analysis between densities of *M. faveolata* and incidence of CYBD on the different reefs ($\rho = 0.70$; $P = 0.01$).

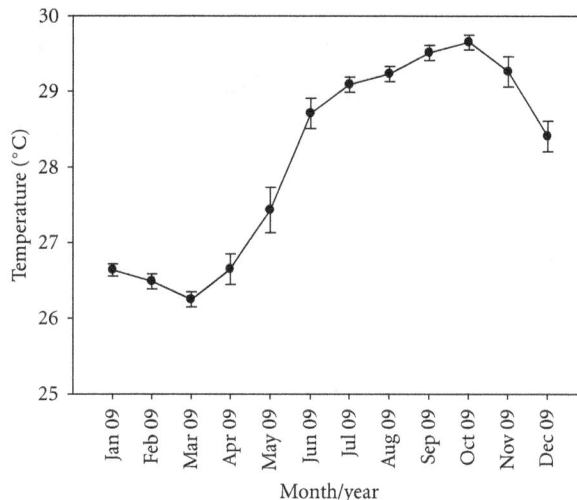

FIGURE 4: Line graph of monthly average ($\pm$SD) seawater temperatures from January–December 2009 in La Parguera, Puerto Rico.

Proportion of monthly CYBD incidence in *M. faveolata* was higher on mid-shelf reefs (0.58 newly infected colonies/month) than on inner (0.41 newly infected colonies/month) or on shelf-edge reefs (0.16 newly infected colonies/month) throughout 2009 ($H = 9.74$; $df = 2$; $P = 0.04$). There was a significant positive correlation (Spearman correlation analysis; $\rho = 0.70$, $P = 0.01$) between CYBD incidence and densities of *M. faveolata* in each of the three $100\,\text{m}^2$ quadrants per site (Figure 3).

CYBD incidence (median $\pm$ SE%) on *M. faveolata* increased significantly $0.7 \pm 0.8\%$ to $1.5 \pm 1.1\%$ ($n = 23$) from winter to summer (Sign Test; $Z = 2.40$; $P = 0.01$). Incidence in *M. faveolata* ranged from 2.00 newly infected colonies/month during the cold season (winter or January–April) to 3.75 newly infected colonies/month during the warm water season (summer or June-September) of 2009 (Table 2). Monthly average temperatures in 2009 increased from $26.6°C$ in the winter to $29.6°C$ in summer (Figure 4). During February ($26.5°C$) and June 2009 ($28.8°C$), CYBD incidence was highest (Table 2). February finished with 3 new colonies infected in Media Luna, 1 in Turrumote and 2 in Buoy. In June, 1 newly infected colony was found in Enrique, 2 in Media Luna and 4 in Turrumote.

Analysis of the spatial distribution patterns of CYBD in *M. faveolata* showed an aggregated pattern for diseased colonies and for all colonies (pooled diseased + healthy) analyzed together. This aggregated pattern was found in all of the $100\,\text{m}^2$ areas surveyed (Table 3) and was significantly different from randomness (Table 4). There were no CYBD-infected *M. faveolata* colonies in the $100\,\text{m}^2$ areas surveyed in Pelotas (i.e., inner shelf reef) so these were excluded from the analyses.

4. Discussion

In La Parguera and on the west coast of Puerto Rico, winter average temperatures used to drop below $25.5°C$. In the last 10 years, however, winter mean temperatures

have not dropped below $26.5°C$ potentially compromising the immune response of corals or increasing virulence of pathogens [3]. In the present study, highest incidence levels were found in June and September (summer). However, high incidence levels were also found in February (winter). *M. faveolata* colonies are more susceptible to any pathogen in the environment due to a weakened state when temperature changes. Susceptibility to opportunistic pathogens increases significantly in corals that are harmed from climatic and physical perturbations [2]. Although previous studies have shown that prevalence of CYBD increases in *Montastraea* species when temperature increases [3, 30–32], a decrease in temperature may cause harm to these colonies as well. Cold temperatures can be detrimental to corals by causing coral bleaching [33, 34]. Further research on how low temperatures may trigger coral diseases, in this case CYBD, should provide interesting information on the dynamics of these stressors.

CYBD distribution found in the different reefs through the Caribbean is due to the combination of host population abundance and distribution, presence of potential pathogens and different environmental factors [24, 35], a pattern similar to most infectious diseases. The significant positive correlation between densities of *M. faveolata* and CYBD incidence found in the present study suggests that this disease is affecting abundant and susceptible hosts mainly at mid-shelf reefs (i.e., Turrumote and Media Luna). These reefs showed the highest colony densities and CYBD incidence levels compared to inner and shelf-edge reefs. Our results concur with other studies where the number of diseased colonies was positively correlated with the abundance of susceptible species [36, 37].

CYBD infected colonies showed an aggregated pattern in all of the $100\,\text{m}^2$ areas surveyed. The same spatial dispersion pattern was found for the whole population (i.e., healthy + diseased) as well. This could mean that *M. faveolata* colonies are harboring at all times the *Vibrio* pathogens that cause

TABLE 2: Number of newly infected *M. faveolata* colonies per month in 2009 at the different reef sites. (—): no newly infected colonies. Winter included the months from January–April and summer included the months from June–September.

	Jan.	Feb.	Mar.	Apr.	May	Jun.	Jul.	Aug.	Sep.	Oct.	Nov.	Dec.
Enrique	—	—	—	2	—	1	—	—	2	—	—	—
M. Luna	—	3	—	—	—	2	—	2	—	—	—	—
Turrumote	—	1	—	—	—	4	—	—	2	—	—	—
Buoy	—	2	—	—	—	—	—	—	—	—	—	—
Weinberg	—	—	—	—	—	—	—	—	2	—	—	—

TABLE 3: Index of aggregation, *R* values, with sample size (in parenthesis) for the three 100 m² quadrants (Q) in the different sites. (—): no data was found. (*): not significant data.

Reef	Q 1		Q 2		Q 3	
	Healthy + CYBD	CYBD	Healthy + CYBD	CYBD	Healthy + CYBD	CYBD
Pelotas	—	—	—	—	—	—
Enrique	*	*	0.63 (26)	0.5 (2)	0.70 (16)	0.48 (2)
Media Luna	0.55 (80)	0.66 (15)	0.48 (124)	0.7 (16)	0.46 (150)	0.60 (28)
Turrumote	0.68 (70)	0.86 (40)	0.65 (61)	0.69 (13)	0.58 (27)	*
Buoy	0.64 (11)	*	0.84 (14)	*	0.75 (36)	0.92 (5)
Weinberg	0.89 (17)	0.84 (2)	0.66 (20)	1.01 (5)	0.66 (51)	0.48 (6)

TABLE 4: Colony distribution departure from randomness; *z* and *P* values (in parenthesis) from the test of significance for the three 100 m² quadrants in the different sites. (—): no data was found. (*): not significant data.

Reef	Q 1		Q 2		Q 3	
	Healthy + CYBD	CYBD	Healthy + CYBD	CYBD	Healthy + CYBD	CYBD
Pelotas	—	—	—	—	—	—
Enrique	*	*	0.63 (26)	0.5 (2)	0.70 (16)	0.48 (2)
Media Luna	0.55 (80)	0.66 (15)	0.48 (124)	0.7 (16)	0.46 (150)	0.60 (28)
Turrumote	0.68 (70)	0.86 (40)	0.65 (61)	0.69 (13)	0.58 (27)	*
Buoy	0.64 (11)	*	0.84 (14)	*	0.75 (36)	0.92 (5)
Weinberg	0.89 (17)	0.84 (2)	0.66 (20)	1.01 (5)	0.66 (51)	0.48 (6)

CYBD, and that these trigger the disease when colonies are more vulnerable due to temperature changes such as those that occur in the summer or during warmer winters. Further research in the bacterial strains within *M. faveolata* tissues may help to find if these colonies are harboring the *Vibrio* consortium that is associated to CYBD. Foley et al. [22], studying spatial dispersion patterns of CYBD in *Montastraea annularis* in Akumal, Mexico, found contrasting results. They found that CYBD-affected colonies showed a more regular (less clustered) spatial pattern compared to the whole population (healthy + diseased) and concluded that proximity to other *M. annularis* may offer protection from CYBD perhaps in the form of barriers to disease agents or genetic resistance of some coral colonies. Foley et al. [22] defined subcolonies of *M. annularis* "as distinct lobes of aggregated clonal corallites that form a recognizable unit of a main corallum (boulder)." It is imperative to carefully inspect this aspect of the analysis due to a high probability of confusion and error if one selects ramets of a different genet pooled as a single colony. In our study, CYBD-infected *M. faveolata* showed highly aggregated

spatial patterns (similar to the whole population) in all reefs in La Parguera. Distribution patterns depend on habitat characteristics, substrate availability, and current regimes. Differences in species distribution and spatial scale used can explain the contrasting results between our study and the one from Foley et al. [22]. In our study, we used a spatial scale of 100 m² quadrants, but Foley et al. [22] studied a spatial scale between 10–30 m. Spatial patterns can change at different scales of observation and this is highly important when studying spatial distribution patterns of coral diseases [22]. Jolles et al. [19] found that clusters of infected *G. ventalina* were larger than expected under the direct contact hypothesis. It is important to recognize that prevalence (proportion of infected colonies in a population) is not the same epidemiological measure as incidence (new cases of diseased colonies over time). However, both parameters can indicate the level at which a disease is affecting a population. Incidence indicates the risk of contracting the disease, whereas prevalence indicates how extensive the disease is. Although we did not assess CYBD prevalence in the 100 m² quadrants on each site, prevalence from other areas in these same sites during 2009 showed a similar pattern to the incidence data we gathered by having more infected colonies in mid-shelf reefs where host abundances were higher (Soto-Santiago and Weil, in preparation). Some of the index of aggregation data showed a low sample size (Table 3). This could have affected the patterns of aggregation since lower number of colonies may result in less power when applying the nearest neighbor method. However, we only included in the analysis data that showed significant differences from randomness (Tables 3 and 4).

If the pattern of warmer winters continues spreading worldwide, this could be highly detrimental to coral reefs around the world. Furthermore, all the ecosystem services that these reefs provide can be lost in the near future. By having warmer winters, corals will not be able to adapt to temperature regimes and susceptibility to any pathogens in the surrounding environment will affect more dramatically these organisms. It will be more difficult for corals to recover from diseases and bleaching as they will be vulnerable all year long. Coral diseases have shown associations to different environmental variables at different spatial scales [21]. Although baseline spatiotemporal assessments such as the study presented here are useful, ultimately this information needs to be combined with covarying environmental measurements over long time periods with high spatial replication and the results coupled and investigated for relationships using predictive modeling and other spatial analyses. Our results have shown that the study of incidence and spatial distribution patterns can highlight important information on the etiology and potential mechanisms of CYBD affecting important reef framework species such as *M. faveolata*. These types of studies can elucidate the dynamics of spread and infection of emerging coral diseases which are threatening to wipe out coral reefs and its ecosystem services in Puerto Rico and the wider Caribbean.

Acknowledgments

Funding for this project was provided by NOAA-CRES grant no. NA170P2919, the GEF-World Bank Coral Reef Targeted Research and Capacity Building Project through the Coral Disease Group, and the Department of Marine Sciences, University of Puerto Rico. Thanks to Jorge Casillas, Emmanuel Irizarry, and employees of the Department of Marine Sciences, University of Puerto Rico for help in the field and support with logistics. Thanks to Michael Nemeth for providing the map of the study sites. Brian Bingham and three anonymous reviewers provided critical comments to improve this manuscript.

References

[1] T. J. Goreau, J. Cervino, M. Goreau et al., "Rapid spread of diseases in Caribbean coral reefs," *Revista de Biologia Tropical*, vol. 46, supplement 5, pp. 157–171, 1998.

[2] L. L. Richardson, "Coral diseases: what is really known?" *Trends in Ecology and Evolution*, vol. 13, no. 11, pp. 438–443, 1998.

[3] D. Harvell, S. Altizer, I. M. Cattadori, L. Harrington, and E. Weil, "Climate change and wildlife diseases: when does the host matter the most?" *Ecology*, vol. 90, no. 4, pp. 912–920, 2009.

[4] E. P. Green and A. W. Bruckner, "The significance of coral disease epizootiology for coral reef conservation," *Biological Conservation*, vol. 96, no. 3, pp. 347–361, 2000.

[5] E. Weil, I. Urreiztieta, and J. Garzón-Ferreira, "Geographic variability in the incidence of coral and octocoral diseases in the wider Caribbean," in *Proceedings of the 9th International Coral Reef Symposium*, vol. 2, pp. 1231–1238, 2002.

[6] E. Weil, "Coral reef diseases in the wider Caribbean," in *Coral Health and Disease*, E. Rosenberg and Y. Loya, Eds., pp. 35–68, Springer, New York, NY, USA, 2004.

[7] J. Miller, R. Waara, E. Muller, and C. S. Rogers, "Coral bleaching and disease combine to cause extensive mortality on reefs in US Virgin Islands," *Coral Reefs*, vol. 25, no. 3, p. 418, 2006.

[8] A. W. Bruckner and R. L. Hill, "Ten years of change to coral communities off Mona and Desecheo Islands, Puerto Rico, from disease and bleaching," *Diseases of Aquatic Organisms*, vol. 87, no. 1-2, pp. 19–31, 2009.

[9] E. Weil, A. Croquer, and I. Urreiztieta, "Temporal variability and impact of coral diseases and bleaching in La Parguera, Puerto Rico from 2003–2007," *Caribbean Journal of Science*, vol. 45, no. 2-3, pp. 221–246, 2009.

[10] C. D. Harvell, E. Jordán-Dahlgren, S. Merkel et al., "Coral diseases, environmental drivers, and the balance between coral and microbial associates," *Oceanography*, vol. 20, no. 1, pp. 36–59, 2007.

[11] J. F. Bruno, L. E. Petes, C. D. Harvell, and A. Hettinger, "Nutrient enrichment can increase the severity of coral diseases," *Ecology Letters*, vol. 6, no. 12, pp. 1056–1061, 2003.

[12] J. D. Voss and L. L. Richardson, "Nutrient enrichment enhances black band disease progression in corals," *Coral Reefs*, vol. 25, no. 4, pp. 569–576, 2006.

[13] G. S. Aeby, G. J. Williams, E. C. Franklin et al., "Growth anomalies on the coral genera Acropora and Porites are strongly associated with host density and human population size across the Indo-Pacific," *PLoS ONE*, vol. 6, no. 2, Article ID e16887, 2011.

[14] G. S. Aeby, G. J. Williams, E. C. Franklin et al., "Patterns of coral disease across the Hawaiian Archipelago: relating disease to environment," *PLoS ONE*, vol. 6, no. 5, Article ID e20370, 2011.

[15] E. A. Dinsdale, O. Pantos, S. Smriga et al., "Microbial ecology of four coral atolls in the Northern Line Islands," *PLoS ONE*, vol. 3, no. 2, Article ID e1584, 2008.

[16] E. Weil, G. Smith, and D. L. Gil-Agudelo, "Status and progress in coral reef disease research," *Diseases of Aquatic Organisms*, vol. 69, no. 1, pp. 1–7, 2006.

[17] A. W. Bruckner and R. J. Bruckner, "Consequences of yellow band disease (YBD) on Montastraea annularis (species complex) populations on remote reefs off Mona Island, Puerto Rico," *Diseases of Aquatic Organisms*, vol. 69, no. 1, pp. 67–73, 2006.

[18] D. L. Ballantine, R. S. Appeldoorn, and P. Yoshioka, "Biology and ecology of puerto rican coral reefs," in *Coral Reefs of the USA*, B. M. Riegl and R. E. Dodge, Eds., chapter 9, pp. 375–406, 2008.

[19] A. E. Jolles, P. Sullivan, A. P. Alker, and C. D. Harvell, "Disease transmission of aspergillosis in sea fans: inferring process from spatial pattern," *Ecology*, vol. 83, no. 9, pp. 2373–2378, 2002.

[20] S. H. Sokolow, P. Foley, J. E. Foley, A. Hastings, and L. L. Richardson, "Disease dynamics in marine metapopulations: modelling infectious diseases on coral reefs," *Journal of Applied Ecology*, vol. 46, no. 3, pp. 621–631, 2009.

[21] G. J. Williams, G. S. Aeby, R. O. M. Cowie, and S. K. Davy, "Predictive modeling of coral disease distribution within a reef system," *PLoS ONE*, vol. 5, no. 2, Article ID e9264, 2010.

[22] J. E. Foley, S. H. Sokolow, E. Girvetz, C. W. Foley, and P. Foley, "Spatial epidemiology of Caribbean yellow band syndrome in *Montastrea* spp. coral in the eastern Yucatan, Mexico," *Hydrobiologia*, vol. 548, no. 1, pp. 33–40, 2005.

[23] T. M. Work, L. L. Richardson, T. L. Reynolds, and B. L. Willis, "Biomedical and veterinary science can increase our understanding of coral disease," *Journal of Experimental Marine Biology and Ecology*, vol. 362, no. 2, pp. 63–70, 2008.

[24] J. M. Cervino, R. Hayes, T. J. Goreau, and G. W. Smith, "Zooxanthellae regulation in yellow blotch/band and other coral diseases contrasted with temperature related bleaching: *In Situ* destruction vs expulsion," *Symbiosis*, vol. 37, no. 1–3, pp. 63–85, 2004.

[25] J. M. Cervino, R. L. Hayes, S. W. Polson et al., "Relationship of *Vibrio* species infection and elevated temperatures to yellow blotch/band disease in caribbean corals," *Applied and Environmental Microbiology*, vol. 70, no. 11, pp. 6855–6864, 2004.

[26] E. . Weil, G. W. Smith, K. B. Ritchie, and A. Croquer, "Inoculation of *Vibrio* spp. onto *Montastraea faveolata* fragments to determine potential pathogenicity," in *Proceedings of 11th International Coral Reef Symposium Session*, vol. 7, pp. 202–205, Ft. Lauderdale, Fla, USA, 2009.

[27] K. Flynn and E. Weil, "Variability of aspergillosis in *Gorgonia ventalina* in La Parguera, Puerto Rico," *Caribbean Journal of Science*, vol. 45, no. 2-3, pp. 215–220, 2009.

[28] P. J. Clark and F. C. Evans, "Distance to nearest neighbor as a measure of spatial relationships in populations," *Ecology*, vol. 35, pp. 445–453, 1954.

[29] C. J. Krebs, *Ecological Methodology*, Addison-Wesley, Reading, Mass, USA, 2nd edition, 1999.

[30] E. Weil, A. Croquer, and I. Urreiztieta, "Temporal variability and impact of coral diseases and bleaching in La Parguera, Puerto Rico from 2003–2007," *Caribbean Journal of Science*, vol. 45, no. 2-3, pp. 221–246, 2009.

[31] J. M. Cervino, F. L. Thompson, B. Gomez-Gil et al., "The *Vibrio* core group induces yellow band disease in Caribbean and Indo-Pacific reef-building corals," *Journal of Applied Microbiology*, vol. 105, no. 5, pp. 1658–1671, 2008.

[32] A. W. Bruckner and R. L. Hill, "Ten years of change to coral communities off Mona and Desecheo Islands, Puerto Rico, from disease and bleaching," *Diseases of Aquatic Organisms*, vol. 87, no. 1-2, pp. 19–31, 2009.

[33] T. Saxby, W. C. Dennison, and O. Hoegh-Guldberg, "Photosynthetic responses of the coral *Montipora digitata* to cold temperature stress," *Marine Ecology Progress Series*, vol. 248, pp. 85–97, 2003.

[34] T. C. LaJeunesse, H. Reyes-Bonilla, and M. E. Warner, "Spring "bleaching" among *Pocillopora* in the Sea of Cortez, Eastern Pacific," *Coral Reefs*, vol. 26, no. 2, pp. 265–270, 2007.

[35] E. Weil and A. Cróquer, "Spatial variability in distribution and prevalence of Caribbean scleractinian coral and octocoral diseases. I. Community-level analysis," *Diseases of Aquatic Organisms*, vol. 83, no. 3, pp. 195–208, 2009.

[36] J. L. Borger and S. C. C. Steiner, "The spatial and temporal dynamics of coral diseases in Dominica, West Indies," *Bulletin of Marine Science*, vol. 77, no. 1, pp. 137–154, 2005.

[37] J. L. Borger, "Dark spot syndrome: a scleractinian coral disease or a general stress response?" *Coral Reefs*, vol. 24, no. 1, pp. 139–144, 2005.

Permissions

The contributors of this book come from diverse backgrounds, making this book a truly international effort. This book will bring forth new frontiers with its revolutionizing research information and detailed analysis of the nascent developments around the world.

We would like to thank all the contributing authors for lending their expertise to make the book truly unique. They have played a crucial role in the development of this book. Without their invaluable contributions this book wouldn't have been possible. They have made vital efforts to compile up to date information on the varied aspects of this subject to make this book a valuable addition to the collection of many professionals and students.

This book was conceptualized with the vision of imparting up-to-date information and advanced data in this field. To ensure the same, a matchless editorial board was set up. Every individual on the board went through rigorous rounds of assessment to prove their worth. After which they invested a large part of their time researching and compiling the most relevant data for our readers. Conferences and sessions were held from time to time between the editorial board and the contributing authors to present the data in the most comprehensible form. The editorial team has worked tirelessly to provide valuable and valid information to help people across the globe.

Every chapter published in this book has been scrutinized by our experts. Their significance has been extensively debated. The topics covered herein carry significant findings which will fuel the growth of the discipline. They may even be implemented as practical applications or may be referred to as a beginning point for another development. Chapters in this book were first published by Hindawi Publishing Corporation; hereby published with permission under the Creative Commons Attribution License or equivalent.

The editorial board has been involved in producing this book since its inception. They have spent rigorous hours researching and exploring the diverse topics which have resulted in the successful publishing of this book. They have passed on their knowledge of decades through this book. To expedite this challenging task, the publisher supported the team at every step. A small team of assistant editors was also appointed to further simplify the editing procedure and attain best results for the readers.

Our editorial team has been hand-picked from every corner of the world. Their multi-ethnicity adds dynamic inputs to the discussions which result in innovative outcomes. These outcomes are then further discussed with the researchers and contributors who give their valuable feedback and opinion regarding the same. The feedback is then collaborated with the researches and they are edited in a comprehensive manner to aid the understanding of the subject.

Apart from the editorial board, the designing team has also invested a significant amount of their time in understanding the subject and creating the most relevant covers. They scrutinized every image to scout for the most suitable representation of the subject and create an appropriate cover for the book.

The publishing team has been involved in this book since its early stages. They were actively engaged in every process, be it collecting the data, connecting with the contributors or procuring relevant information. The team has been an ardent support to the editorial, designing and production team. Their endless efforts to recruit the best for this project, has resulted in the accomplishment of this book. They are a veteran in the field of academics and their pool of knowledge is as vast as their experience in printing. Their expertise and guidance has proved useful at every step. Their uncompromising quality standards have made this book an exceptional effort. Their encouragement from time to time has been an inspiration for everyone.

The publisher and the editorial board hope that this book will prove to be a valuable piece of knowledge for researchers, students, practitioners and scholars across the globe.

List of Contributors

Janet L. Neilson and Christine M. Gabriele
Division of Resource Management, Glacier Bay National Park and Preserve, P.O. Box 140, Gustavus, AK 99826, USA

Aleria S. Jensen and Kaili Jackson
Office of Protected Resources, National Marine Fisheries Service, P.O. Box 21668, Juneau, AK 99802, USA

Janice M. Straley
Department of Biology, University of Alaska Southeast Sitka Campus, 1332 Seward Avenue, Sitka, AK 99835, USA

Siamak Bagheri, Jalil Sabkara, Seyed Hojat Khodaparast and Esmaeil Yosefzad
Inland Waters Aquaculture Institute, Iranian Fisheries Research Organization (IFRO), Anzali 66, Iran

Alireza Mirzajani
Inland Waters Aquaculture Institute, Iranian Fisheries Research Organization (IFRO), Anzali 66, Iran
Faculty of Natural Resources, University of Tehran, P.O. Box 4314, Karaj 31587-77878, Iran

Foong Swee Yeok
School of Biological Sciences, Universiti Sains Malaysia, 11800 Penang, Malaysia

Paul Q. Sims
Department of Biology, McGill University, 1205 Avenue Docteur Penfield, Montreal, QC, Canada

Samuel K. Hung
Hong Kong Cetacean Research Project, Flat C 22/F., Block 13, Sceneway Garden, Lam Tin, Kowloon, Hong Kong

Bernd Wursig
Department of Marine Biology, Texas A&M University at Galveston, 200 SeaWolf Pkwy, Galveston, TX 77553, USA

Enrique Godínez-Domínguez
Departamento de Estudios para el Desarrollo Sustentable de la Zona Costera, Centro Universitario de la Costa Sur (CUCSUR), Universidad de Guadalajara, Avendia V. Gomez Farıas 82, 48980 San Patricio Melaque, JAL, Mexico

Gerardo Aceves-Medina
Instituto Politecnico Nacional, CICIMAR-IPN, Departamento de Plancton y Ecologıa Marina, 23096 La Paz, BCS, Mexico

Eduardo González-Rodríguez and Armando Trasviña
Centro de Investigacion Cientıfica y de Educacion Superior de Ensenada, Unidad La Paz, 23050 La Paz, BCS, Mexico

Raymundo Avendaño-Ibarra
Departamento de Estudios para el Desarrollo Sustentable de la Zona Costera, Centro Universitario de la Costa Sur (CUCSUR), Universidad de Guadalajara, Avendia V. Gomez Farıas 82, 48980 San Patricio Melaque, JAL, Mexico
Instituto Politecnico Nacional, CICIMAR-IPN, Departamento de Plancton y Ecologıa Marina, 23096 La Paz, BCS, Mexico

Bastiaan Knoppers
Programa de Pos-Graduacao em Geoquımica, Instituto de Quımica, Universidade Federal Fluminense, Outeiro de Sao Joao Batista, s/n, 24020-141, Niteroi, RJ, Brazil

Adriana C. Fonseca
Instituto Chico Mendes de Conservacao da Biodiversidade (ICMBIO), Rodovia Mauricio S. Sobrinho s/n, km 2, 88053-700, Florianopolis, SC, Brazil
Programa de Pos-Graduacao em Geoquımica, Instituto de Quımica, Universidade Federal Fluminense, Outeiro de Sao Joao Batista, s/n, 24020-141, Niteroi, RJ, Brazil

Roberto Villaca
Programa de Pos-Graduacao em Biologia Marinha, Instituto de Biologia, Universidade Federal Fluminense, Outeiro de Sao Joao Batista, s/n, 24001-970, Niteroi, RJ, Brazil

H. M.Murray and G. L. Sheppard
Fisheries and Oceans Canada, 80 White Hills Road, P.O. Box 5667, St. John's, NL, Canada

D. Gallardi
Fisheries and Oceans Canada, 80 White Hills Road, P.O. Box 5667, St. John's, NL, Canada
School of Fisheries, Marine Institute of Memorial University of Newfoundland, St. John's, NL, Canada

Y. S. Gidge
Fisheries and Oceans Canada, 80 White Hills Road, P.O. Box 5667, St. John's, NL, Canada
Department of Environmental Science, Memorial University of Newfoundland, St. John's, NL, Canada

L. Charpy and M. J. Langlade
Mediterranean Institute of Oceanography (MIO), IRD, UR235 Center of Tahiti, BP 529, 98713 Papeete, French Polynesia

B. E. Casareto and Y. Suzuki
Graduate School of Science and Technology, Shizuoka University, 836 Ohya, Suruga-ku, Shizuoka 422-8529, Japan

G. D Onghia, A. Giove, P.Maiorano, R. Carlucci, M. Minerva, F. Capezzuto, L. Sion and A. Tursi
Department of Biology, University of Bari "Aldo Moro", Via E. Orabona 4, 70125 Bari, Italy

Nathalie Caill-Milly and Catherine Borie
Laboratoire Ressources Halieutiques Aquitaine, IFREMER, FED 4155 MIRA, 1 allee du Parc Montaury, 64600 Anglet, France

Noelle Bru
Laboratoire de Mathematiques et de leurs Applications de Pau, UMR CNRS 5142, FED 4155 MIRA, UNIV PAU & PAYS ADOUR, 64000 Pau, France

Kelig Mahe
Pole de Sclerochronologie, IFREMER Centre Manche-Mer du Nord, 150 Quai Gambetta, 62200 Boulogne-sur-Mer, France

Frank D Amico
UMR ECOBIOP, FED 4155 MIRA, UNIV PAU & PAYS ADOUR, Campus Montaury, 64600 Anglet, France

Anne Klöppel and Franz Brümmer
Department of Zoology, Biological Institute, University of Stuttgart, Pfaffenwaldring 57, 70569 Stuttgart, Germany

Denise Schwabe
Animal Evolutionary Ecology, Zoological Institute, University of Tubingen, Auf der Morgenstelle 28, 72076 Tubingen, Germany

Gertrud Morlock
Institute of Food Chemistry, University of Hohenheim, Garbenstraße 28, 70599 Stuttgart, Germany

Abdulmumin A. Nuhu
Department of Chemistry, Ahmadu Bello University, P.M.B. 1069, Zaria, Kaduna, Nigeria

Z. T. Richards, M. Bryce and C. Bryce
Aquatic Zoology, West Australian Museum, Locked Bag 49, Welshpool DC, WA 6986, Australia

Jenna R. Lueg, Alison L. Moulding and David S. Gilliam
Oceanographic Center, Nova Southeastern University, 8000 N. Ocean Drive, Dania Beach, FL 33004, USA

Vladimir N. Kosmynin
Bureau of Beaches and Coastal Systems, Florida Department of Environmental Protection, 3900 Commonwealth Boulevard, MS 49, Tallahassee, FL 32399, USA

Jeffrey S. Prince
Dauer Electron Microscopy Laboratory, Department of Biology, The University of Miami, P.O. Box 249118, Coral Gables, FL 33124, USA

Melissa R. Romero, Kimberly M. Walker, Kimberly J. Nelson, Daisha C. Ortega, Serra L. Smick, William J. Hoese and Danielle C. Zacherl
Department of Biological Science, California State University Fullerton, P.O. Box 6850, Fullerton, CA 92834-6850, USA

Carmen J. Cortez
Graduate Group in Ecology, University of California Davis, One Shields Avenue, Davis, CA 95616, USA

Yareli Sanchez
Global Change Research Group, Department of Biology, San Diego State University, 5500 Campanile Drive, San Diego, CA 92182, USA

Karin Schiefenhovel and Andreas Kunzmann
Department of WG Ecophysiology, Leibniz Center for Tropical Marine Ecology, Fahrenheitstraße 6, 28359 Bremen, Germany

Francisco Javier Murillo
Instituto Espanol de Oceanografía, 36280 Vigo, Spain

Konstantin R. Tabachnick
Department of Benthic Fauna, P. P. Shirshov Institute of Oceanology, Russian Academy of Sciences, Nakhimovsky Prospekt 36, Moscow 117997, Russia

Larisa L. Menshenina
Physical Department, Moscow State University, Moscow 119991, Russia

Yun Li
East China Sea Fisheries Research Institute, Chinese Academy of Fishery Sciences, Shanghai 200090, China
State Key Laboratory of Estuarine and Coastal Research, East China Normal University, Shanghai 200062, China

Daoji Li
State Key Laboratory of Estuarine and Coastal Research, East China Normal University, Shanghai 200062, China

Nathan L. Stewart
School of Fisheries and Ocean Sciences, University of Alaska Fairbanks, 905 N. Koyukuk Drive, 245 O Neill Building, Fairbanks, AK 99775, USA

Brenda Konar
Global Undersea Research Unit, University of Alaska Fairbanks, 905 N. Koyukuk Drive, 217 O Neill Building, Fairbanks, AK 99775, USA

Ernesto Weil
Department of Marine Sciences, University of Puerto Rico, Mayaguez, P.O. Box 9000, San Juan P.R. 00681, Puerto Rico

Francisco J. Soto-Santiago
Department of Marine Sciences, University of Puerto Rico, Mayaguez, P.O. Box 9000, San Juan P.R. 00681, Puerto Rico
Department of Environmental Sciences, University of Puerto Rico, Rio Piedras Campus, P.O. Box 70377, San Juan P.R. 00936, Puerto Rico